PURE AND APPLIED MATHEMATICS

GEOMETRY OF REFLECTING RAYS AND INVERSE SPECTRAL PROBLEMS

GEOMETRY OF REFLECTING RAYS AND INVERSE SPECTRAL PROBLEMS

Vesselin M. Petkov

and

Luchezar N. Stoyanov

Institute of Mathematics,
Bulgarian Academy of Sciences

JOHN WILEY & SONS

Chichester • New York • Brisbane • Toronto • Singapore

Other Wiley Editorial Offices

John Wiley & Sons, Inc., 605 Third Avenue,
New York, NY 10158-0012, USA.

Jacaranda Wiley Ltd, G.P.O. Box 859, Brisbane,
Queensland 4001, Australia

John Wiley & Sons (Canada) Ltd, 22 Worcester Road,
Rexdale, Ontario M9W 1L1, Canada

John Wiley & Sons (SEA) Pte Ltd, 37 Jalan Pemimpin 05-04,
Block B, Union Industrial Building, Singapore 2057

Library of Congress Cataloging-in-Publication Data:

Petkov, Vesselin.
    Geometry of reflecting rays and inverse spectral problems /
Vesselin M. Petkov and Luchezar N. Stoyanov.
    p.   cm. — (Pure and applied mathematics)
Includes bibliographical references and indexes.
ISBN 0 471 93174 8
1. Spectral theory (Mathematics)   2. Inverse problems
(Differential equations)   3. Geometry, Differential.   I. Stoyanov,
Luchezar N., 1954– .   II. Title.   III. Series: Pure and applied
mathematics (John Wiley & Sons : Unnumbered)
QA320.P435   1992
515'.7222 — dc20
                                        91-38230
                                        CIP

*A catalogue record for this book is
available from the British Library.*

ISBN 0 471 93174 8

Typeset by Laser Words, Madras, India
Printed in Great Britain by Biddles Ltd, Guildford and Kings Lynn

CONTENTS

INTRODUCTION

This monograph is devoted to the analysis of some inverse problems concerning the spectrum of the Laplace operator in a bounded domain $\Omega \subset \mathbf{R}^n$, $n \geq 2$, and of the scattering operator in the exterior of a bounded obstacle $K \subset \mathbf{R}^n$, $n \geq 3$. In both cases our aim is to obtain geometrical information on Ω (resp. K) from the spectral (resp. scattering) data. We treat both inverse problems using similar techniques based on certain properties of the billiard and geodesic flows in the generic case and on a microlocal analysis of mixed problems.

Let $\Omega \subset \mathbf{R}^n$, $n \geq 2$, be a bounded domain with C^∞ smooth boundary $\partial\Omega$ and let A be the self-adjoint operator in $L^2(\Omega)$ related to the *Laplacian*

$$-\Delta = -\sum_{j=1}^{n} \partial_{x_j}^2,$$

in Ω with Dirichlet boundary condition on $\partial\Omega$. The *spectrum* of A is given by a sequence

$$0 \leq \lambda_1^2 \leq \lambda_2^2 \leq \ldots \lambda_m^2 \leq \ldots \tag{0.1}$$

of eigenvalues λ_j^2 for which the problem

$$\begin{cases} -\Delta\varphi_j = \lambda_j^2\varphi_j \text{ in } \Omega, \\ \varphi_j = 0 \text{ on } \partial\Omega \end{cases}$$

has a non-trivial solution $\varphi_j \in C^\infty(\Omega)$. The *counting function*

$$N(\lambda) = \#\{j : \lambda_j^2 \leq \lambda^2\}$$

admits a polynomial bound

$$N(\lambda) \leq C\lambda^n, \quad \lambda \geq 0. \tag{0.2}$$

This function has a Weyl type asymptotic

$$N(\lambda) = \frac{(4\pi)^{-n/2}}{\Gamma(n/2+1)} \text{ vol } (\Omega)\lambda^n + O(\lambda^{n-1}), \quad \lambda \to \infty. \tag{0.3}$$

Thus, from the spectrum (0.1) we can recover the volume of Ω. Around 1912 Weyl conjectured that for every bounded domain Ω in $\mathbf{R}^n$ with smooth boundary $\partial\Omega$ we have

$$N(\lambda) = \frac{(4\pi)^{-n/2}}{\Gamma(n/2+1)} \text{vol}\,(\Omega)\lambda^n - \frac{(4\pi)^{-(n-1)/2}}{4\Gamma((n-1)/2+1)} \text{vol}\,(\partial\Omega)\lambda^{n-1}$$
$$+ o(\lambda^{n-1}), \quad \lambda \to \infty. \tag{0.4}$$

Ivrii [Iv1] proved that if the periodic billiard trajectories in Ω form a subset of Lebesgue measure zero in the phase space of the billiard in Ω, then the asymptotic (0.4) holds. Hence for such domains vol $(\partial\Omega)$ becomes another spectral invariant. It is not known so far if the assumption in Ivrii's result is always satisfied.

To obtain more information in this direction, it is convenient to examine some function determined by the sequence (0.1). The distribution

$$\tau(t) = \sum_j e^{-\lambda_j^2 t} \in \mathcal{D}'(\bar{\mathbf{R}}_+)$$

has the asymptotic

$$\tau(t) \sim \sum_{j=0}^{\infty} c_j t^{-(n/2)+j/2} \text{ as } t \searrow 0, \tag{0.5}$$

and the constants c_j are spectral invariants. Moreover, one can recover vol (Ω) and vol $(\partial\Omega)$ from c_0 and c_1.

In his classical work [Kac] Kac posed the problem of recovering the form of a strictly convex domain $\Omega \subset \mathbf{R}^2$ from the spectrum (0.1). This article influenced the investigations of different inverse spectral problems for manifolds with and without boundary as well as the analysis of the so called isospectral manifolds, that is manifolds for which the spectrums of the corresponding Laplace–Beltrami operators coincide.

To determine a strictly convex planar domain Ω, modulo Euclidean transformations, it suffices to know the curvature $\kappa(x)$ of $\partial\Omega$ at each point $x \in \partial\Omega$. In general, the spectral data $\{c_j\}$, given by (0.5), is not sufficient to determine the function $\kappa(x)$. Let us mention that the distribution $\tau(t)$ is singular only at $t = 0$. To obtain a distribution with a larger singular set, consider

$$\sigma(t) = \sum_{j=1}^{\infty} \cos(\lambda_j t) \in \mathcal{S}'(\mathbf{R}). \tag{0.6}$$

Then $\sigma(t)$ is singular at 0 and

$$\sigma(t) \sim \sum_{j=0}^{\infty} d_j t^{-n+j}$$

(see [Me3], [Iv2]). The constants d_j provide other spectral invariants, and the first two determine vol (Ω) and vol $(\partial\Omega)$ again.

It turns out that the set of singularities of $\sigma(t)$ is related to the *length spectrum* L_Ω of Ω. By definition, L_Ω is the set of periods (lengths) of all *periodic generalized geodesics* in Ω. Let us mention that the generalized geodesics are projections on Ω of generalized bicharacteristics of the wave operator $\partial_t^2 - \Delta_x$ in $T^*(\mathbf{R} \times \Omega)$ defined by Melrose and Sjöstrand [MS1], [MS2]. We refer the reader to Chapter 1 for the precise definitions. The so-called *Poisson relation for manifolds with boundary* has the form

$$\text{sing supp}\,\sigma(t) \subset \{0\} \cup \{T \in \mathbf{R} : |T| \in L_\Omega\}. \tag{0.7}$$

For strictly convex (concave) domains this relation has been established by Anderson and Melrose [AM]. Its proof for general domains is based on the results in [MS2] on the propagation of C^∞ singularities. In fact, a relation similar to (0.6), was first established for Riemannian manifolds without boundary. This was achieved independently by Chazarain [Ch2] and Duistermaat and Guillemin [DG]. Moreover, under certain assumptions on $T \in \text{sing supp}\,\sigma(t)$, the leading singularity of $\sigma(t)$ at T was examined in [DG].

It is natural to investigate the inverse inclusion in (0.7), however in the general case very little is known so far. For certain strictly convex planar domains Ω, Marvizi and Melrose [MM] found a sequence of closed billiard trajectories in Ω the lengths of which belong to sing supp $\sigma(t)$. It was expected ([C1, GM3]) that for generic strictly convex domains in $\mathbf{R}^2$ the inclusion (0.7) could become an equality. Such a result was established in [PS2] (see also [PS1]) not only for strictly convex but for all generic domains as well. Its analogue in the case $n > 2$ is proved only for strictly convex domains ([S3]). The results, just mentioned, form one of the main topics in this book.

If the equality

$$\text{sing supp}\,\sigma(t) = \{0\} \cup \{T : |T| \in L_\Omega\} \tag{0.8}$$

holds for some domain Ω, then the lengths of periodic geodesics in Ω can be considered as spectral invariants. From them one can determine various spectral invariants of Ω. The reader may consult [MM, C1, KP, P1, P2, P3] for more information and further results in this direction.

Let $\mathcal{L}_\Omega$ be the set of all periodic geodesics in Ω. For $\gamma \in \mathcal{L}_\Omega$ we denote by T_γ the *period (length)* of γ. There are three types of elements of $\mathcal{L}_\Omega$: periodic reflecting rays (i.e. closed billiard trajectories in Ω), closed geodesics on $\partial\Omega$, and *periodic geodesics of mixed type*, containing both linear segments in Ω and geodesic segments on $\partial\Omega$. Among the periodic reflecting rays we shall distinguish those without segments tangent to the boundary $\partial\Omega$; such rays will be called *ordinary*. As for the closed geodesics on $\partial\Omega$, for each ordinary periodic reflecting ray γ one can naturally define a *Poincaré map* $\mathcal{P}_\gamma$, such that the spectrum spec P_γ of the linearization P_γ of $\mathcal{P}_\gamma$ contains certain information about the behaviour of the billiard flow along γ. Such a ray γ will be called *non-degenerate* if $1 \notin \operatorname{spec} P_\gamma$. The Poincaré map for periodic reflecting rays is defined and studied in Chapter 2.

Given a smooth submanifold X of $\mathbf{R}^n$, we denote by $C^\infty(X, \mathbf{R}^n)$ the *space of all smooth maps* $f : X \to \mathbf{R}^n$, endowed with the Whitney C^∞ topology (cf. Chapter 1). Let $\mathbf{C}(X) = C^\infty_{\mathrm{emb}}(X, \mathbf{R}^n)$ be its subspace consisting of all smooth embeddings of X into $\mathbf{R}^n$. Being open in $C^\infty(X, \mathbf{R}^n)$, $\mathbf{C}(X)$ is a Baire space, so every residual subset of $\mathbf{C}(X)$ is dense in it. Recall that a subset R of a topological space Z is called *residual* if R is a countable intersection of open dense subsets of Z.

Throughout the book we shall consider very often the situation when Ω is a compact domain with smooth boundary $\partial\Omega$ and $X = \partial\Omega$. Then, for every $f \in \mathbf{C}(X)$ there exists a unique compact domain Ω_f in $\mathbf{R}^n$ with boundary $\partial\Omega_f = f(X) = f(\partial\Omega)$. Let us note that if Ω is strictly convex, the set $\mathcal{O}(\Omega)$ of those $f \in \mathbf{C}(X)$ such that Ω_f is strictly convex, is open in $\mathbf{C}(X)$, and so it is a Baire topological space, too. If Ω is a domain in $\mathbf{R}^n$ with a bounded complement, for $f \in \mathbf{C}(X)$ we denote by Ω_f the unbounded domain in $\mathbf{R}^n$ with $\partial\Omega_f = f(X)$. In the following we sometimes say that a property is generically satisfied (briefly, *generic property*) in some class of objects, say for the compact domains in $\mathbf{R}^n$ with smooth boundaries. By this we mean a property S such that for every bounded domain Ω with smooth boundary $X = \partial\Omega$ there exists a residual subset R of $\mathbf{C}(X)$ such that Ω_f has the property S for every $f \in R$. In the same way, considering residual subsets of $\mathcal{O}(\Omega)$, one can speak about generic properties of the strictly convex domains, etc.

Let us note that in the whole book 'smooth' means C^∞. By a domain we always mean a domain with smooth boundary.

Exploiting the multijet transversality theorem, we establish that the following properties of the compact domains in $\mathbf{R}^n$ are generic:

(I) $T_\gamma / T_\delta \notin \mathbf{Q}$ for all periodic reflecting rays γ and δ such that neither of them is a multiple of the other;

(II) every periodic reflecting ray in Ω is ordinary and non-degenerate.

As a consequence of this, it is established that the asymptotic (0.4) holds for generic domains $\Omega \subset \mathbf{R}^n$. Using (I) and (II), we prove (0.8) for generic strictly convex domains in the plane. In fact, if Ω has the properties (I) and (II), then each periodic reflecting ray γ in Ω has a period T_γ which is an isolated point in L_Ω. For the kernel $\mathcal{E}(t, x, y)$ of the operator $\cos(t\sqrt{A})$ we have

$$\sigma(t) = \int_\Omega \mathcal{E}(t, x, x)\, \mathrm{d}x.$$

One can compute the leading singularity of $\sigma(t)$ for t close to T_γ by the Poisson summation formula discussed in Chapter 6. This leads to (0.8), since by (I) the singularities, related to different periodic rays, cannot be cancelled.

In general, a domain $\Omega \subset \mathbf{R}^2$ might admit periodic geodesics of mixed type. The analysis of the singularities of $\sigma(t)$, related to the periods of such geodesics, imposes a rather difficult problem. We overcome this difficulty by showing that the following property is generic for domains $\Omega \subset \mathbf{R}^2$.

(III) there are no periodic geodesics of mixed type in Ω.

The analysis of generic properties, like (I)–(III), is the second main topic of this book. To establish (0.8) for generic strictly convex domains in $\mathbf{R}^n$, $n \geq 3$, in Chapter 4 we prove an analogue of the classical bumpy metric theorem of Abraham–Klingenberg–Takens–Anosov, considering Riemannian metrics on $X \subset \mathbf{R}^n$, induced by smooth embeddings of X into $\mathbf{R}^n$.

Our third topic is connected with inverse problems for the kernel $s(t - t', \theta, \omega)$ of the operator

$$S - \text{Id} : L^2(\mathbf{R} \times S^{n-1}) \rightarrow L^2(\mathbf{R} \times S^{n-1}).$$

Here $\theta, \omega \in S^{n-1}$, $t, t' \in \mathbf{R}$, and S is the scattering operator related to the Dirichlet problem for the wave operator

$$\square = \partial_t^2 - \Delta$$

in the exterior Ω of a bounded obstacle K with smooth boundary $\partial\Omega = \partial K$ (see [LPh]). For fixed $\theta, \omega \in S^{n-1}$ the *scattering kernel* $s(t, \theta, \omega)$ is a tempered distribution in $\mathcal{S}'(\mathbf{R})$. It is well known that the distribution $s(t, \theta, \omega)$ determines uniquely the obstacle K ([LPh]). On the other hand, in the applications, for given directions ω and θ, one can measure only the singularities of $s(t, \theta, \omega)$. It turns out that these singularities are connected with the sojourn times of the (generalized) (ω, θ)-rays in Ω. These rays are generalized geodesics in Ω, incoming with direction ω and outgoing with direction θ. For such a ray γ, the sojourn time T_γ was defined by Guillemin [G1] as an analogue of the notion of period of a periodic geodesic. The sojourn time measures the time which a point, moving along γ with a unit speed, spends near the obstacle K. For strictly convex obstacles K and fixed $\theta \neq \omega$ we have

$$\text{sing supp } s(t, \theta, \omega) = \{-T_\gamma\},$$

γ being the only (ω, θ)-ray in Ω ([Ma2]). In general, the *set $\mathcal{L}_{\omega,\theta}(\Omega)$ of all (ω, θ)-rays* in Ω could contain more than one element. Assuming that for every point ρ in $T^*(\mathbf{R} \times \Omega)$ there is only one generalized bicharacteristic of $\square$ passing through ρ, for $\theta \neq \omega$ we prove the inclusion

$$\text{sing supp } s(t, \theta, \omega) \subset \{-T_\gamma : \gamma \in \mathcal{L}_{\omega,\theta}(\Omega)\}, \tag{0.9}$$

which is called *Poisson relation for the scattering kernel*. It is shown also that for generic obstacles in $\mathbf{R}^3$ this relation becomes an equality, and for t close to $-T_\gamma, \gamma \in \mathcal{L}_{\omega,\theta}(\Omega)$, the leading singularity of $s(t, \theta, \omega)$ is described. In order to do this, we study generic properties of (ω, θ)-rays, similar to (I)–(III). Here the analogue of a periodic reflecting ray is a reflecting (ω, θ)-ray, and that of a Poincaré map is the so-called differential cross section dJ_γ of an ordinary periodic

(ω, θ)-ray. Correspondingly, the non-degeneracy of such a ray γ means that det $\mathrm{d}J_\gamma \neq 0$. The analogue of (III) says that, given ω and θ, for generic obstacles in $\mathbf{R}^3$ there are no (ω, θ)-rays of mixed type in Ω.

As the length spectrum for bounded domains, the right-hand side of (0.9) contains certain information about the geometry of obstacle K, and might be called *scattering length spectrum* (with respect to ω and θ). In Chapter 10 we treat some inverse spectral problems involving the scattering length spectrum and the set of the corresponding singularities of the scattering kernel $s(t, \theta, \omega)$.

Below we present a brief summary of the contents of the book. Chapter 1 has a preliminary character and contains some facts from differential topology and microlocal analysis, including some results on wave fronts of distributions and operators. The main reference book on the latter subject is [H1]. In this chapter we also introduce the generalized bicharacteristics of the wave operator and list some of its properties which will be used later.

In Chapter 2 periodic reflecting rays and (ω, θ)-rays are defined and studied. The central moments here are the matrix representations of the Poincaré map and the differential cross section, respectively. For an obstacle K, which is a finite union of disjoint strictly convex domains in $\mathbf{R}^n$, the periodic reflecting rays in the exterior of K are examined and an estimate is obtained of the number of these rays with a given number of reflection points. In the last section we consider a class of star-shaped non-convex domains in $\mathbf{R}^3$, which admit more than one (ω, θ)-ray for appropriately fixed ω and θ, and have the generic properties discussed later in Chapters 3 and 9.

In Chapter 3 various generic properties of reflecting rays are established. The proofs use a general idea, described in the first two sections, and make use of the multijet transversality theorem.

The bumpy metric theorem for compact smooth hypersurfaces X in $\mathbf{R}^n$ is proved in Chapter 4. It says that for a residual set of embeddings $f \in \mathbf{C}(X)$ all closed geodesics on $f(X)$ with respect to the standard metric on $f(X)$ inherited from the Euclidean structure of $\mathbf{R}^n$, are non-degenerate.

In Chapter 5 we prove the Poisson relation (0.7) for bounded domains $\Omega \subset \mathbf{R}^n$. For convenience of the reader, we give a separate proof for convex domains, since this case is much simpler than the general one. The rest of the chapter is devoted to the prove of (0.7) for all domains, and can be omitted if one is only interested in the strictly convex case.

The leading singularity of the distribution $\sigma(t)$ near an isolated ordinary periodic reflecting ray γ is studied in Chapter 6. To this end we use global Fourier integral operators and the calculus for such operators developed in [H4]. The central result here is Theorem 6.3.1, which is often used in the subsequent chapters.

In Chapter 7 the equality (0.8) is established for generic domains in $\mathbf{R}^2$ and for generic strictly convex domains in $\mathbf{R}^n$, $n \geq 3$. Modulo the results in the previous chapters, the central moment in the latter case is the proof of Magnuson's theorem about approximation of closed geodesics on $\partial\Omega$ by periodic reflecting rays in Ω. The main tools here are the so called interpolating Hamiltonians for the billiard ball map, the existence of which follows by a general result of Melrose [M1]. For the sake of space, we only give a brief presentation of the latter result; the reader is referred to [M1] or [H3] for a proof.

Chapters 8 and 9 are devoted to the singularities of the scattering kernel. To obtain (0.9), we exploit certain representations of the Fourier transform of $s(t, \theta, \omega)$ involving the Green function. In this way we are able to eliminate some singularities which are not related to (ω, θ)-rays. The leading singularity near an isolated ordinary reflecting (ω, θ)-ray is examined by using the techniques developed in Chapter 6. Finally, the fact that for generic obstacles in $\mathbf{R}^3$ the (ω, θ)-rays of mixed type disappear is established, making use of the genericity argument from Chapter 3.

Inverse scattering problems for several strictly convex disjoint obstacles are examined in Chapter 10. Under a geometric condition (H), introduced by Ikawa, first, a hyperbolic property of the billiard trajectories in the exterior Ω of the obstacles is studied. This allows us to show that all periodic reflecting rays in Ω can be approximated by (ω, θ)-rays for appropriately fixed ω and θ, and that their periods can be discovered from the scattering length spectrum of Ω, related to ω and θ. Finally, we find the asymptotic of the coefficients in front of the leading singularities of the scattering kernel, corresponding to the sojourn times of the approximating (ω, θ)-rays.

Hereafter we assume some knowledge of differential geometry (including basic facts in symplectic geometry) and differential topology. Several profound results from microlocal analysis and global Fourier integral operators are used in Chapters 5 and 6. The reader may choose to read these results informally, omitting their proofs, and then go to Chapters 7, 9 and 10.

In the References we have not even attempted to cover the immense range of works devoted to inverse problems in general. Most of the publications cited there concern inverse spectral results for manifolds with boundary and inverse scattering problems for the scattering kernel.

Acknowledgements: We are indebted to numerous colleagues for their assistance during the work on this book. Special thanks are due to V. Kovachev, G. Popov, Yu. Safarov, J. Sjöstrand and Pl. Stefanov.

Most of the material in this book was written while the first author was a visiting professor at the University of Bordeaux I and the second author was a fellow of the Alexander von Humboldt Foundation at TH Darmstadt, Germany. Thanks are due to Prof. Dr H.-D. Alber and Prof. Dr M. Fuchs for their attention and useful discussions, and to Mrs M. Tabbert, Mrs G. Gehring and Mrs A. Polzin for the consultations during the technical preparations of the text.

We are very grateful to the following publishing companies and societies for permitting us to include parts of papers of ours in this book: the American Mathematical Society for [PS5] and [S4]; Cambridge University Press for [PS4] and [S1] as well as for Figures 1, 2, 3, 4 in [S1]; the Johns Hopkins University Press for [PS2]; Springer-Verlag for [PS3], [S2] and [S3].

1 PRELIMINARIES FROM DIFFERENTIAL TOPOLOGY AND MICROLOCAL ANALYSIS

Here we collect some facts concerning manifolds of jets, spaces of smooth maps and transversality, as well as some material from microlocal analysis. A special emphasis is given to the definition and main properties of the generalized bicharacteristics of the wave operator and the corresponding generalized geodesics.

1.1. Spaces of jets and transversality theorems

We begin with the notion of transversality, manifolds of jets and spaces of smooth maps. The reader is referred to Golubitsky and Guillemin [GG] or Hirsch [Hir] for a detailed presentation of this material.

In this book **smooth** means C^∞.

Let X and Y be smooth manifolds and let

$$f : X \to Y$$

be a smooth map. Given $x \in X$, we shall denote by $T_x f$ the *tangential map* of f at x. Sometimes we shall use the notation $d_x f = T_x f$. If rank $T_x f = \dim X \leq \dim Y$ (resp. rank $T_x f = \dim Y \leq \dim X$), then f is called *immersion* (resp. *submersion*) at x. Let W be a smooth submanifold of Y. We shall say that f is *transversal* to W at $x \in X$ and shall denote this by $f \pitchfork_x W$, if either $f(x) \notin W$ or $f(x) \in W$ and $\operatorname{Im} T_x f + T_{f(x)} W = T_{f(x)} Y$. Here for every $y \in W$ we identify $T_y W$ with its image under the map $T_y i : T_y W \to T_y Y$, where $i : W \to Y$ is the identical inclusion. Clearly, if f is a submersion at x, then $f \pitchfork_x W$ for every submanifold W of Y. If $Z \subset X$ and $f \pitchfork_x W$ for every $x \in Z$, we say that f is transversal to W on Z. Finally, if f is transversal to W on the whole X, we say simply that f is transversal to W and write $f \pitchfork W$.

The next proposition contains a basic property of the notion of transversality, which will be used many times in the following.

Proposition 1.1.1: *Let* $f : X \to Y$ *be a smooth map, and let* W *be a smooth submanifold of* Y *such that* $f \pitchfork W$. *Then* $f^{-1}(W)$ *is a smooth submanifold of* X *with*

$$\operatorname{codim} f^{-1}(W) = \operatorname{codim} W. \tag{1.1}$$

In particular:

(a) *if* $\dim X < \operatorname{codim} W$, *then* $f^{-1}(W) = \emptyset$, *i.e.* $f(X) \cap W = \emptyset$;

(b) *if* $\dim X = \operatorname{codim} W$, *then* $f^{-1}(W)$ *consists of isolated points in* X. ♠

Consequently, if f is a submersion, then for every submanifold W of Y, $f^{-1}(W)$ is a submanifold of X with (1.1). Thus, in this case $f^{-1}(y)$ is a submanifold of X of codimension equal to $\dim Y$ for every $y \in Y$.

Let again X and Y be smooth manifolds and $x \in X$. Given two smooth maps $f, g : X \to Y$, we write $f \sim_x g$ if $d_x f = d_x g$. For an integer $k \geq 2$ write $f \sim_x^k g$ if for the smooth maps

$$df, dg : TX \to TY$$

we have $df \sim_\xi^{k-1} dg$ for every $\xi \in T_x X$. Thus, by induction one defines the relation $f \sim_x^k g$ for all integers $k \geq 1$. Fix for a moment $x \in X$ and $y \in Y$. Denote by $J^k(X,Y)_{x,y}$ the family of all equivalence classes with respect to the relation $\sim_x^k$ in the set of all smooth maps $f : X \to Y$ with $f(x) = y$. Define the *space of k-jets* by

$$J^k(X,Y) = \cup_{(x,y) \in X \times Y} J^k(X,Y)_{x,y}.$$

So, for each k-jet $\sigma \in J^k(X,Y)$ there exist $x \in X$ and $y \in Y$ with $\sigma \in J^k(X,Y)_{x,y}$. We set $\alpha(\sigma) = x$, $\beta(\sigma) = y$, thus obtaining two maps

$$\alpha : J^k(X,Y) \to X, \quad \beta : J^k(X,Y) \to Y, \tag{1.2}$$

called *source and target map*, respectively. Given an arbitrary smooth $f : X \to Y$, define

$$j^k f : X \to J^k(X,Y), \tag{1.3}$$

assigning to each $x \in X$ the equivalence class $j^k f(x)$ of f in $J^k(X,Y)_{x,f(x)}$.

There is a natural structure of a smooth manifold on $J^k(X,Y)$ for every k. We refer the reader to [GG] or [Hir] for its description and main properties. Let us only mention that with respect to this structure for each f the maps (1.2) and (1.3) are smooth.

For a set A and an integer $s \geq 1$ define

$$A^{(s)} = \{(a_1, \ldots, a_s) \in A^s : a_i \neq a_j, \quad 1 \leq i < j \leq s\}.$$

Note that if A is a topological space, then $A^{(s)}$ is an open (dense) subset of the product-space A^s. If $f : A \to B$ is an arbitrary map, define $f^s : A^s \to B^s$ by

$$f^s(a_1, \ldots, a_s) = (f(a_1), \ldots, f(a_s)).$$

Let X and Y be smooth manifolds, and s and k be natural numbers. Consider the map

$$\alpha^s : (J^k(X, Y))^s \to X^s.$$

The open submanifold

$$J^k_s(X, Y) = (\alpha^s)^{-1}(X^{(s)})$$

of $(J^k(X, Y))^s$ is called *s-fold k-jet bundle*. For a smooth $f : X \to Y$, determine the smooth map

$$j^k_s f : X^{(s)} \to J^k_s(X, Y)$$

by

$$j^k_s f(x_1, \ldots, x_s) = (j^k f(x_1), \ldots, j^k f(x_s)).$$

We are now going to define the Whitney C^k topology on the *space* $C^\infty(X, Y)$ *of all smooth maps* of X into Y. Let $k \geq 0$ be an integer and let U be an open subset of $J^k(X, Y)$. Set

$$M(U) = \{ f \in C^\infty(X, Y) : j^k f(X) \subset U \}.$$

The family $\{M(U)\}_U$, where U runs over the open subsets of $J^k(X, Y)$, is a basis for a topology on $C^\infty(X, Y)$, called *the Whitney C^k topology*. The supremum of all Whitney C^k topologies, $k = 0, 1, 2, \ldots$, is called *the Whitney C^∞ topology*. It follows by the latter definition that $f_\lambda \to f$ in the C^∞ topology if and only if $f_\lambda \to f$ in the C^k topology for all $k = 0, 1, 2, \ldots$. Note that if X is not compact (and dim $Y > 0$), then all C^k topologies (including $k = \infty$) do not satisfy the first axiom of countability, and therefore are not metrizable. On the other hand, if X is compact, then all C^k topologies are metrizable with complete metrics.

In this book we always consider $C^\infty(X, Y)$ with the Whitney C^∞ topology. One of the basic facts concerning these spaces, which will be often used, is that for all smooth manifolds X and Y the space $C^\infty(X, Y)$ is a Baire topological space. Recall that a subset R of a topological space Z is called *residual in Z* if R is a countable intersection of open dense subsets of Z. If every residual subset of Z is dense in it, Z is called a *Baire space*.

In the next chapters we consider very often spaces of the form $C^\infty(X, R^n)$, X being a smooth submanifold of R^n for some $n \geq 2$. Let us note that these spaces have a natural structure of Frechet spaces. Moreover, if X is compact, then $C^\infty(X, R^n)$ has a natural structure of a Banach space. Denote by

$$\mathbf{C}(X) = C^\infty_{\text{emb}}(X, \mathbf{R}^n)$$

the subset of $C^\infty(X,\mathbf{R}^n)$ consisting of all smooth embeddings $X \to \mathbf{R}^n$. Then $\mathbf{C}(X)$ is open in $C^\infty(X,\mathbf{R}^n)$ (cf. Chapter II in [Hir]), and therefore it is a Baire topological space with respect to the topology induced by $C^\infty(X,\mathbf{R}^n)$. Finally, notice that for compact X, the space $\mathbf{C}(X)$ has natural structure of a *Banach manifold*. We refer the reader to [Lang] for the definition of Banach manifolds and their main properties.

The following theorem is known as the *multijet transversality theorem*, and will be used many times in the next chapters.

Theorem 1.1.2: *Let X and Y be smooth manifolds, k and s be natural numbers, and let W be a smooth submanifold of $J_s^k(X,Y)$. Then*

$$T_W = \{f \in C^\infty(X,Y) : j_s^k \pitchfork W\}$$

is a residual subset of $C^\infty(X,Y)$. Moreover, if W is compact, then T_W is open in $C^\infty(X,Y)$. ♠

For $s = 1$ this theorem coincides with the Thom transversality theorem.

We conclude this section with a consequence of the Abraham transversality theorem, which will be used in Chapter 4. The main reference here is the book [AbR] of Abraham and Robbin. Now by a smooth manifold we mean a smooth Banach manifold with finite or infinite dimension (cf. [Lang]).

Let $\mathcal{A}$, X and Y be smooth manifolds, and let

$$\rho : \mathcal{A} \to C^\infty(X,Y) \tag{1.4}$$

be a map. The image of $a \in \mathcal{A}$ with respect to ρ will be denoted by ρ_a. Define

$$ev_\rho : \mathcal{A} \times X \to Y \tag{1.5}$$

by $ev_\rho(a,x) = \rho_a(x)$.

The next theorem is a proper case of the *Abraham transversality theorem*.

Theorem 1.1.3: *Let ρ have the form (1.4) and let W be a smooth submanifold of Y.*

(a) *If the map (1.5) is C^1-differentiable and K is a compact subset of X, then*

$$\mathcal{A}_{K,W} = \{a \in \mathcal{A} : \rho_a \pitchfork_x W, \quad x \in K\}$$

is an open subset of $\mathcal{A}$;

(b) *Let $\dim X = n < \infty$, $\operatorname{codim} W = q < \infty$ and let r be a natural number with $r > n - q$. Suppose that the manifolds $\mathcal{A}$, X and Y satisfy the second axiom of countability, the map (1.5) is C^r differentiable and $ev_\rho \pitchfork W$. Then*

$$\mathcal{A}_W = \{a \in \mathcal{A} : \rho_a \pitchfork W\}$$

is a residual subset of $\mathcal{A}$. ♠

1.2. Generalized bicharacteristics

Our aim in this section is to define the generalized bicharacteristics of the *wave operator*

$$\Box = \partial_t^2 - \Delta_x$$

and to present their main properties which will be used throughout the book. Here we use the notation from Section 24 in [H3]. In what follows $\Omega \subset \mathbf{R}^{n+1}$ is a closed domain with smooth boundary $\partial\Omega$.

Given a point on $\partial\Omega$, we choose local coordinates

$$x = (x_1, \ldots, x_{n+1}), \quad \xi = (\xi_1, \ldots, \xi_{n+1})$$

in $T^*\mathbf{R}^{n+1}$ around it such that the boundary $\partial\Omega$ is given by $x_1 = 0$ and Ω is locally defined by $x_1 \geq 0$. We assume that the coordinates ξ_i are those dual to x_i. The coordinates x, ξ can be chosen in such a way that the *principal symbol* of $\Box$ has the form

$$p(x, \xi) = \xi_1^2 - r(x, \xi'),$$

where

$$x' = (x_2, \ldots, x_{n+1}), \quad \xi' = (\xi_2, \ldots, \xi_{n+1}),$$

and $r(x, \xi')$ is homogeneous of order 2 in ξ'. Introduce the sets

$$\Sigma = \{(x, \xi) \in T^*\mathbf{R}^{n+1} \setminus \{0\} : p(x, \xi) = 0\},$$
$$\Sigma_0 = \{(x, \xi) \in T^*\mathbf{R}^{n+1} : x_1 > 0\} \cap \Sigma,$$
$$H = \{(x, \xi) \in \Sigma : x_1 = 0, \ r(0, x', \xi') > 0\},$$
$$G = \{(x, \xi) \in \Sigma : x_1 = 0, \ r(0, x', \xi') = 0\}.$$

The sets Σ, H and G are called *characteristic, hyperbolic* and *glancing*, respectively. Let

$$r_0(x', \xi') = r(0, x', \xi'), \quad r_1(x', \xi') = \frac{\partial r}{\partial x_1}(0, x', \xi').$$

Define the *diffractive* and the *gliding sets* by

$$G_d = \{(x, \xi) \in G : r_1(x', \xi') > 0\},$$
$$G_g = \{(x, \xi) \in G : r_1(x', \xi') < 0\},$$

respectively.

Further, consider the Hamiltonian vector fields

$$H_p = \sum_{j=1}^{n+1} \left(\frac{\partial p}{\partial \xi_j} \cdot \frac{\partial}{\partial x_j} - \frac{\partial p}{\partial x_j} \cdot \frac{\partial}{\partial \xi_j} \right),$$

$$H_{r_0} = \sum_{j=2}^{n+1} \left(\frac{\partial r_0}{\partial \xi_j} \cdot \frac{\partial}{\partial x_j} - \frac{\partial r_0}{\partial x_j} \cdot \frac{\partial}{\partial \xi_j} \right).$$

Notice that

$$d_\xi p(x, \xi) \neq 0 \text{ on } \Sigma,$$
$$d_{\xi'} r_0(x', \xi') \neq 0 \text{ on } G,$$

so H_p and H_{r_0} are not radial on Σ and G, respectively. Next, introduce the sets

$$G^k = \{((x, \xi) \in G : r_1 = H_{r_0}(r_1) = \ldots = H_{r_0}^{k-3}(r_1) = 0\}, \quad k \geq 3,$$
$$G^\infty = \cap_{k=3}^\infty G^k.$$

The above definitions are independent of the choice of the local coordinates. Let us mention that if $\partial\Omega$ is given locally by $\varphi = 0$ and Ω by $\varphi \geq 0$, φ being a smooth function, then:

$$H = \{(x, \xi) \in T^*(\mathbf{R} \times \Omega) : p(x, \xi) = 0, \ H_p \varphi(x, \xi) \neq 0\},$$
$$G = \{(x, \xi) \in T^*(\mathbf{R} \times \Omega) : p(x, \xi) = 0, \ H_p \varphi(x, \xi) = 0\},$$
$$G_d = \{x, \xi) \in G : H_p^2 \varphi(x, \xi) > 0\},$$
$$G_g = \{(x, \xi) \in G : H_p^2 \varphi(x, \xi) < 0\},$$
$$G^k = \{(x, \xi) \in \mathbf{G} : H_p^j \varphi(x, \xi) = 0, \ 0 \leq j < k\}.$$

We define the generalized bicharacteristics of $\square$ using the special coordinates x, ξ chosen above.

Definition 1.2.1: Let I be an open interval in $\mathbf{R}$. A curve

$$\gamma : I \rightarrow \Sigma \tag{1.6}$$

is called a *generalized bicharacteristic* of $\square$ if there exists a discrete subset B of I such that the following conditions hold.

(i) If $\gamma(t_0) \in \Sigma_0 \cup G_d$ for some $t_0 \in I \backslash B$, then γ is differentiable at t_0 and

$$\frac{d}{dt}\gamma(t_0) = H_p(\gamma(t_0)).$$

(ii) If $\gamma(t_0) \in G \backslash G_d$ for some $t_0 \in I \backslash B$, then

$$\gamma(t) = (x_1(t), x'(t), \xi_1(t), \xi'(t))$$

is differentiable at t_0 and

$$\frac{dx_1}{dt}(t_0) = \frac{d\xi_1}{dt}(t_0) = 0, \quad \frac{d}{dt}(x'(t), \ \xi'(t))_{|t=t_0} = H_{r_0}(\gamma(t_0)).$$

(iii) If $t_0 \in B$, then $\gamma(t) \in \Sigma_0$ for all $t \neq t_0$, $t \in I$ with $|t - t_0|$ sufficiently small. Moreover, for

$$\xi_1^\pm(x', \xi') = \pm\sqrt{r_0(x', \xi')},$$

we have

$$\lim_{t \to t_0, \pm(t-t_0)>0} \gamma(t) = (0, x'(t_0), \xi_1^\pm(x'(t_0)), \xi'(t_0)) \in H.$$

This definition does not depend on the choice of the local coordinates. Note that when $\partial\Omega$ is locally given by $\varphi = 0$ and Ω by $\varphi \geq 0$, then the condition (ii) means that if $\gamma(t_0) \in G \backslash G_d$, then

$$\frac{d\gamma}{dt}(\gamma(t_0)) = H_p^G(\gamma(t_0)),$$

where

$$H_p^G = H_p + \frac{H_p^2 \varphi}{H_\varphi^2 p} H_\varphi$$

is the so-called *glancing vector field* on G.

It follows from the above definition that if (1.6) is a generalized bicharacteristic, then the functions $x(t)$, $\xi'(t)$, $|\xi_1(t)|$ are continuous on I, while $\xi_1(t)$ has jump discontinuities at any $t \in B$. The functions $x'(t)$, $\xi'(t)$ are continuously differentiable on I and

$$\frac{dx'}{dt} = -\frac{\partial r}{\partial \xi'}, \quad \frac{d\xi'}{dt} = \frac{\partial r}{\partial x'}. \tag{1.7}$$

Moreover, for $t \in B$, $x_1(t)$ admits left and right derivatives

$$\frac{d^\pm x_1}{dt}(t) = \lim_{\epsilon \to +0} \pm\frac{x_1(t \pm \epsilon) - x_1(t)}{\epsilon} = 2\xi_1(t \pm 0). \tag{1.8}$$

The function $\xi_1(t)$ also has a left and a right derivative. For $\gamma(t) \notin G_g$ we have

$$\frac{d^\pm \xi_1}{dt}(t) = \lim_{\epsilon \to +0} \pm\frac{\xi_1(t \pm \epsilon) - \xi_1(t)}{\epsilon} = \frac{\partial r}{\partial x_1}(x(t), \xi'(t)), \tag{1.9}$$

while for $\gamma(t) \in G_g$,

$$\frac{d^\pm \xi_1}{dt}(t) = 0.$$

Thus, if $\gamma(t)$ remains in a compact set, then the functions $x(t)$, $\xi'(t)$, $\xi_1^2(t)$ and $x_1(t)\xi_1(t)$ satisfy a uniform Lipschitz condition. For the left and right derivatives of $|\xi_1(t)|$ one gets

$$\left|\frac{d^\pm |\xi_1(t)|}{dt}\right| \leq \left|\frac{\partial r}{\partial x_1}(x(t), \xi'(t))\right|. \tag{1.10}$$

Melrose and Sjöstrand [MS2] (see also Section 24 in [H3]) showed that for each $z_0 \in \Sigma$ there exists a generalized bicharacteristic (1.6) of $\square$ with $\gamma(t_0) = z_0$ for some $t_0 \in I$. Since the vector fields H_p and H_p^G are not radial on Σ and G, respectively, such a bicharacteristic γ can be extended for all $t \in \mathbf{R}$. However, in general γ is not unique. We refer the reader to [Tay] or [H3] for an example of two different bicharacteristics passing through one and the same point.

For $\rho \in \Sigma$ denote by $C_t(\rho)$ the set of those $\mu \in \Sigma$ such that there exists a generalized bicharacteristic (1.6) with $0, t \in I$, $\gamma(0) = \rho$, $\gamma(t) = \mu$. In many cases $C_t(\rho)$ is related to a uniquely determined bicharacteristic γ. In the general case it is convenient to introduce the following notion.

Definition 1.2.2: A generalized bicharacteristic

$$\gamma : \mathbf{R} \to \Sigma$$

of $\square$ is called *uniquely extendible* if for each $t \in \mathbf{R}$ the only generalized bicharacteristic (up to a change of the parameter t) passing through $\gamma(t)$ is γ. That is, for $\gamma(0) = \rho$ we have

$$C_t(\rho) = \{\gamma(t)\}, \quad t \in \mathbf{R}.$$

It was proved by Melrose and Sjöstrand [MS1] that if Im $\gamma \subset \Sigma \backslash G^\infty$, then γ is uniquely extendible. If $z_0 = \gamma(t_0) \in H$, then $\gamma(t)$ transversally meets $\partial\Omega$ at $x(t_0)$, $t_0 \in B$, and (iii) holds. For $z_0 \in \Sigma_0 \cup G_d$ we have $\gamma(t) \in \Sigma_0$ for $|t - t_0|$ small enough, while in the case $z_0 \in G_g$ for small $|t - t_0|$, $\gamma(t)$ coincides with the gliding ray

$$\gamma_0(t) = (0, x'(t), 0, \xi'(t)), \tag{1.11}$$

where $(x'(t), \xi'(t))$ is a null bicharacteristic of the Hamiltonian vector field H_{r_0}.

To discuss the local uniqueness of the generalized bicharacteristics, let $\gamma(t) = (x(t), \xi(t))$ be such a bicharacteristic and let $(y'(t), \eta'(t))$ be the solution of the problem

$$\begin{cases} \dfrac{dy'}{dt}(t) = \dfrac{\partial r_0}{\partial \xi'}(y'(t), \eta'(t)), \\[2mm] \dfrac{d\eta'}{dt}(t) = -\dfrac{\partial r_0}{\partial x'}(y'(t), \eta'(t)), \\[2mm] y'(0) = x'(0), \eta'(0) = \xi'(0). \end{cases} \tag{1.12}$$

Then, setting $e(t) = r_1(y'(t), \eta'(t))$, we have the following local description of the generalized bicharacteristic γ.

Proposition 1.2.3: Let $\gamma(0) \in G^3$. If $e(t)$ *increases for small* $t > 0$, *then for such* t *the bicharacteristic* $\gamma(t)$ *is a trajectory of* H_p. *If* $e(t)$ *decreases for* $0 \le t \le T$, *then for such* t, $\gamma(t)$ *is a gliding ray of the form (1.11).* ♠

The reader may consult Section 24.3 in [H3] for a proof of this proposition and for other properties of the generalized bicharacteristics.

Let us mention that for $k \geq 3$ and $\gamma(0) \in G^k \backslash G^{k+1}$ we have

$$e(t) = \frac{1}{2(k-2)!} H_p^k \varphi(\gamma(0)) t^{k-2} + O(t^{k-1}),$$

therefore the sign of $H_p^k \varphi(\gamma(0))$ determines the local behaviour of $e(t)$.

Corollary 1.2.4: *In each of the following cases every generalized bicharacteristic of $\Box$ is uniquely extendible:*

(a) *the boundary $\partial \Omega$ is a real analytic manifold;*

(b) *there are no points $y \in \partial \Omega$ at which the normal curvature of $\partial \Omega$ vanishes of infinite order with respect some direction $\xi \in T_y \partial \Omega$;*

(c) *$\partial \Omega$ is given locally by $\varphi = 0$ and*

$$H_p^2 \varphi(z) \leq 0 \qquad (1.13)$$

for every $z \in G$. If Ω is locally convex in the domain of φ, then (1.13) holds.

Proof: In case (a) the symbols $r_0(x', \xi')$ and $r_1(x', \xi')$ are real analytic, so the solution $(y'(t), \eta'(t))$ of (1.12) is analytic in t. Consequently, the function $e(t)$ is analytic and we can use its Taylor expansion in order to apply Proposition 1.2.3.

In case (c), working in the special coordinates x, ξ, combining (1.13) with (1.9), we find

$$\frac{d^{\pm} \xi_1}{dt}(t) \geq 0.$$

On the other hand, if $\xi_1(t)$ has a jump at $\gamma(t) \in H$, then this jump is equal to $2r_0(x'(t), \xi'(t)) > 0$. Thus, the function $\xi_1(t)$ is increasing. If $e(t) = 0$ for $0 \leq t \leq t_0$, we get $x_1(t) = \xi_1(t) = 0$ for such t, so $\{\gamma(t) : 0 \leq t \leq t_0\}$ is a gliding ray. Assume there exists a sequence $t_k \searrow 0$ such that $e(t_k) \neq 0$. Then $\xi_1(t) > 0$ for all sufficiently small $t > 0$. Now (1.8) shows that $x_1(t)$ is increasing for such t, therefore there is $t_1 > 0$ such that $\gamma(t)$, $0 \leq t \leq t_1$, coincides with a trajectory of H_p.

Let $p = \sum_{j=1}^n \xi_j^2 - \xi_{n+1}^2$ and let φ depend on $x_1, \ldots, x_n$ only. Then

$$(H_p^2 \varphi)(x, \xi) = 4 \sum_{i,j=1}^n \frac{\partial^2 \varphi}{\partial x_i \partial x_j}(x) \xi_i \xi_j,$$

and if the boundary $\partial \Omega$ is locally convex, we obtain (1.13).

Finally, in the case (b), for each $x \in \partial \Omega$ there exists a multiindex α, depending on x, such that $(\partial_x^\alpha \varphi)(x) \neq 0$. This implies that $G^\infty = \emptyset$, which completes the proof. ♠

According to Lemma 7.1.2, in the generic case, discussed in Chapters 7 and 9, assumption (b) is always satisfied.

Let $Q = \Omega \times \mathbf{R}$. We shall use again the coordinates $x = (x_1, \ldots, x_{n+1})$, denoting this time the last coordinate by t, i.e. we set $t = x_{n+1}$. For $x \in \partial Q =$

$\partial\Omega \times \mathbf{R}$, let $N_x(\partial Q)$ be the space of covectors $\xi \in T_x^* Q$ vanishing on $T_x(\partial Q)$. Define the equivalence relation $\sim$ on $T^* Q$ by $(x, \xi) \sim (y, \eta)$ if and only if either $x = y \in Q \backslash \partial Q$ and $\xi = \eta$ or $x = y \in \partial Q$ and $\xi - \eta \in N_x(\partial Q)$. Then $T^* Q / \sim$ over ∂Q can be naturally identified with $T^*(\partial Q)$. Consider the map

$$\sim: T^* Q \ni (x, \xi) \mapsto (x, \xi_{|T_x(\partial Q)}) \in T^*(\partial Q),$$

defined as the identity on $T^*(Q \backslash \partial Q)$. Then

$$\tilde{\Sigma} = \Sigma_b$$

is called *the compressed characteristic set*, while the image $\tilde{\gamma}$ under $\sim$ of a bicharacteristic γ is called *compressed generalized bicharacteristic*. Clearly, $\tilde{\gamma}$ is a continuous curve in Σ_b.

Given $\rho = (x, \xi)$, $\mu = (y, \eta) \in T^* Q$, denote by $d(\rho, \mu)$ the standard Euclidean distance between ρ and μ. For $\rho, \mu \in \Sigma$ define

$$D(\rho, \mu) = \inf_{\nu', \nu'' \in \Sigma, \nu' \sim \nu''} (\min\{d(\rho, \mu), d(\rho, \nu') + d(\nu'', \mu)\}).$$

Clearly, $D(\rho, \mu) = 0$ if and only if $\rho \sim \mu$ and $D(\rho, \mu) = D(\rho', \mu')$, provided $\rho \sim \rho'$ and $\mu \sim \mu'$. It is easy to check that D is symmetric and satisfies the triangle inequality. Thus, D is a pseudo-metric on Σ which induces a metric on Σ_b.

For the next lemma we assume that I is a closed non-trivial interval in $\mathbf{R}$, $(y_0, \eta_0) \in \Sigma$ and Γ is a neighbourhood of (y_0, η_0) in Q.

Lemma 1.2.5: *There exists a constant $C_0 > 0$, depending only on Γ and I, such that for each generalized bicharacteristic $\gamma : I \to \Sigma \cap \Gamma$ we have*

$$D(\gamma(t), \gamma(s)) \leq C_0 |t - s|, \quad t, s \in I.$$

Proof: It is sufficient to examine only the case when $|t - s|$ is small. Then we can use the local coordinates introduced above. From Equations (1.7), (1.8) and (1.10) we find

$$|x(t) - x(s)| + |\xi'(t) - \xi'(s)| \leq C_1 |t - s|, \quad ||\xi_1(t)| - |\xi_1(s)|| \leq C_1 |t - s|,$$

where $C_1 > 0$ does not depend on t and s. Thus, if $\xi_1(t) = 0$ or $\xi_1(s) = 0$, we get

$$|\xi_1(t) - \xi_1(s)| \leq C_1 |t - s|.$$

The last inequality holds also in the case when $\gamma(t') \notin \partial\Omega$ for $t < t' < s$. Consequently,

$$D(\gamma(t), \gamma(s)) \leq C_2 |t - s|$$

whenever either $\xi_1(t)\xi_1(s) = 0$ or $\gamma(t') \in \partial\Omega$ only for finitely many $t' \in (t, s)$. Assume that there are infinitely many $t' \in (t, s)$ such that $\gamma(t')$ is a reflection point of γ. Then there exists $u \in [t, s]$ with $\gamma(u) \in G$. Hence

$$D(\gamma(t), \gamma(u)) \leq C_2 |t - u|, \quad D(\gamma(u), \gamma(t)) \leq C_2 |u - s|,$$

and, applying the triangle inequality for D, we complete the proof of the assertion. ♠

The next lemma shows that any sequence of generalized bicharacteristics has a subsequence which is convergent on a given compact interval.

Lemma 1.2.6: *Let* $I = [a, b]$ *be a compact interval in* **R**, K *be a compact subset of* Σ *and let for every natural* k,

$$\gamma^{(k)} : I \rightarrow K \subset \Sigma$$

be a generalized bicharacteristic of □. *Then there exist an infinite sequence* $k_1 < k_2 < \ldots$ *of natural numbers and a generalized bicharacteristic*

$$\gamma : [a, b] \rightarrow \Sigma$$

such that

$$\lim_{m \to \infty} D(\gamma^{(k_m)}(t), \gamma(t)) = 0 \tag{1.14}$$

for every $t \in [a, b]$.

Proof: Using local coordinates, we see that the derivatives of $(x^{(k)})'$, $(\xi^{(k)})'$ and the left and right derivatives of $x_1^{(k)}$ and $\xi_1^{(k)}$ are uniformly bounded for $t \in [a, b]$ and $k \in$ **N**. Hence the maps $x^{(k)}(t)$, $(\xi^{(k)})'(t)$ and $(\xi_1^{(k)}(t))^2$ are uniformly Lipschitzian, which implies that there exists an infinite sequence $k_1 < k_2 < \ldots$ of natural numbers such that $x^{(k_m)}(t)$, $(\xi^{(k_m)})'(t)$, $(\xi_1^{(k_m)}(t))^2$, $x_1^{(k_m)}(t)\xi_1^{(k_m)}(t)$ are uniformly convergent for $t \in [a, b]$. According to Proposition 24.3.12 in [H3], there exists a generalized bicharacteristic $\gamma : [a, b] \rightarrow \Sigma$ of □ such that

$$\lim_m \gamma^{(k_m)}(t) = \gamma(t) \tag{1.15}$$

for each $t \in [a, b]$ with $\gamma(t) \notin H$.

Let $t' \in [a, b]$ be such that $\gamma(t')$ is a reflection point of γ. Then there exists a sequence $t_j \rightarrow t'$ with $\gamma(t_j) \in \Sigma_0 \cup G$ for all j. Then

$$D(\gamma^{(k_m)}(t'), \gamma(t')) \leq D(\gamma^{(k_m)}(t'), \gamma^{(k_m)}(t_j))$$
$$+ D(\gamma(t_j), \gamma(t')) + D(\gamma^{(k_m)}(t_j), \gamma(t_j)).$$

By Lemma 1.2.5, the first two terms at the right-hand side can be estimated uniformly with respect to m, while for the third term we can apply (1.15). Taking j and m sufficiently large, we obtain (1.14), which proves the assertion. ♠

In what follows we use local coordinates $(t, x) \in \mathbf{R} \times \Omega$ and the corresponding local coordinates $(t, x; \tau, \xi)$ in $T^*(\mathbf{R} \times \Omega)$. In these coordinates the principal symbol p of $\square$ has the form

$$p(x, \tau, \xi) = \xi_1^2 - q_2(x, \xi') - \tau^2,$$

where $\xi' = (\xi_2, \ldots, \xi_n)$ and $q_2(x, \xi')$ is homogeneous of order 2 in ξ'. Consequently, the vector fields H_p and H_p^G do not involve derivatives with respect to τ, so by Definition 1.2.1, the variable τ remains constant along each generalized bicharacteristic. This makes it possible to parametrize the generalized bicharacteristics by the time t.

Given $(y, \eta) \in T^*(\Omega) \backslash \{0\}$, consider the points

$$\mu_\pm = (0, y, \mp |\eta|, \eta) \in \Sigma.$$

Assume that locally $\partial \Omega$ is given by $x_1 = 0$ and Ω by $x_1 \geq 0$. Let μ_+ be a hyperbolic point and let $\xi_1^\pm(y', \eta)$ be the different real roots of the equation

$$p(0, y', |\eta|, z, \eta') = 0$$

with respect to z. Denote by γ the generalized bicharacteristic, parametrized by a parameter s, such that

$$\lim_{s \searrow 0} \gamma(s) = \mu_+.$$

Then $\tau = -|\eta| < 0$ along γ and the time t increases when s increases. Such bicharacteristics will be called *forward*. For the right derivative of $x_1(t)$ we get

$$\frac{d^+ x_1}{dt} = \frac{d^+ x_1/ds}{dt/ds} = \frac{\xi_1(+0)}{-\tau} > 0,$$

since for small $t > 0$, $\gamma(t)$ enters in Ω and $x_1(t) > 0$. Therefore, setting

$$\xi_1^\pm(y', \eta) = \pm \sqrt{|\eta|^2 + q_2(0, y', \eta')},$$

we find

$$\lim_{s \searrow 0} \xi_1(s) = \xi_1^+(y', \eta).$$

In the case $\mu_+ \in G$, it may happen that there exist several forward bicharacteristics passing through μ_+. This gives rise to the following notation. Denote by C_+ the set of those

$$(t, x, y; \tau, \xi, \eta) \in T^*(\mathbf{R} \times \Omega \times \Omega) \backslash \{0\}$$

such that $\tau = -|\xi| = -|\eta|$ and (t, x, τ, ξ) and $(0, y, \tau, \eta)$ lie on a forward generalized bicharacteristic of $\square$. In a similar way we define C_- using *backward bicharacteristics*, determined as the forward ones replacing μ_+ by μ_-. The set $C = C_+ \cup C_-$ is called *the bicharacteristic relation* of $\square$. If $\mu = (0, y, \tau, \eta) \in H$ and $\tau < 0$ (resp. $\tau > 0$), we shall say that μ is a reflection point of a forward (resp. backward) generalized bicharacteristic. Similarly, if $\rho = (t, x, \tau, \xi) \in H$, then ρ is a reflection point of a generalized bicharacteristic passing through $(0, y, \tau, \eta)$, and, working in local coordinates as above, the sign of τ determines uniquely $\xi_1(t+0)$. The sets $C_\pm$ and C are homogeneous with respect to (τ, ξ, η), i.e. $(t, x, y, \tau, \xi, \eta) \in C_\pm$ implies $(t, x, y, s\tau, s\xi, s\eta) \in C_\pm$ for all $s \in \mathbf{R}^+$.

Lemma 1.2.7: $C_\pm$ *are closed in* $T^*(\mathbf{R} \times \Omega \times \Omega) \backslash \{0\}$.

Proof: Since C_+ is homogeneous, it is sufficient to show that if

$$C_+ \ni z_k = (t_k, x_k, y_k, -1, \xi_k, \eta_k), \quad |\xi_k| = |\eta_k| = 1$$

for every integer $k \geq 1$ and if there exists

$$\lim_{k \to \infty} z_k = z_0 = (t_0, x_0, y_0, -1, \xi_0, \eta_0),$$

then $z_0 \in C_+$. Let $\gamma^{(k)}(t)$ be a generalized bicharacteristic of $\Box$ such that $(t_k, x_k, -1, \xi_k)$ and $(0, y_k, -1, \eta_k)$ lie on Im $\gamma^{(k)}$. If one of these points belongs to H, we consider it as a reflection point of $\gamma^{(k)}$, according to the above convention choosing suitably $\xi_1^{(k)}(t)$. Assume $|t_k| \leq T$. Then there exists a compact $K \subset \Sigma$ such that $\gamma^{(k)}(t) \in K$ for all $|t| \leq T$, so we can apply the argument in the proof of Lemma 1.2.6. Consequently, there exists a sequence $k_1 < k_2 < \ldots$ of natural numbers and a generalized bicharacteristic γ satisfying (1.14) and (1.15). Then for the Euclidean distance d we find

$$d(\gamma^{(k_m)}(t_{k_m}), \gamma(t_0)) \leq d(\gamma^{(k_m)}(t_{k_m}), \gamma^{(k_m)}(t_0)) + d(\gamma^{(k_m)}(t_0), \gamma(t_0)).$$

If $\gamma(t_0) \in \Sigma_0 \cup G$, according to (1.15) and the continuity of $x(t)$, $\xi'(t)$, $|\xi_1(t)|$, we get

$$d(\gamma^{(k_m)}(t_{k_m}), \gamma(t_0)) \to_{m \to \infty} 0, \qquad (1.16)$$

which shows that $z_0 \in C_+$. Let $\gamma(t_0) \in H$, it follows by our convention that $\xi_1(t+0)$ and $\xi_1^{k_m}(t+0)$ have the same sign for large m, which implies $z_0 \in C_+$.

Therefore C_+ is closed. In the same way one proves that C is closed, too. ♠

Using C_+, we now define the so called *generalized Hamiltonian flow* Φ^t of $\Box$; it is sometimes called *broken Hamiltonian flow*. Given $(y, \eta) \in T^*\Omega \backslash \{0\}$, set

$$\Phi^t(y, \eta) = \{(x, \xi) \in T^*\Omega \backslash \{0\} : (t, x, y, -|\eta|, \xi, \eta) \in C_+\}.$$

In general, $\Phi^t(y, \eta)$ is not a one-point set. Nevertheless, setting

$$\Phi^t(V) = \{\Phi^t(y, \eta) : (y, \eta) \in V\}$$

for $V \subset T^*\Omega \backslash \{0\}$, we have that the group property

$$\Phi^{t+s}(y, \eta) = \Phi^t(\Phi^s(y, \eta))$$

holds. The flow, generated by C_-, is in fact $\Phi^t(y, -\eta)$.

Let $\partial\Omega$ be locally given by $x_1 = 0$ and let

$$p(x, \tau, \xi) = \xi_1^2 - q_2(x, \xi') - \tau^2$$

be the principal symbol of $\square$. A point

$$\sigma = (t, x', \tau, \xi') \in T^*(\mathbf{R} \times \partial\Omega)\backslash\{0\}$$

is called *hyperbolic* (resp. *glancing*) for $\square$, if the equation

$$p(0, x', \tau, \xi_1, \xi') = 0 \tag{1.17}$$

with respect to ξ_1 has two different real roots (resp. a double real root). These definitions are invariant with respect to the choice of the local coordinates. If (1.17) has no real roots, the point σ is called *elliptic*. Clearly, the set of the hyperbolic points is open in $T^*(\mathbf{R} \times \partial\Omega)$, while that of the glancing points is closed.

Let

$$\pi : T^*(\mathbf{R} \times \Omega) \rightarrow \Omega$$

be the *natural projection*, $\pi(t, x, \tau, \xi) = x$.

Definition 1.2.8: A continuous curve

$$g : [a, b] \rightarrow \Omega$$

is called a *generalized geodesic* in Ω if there exists a generalized bicharacteristic

$$\gamma : [a, b] \rightarrow \Sigma$$

such that

$$g(t) = \pi(\gamma(t)), \quad t \in [a, b]. \tag{1.18}$$

Notice that in general a generalized geodesic is not uniquely determined by a point on it and the corresponding direction. If the generalized bicharacteristic γ with (1.18) satisfies

$$\gamma(t) \in \Sigma_0 \cup H, \quad t \in [a, b],$$

we shall say that g (or Im g) is a *reflecting ray* in Ω. Two special kinds of such rays will be studied in detail in Chapter 2. One of them is defined as follows.

Definition 1.2.9: A point $(x, \xi) \in T^*\Omega\backslash\{0\}$ is called *periodic* with *period* $T \neq 0$ if

$$(T, x, x, \pm|\xi|, \xi, \xi) \in C.$$

A generalized bicharacteristic $\gamma(t) = (t, x(t), \tau, \xi(t)) \in \Sigma, t \in \mathbf{R}$, will be called *periodic* with *period* $T \neq 0$ if for each $t \in \mathbf{R}$ the point $(x(t), \xi(t))$ is periodic with period T. The projections on Ω of the periodic generalized bicharacteristics of $\square$ are called *periodic generalized geodesics*.

Notice that if $(T, x, x, -|\xi|, \xi, \xi) \in C_+$, then $(T, x, x, |\xi|, -\xi, -\xi) \in C_-$, since we may change the orientation on the bicharacteristic, passing through $(0, x, -|\xi|, \xi)$. A uniquely extendible bicharacteristic γ is periodic provided $Im\gamma$ contains a periodic point. If T is the period of a generalized geodesic g, then $|T|$ coincides with the standard length of the curve Im g.

Let $\mathcal{L}_\Omega$ be the set of *all periodic generalized geodesics* in Ω. For $g \in \mathcal{L}_\Omega$ we denote by T_g the length of Im g. Set

$$L_\Omega = \{T_g : g \in \mathcal{L}_\Omega\}.$$

Lemma 1.2.10:   *The set L_Ω is closed in* **R**.

Proof: Consider a sequence

$$L_\Omega \ni T_k \underset{k \to \infty}{\longrightarrow} T_0.$$

Then for every k there exists a generalized bicharacteristic $\gamma^{(k)}$ of $\square$ with period T_k, passing through a point of the form $(0, x_k, -1, \xi_k)$. If $T_0 \neq 0$, choosing a subsequence as in the proof of Lemma 1.2.7, we obtain $T_0 \in L_\Omega$.

It remains to show that the case $T_0 = 0$ is impossible. Assume $T_0 = 0$. Passing to an appropriate subsequence, we may assume that there exists $\lim_k (x_k, \xi_k) = (x_0, \xi_0)$ and for every t there exists

$$\lim_k \gamma^{(k)}(t) = \lim_k (t, x^{(k)}(t), -1, \xi^{(k)}(t)) = \gamma_0(t) = (t, x_0(t), -1, \xi_0(t)),$$

provided $\gamma_0(t) \notin H$ and $|t| \leq T$. If $x_0 \in \Omega^\circ$, then $x_k \in \Omega^\circ$ for large k. For such a k, then $x^{(k)}(t) \in \Omega^\circ$ for all sufficiently small $t > 0$, which is a contradiction. If there exists t' with $|t'| \leq T$ and $x_0(t') \in \Omega^\circ$ we obtain a contradiction by the same argument.

It remains to consider the case when $\gamma_0(t) \in G$ for all $t \in [-T, T]$. Then for such t, $\gamma_0(t)$ is an integral curve of the glancing vector field H_p^G. Since the latter is not radial, γ_0 has no stationary points for $t \in [-T, T]$. Given a small neighbourhood U of x_0 in $\partial\Omega$, there exist δ_0, δ_1 such that $0 < \delta_0 < \delta_1 \leq T$ and $x_0(t) \notin U$ for $\delta_0 \leq |t| \leq \delta_1$. Since $x^{(k)}(t) \underset{k \to \infty}{\longrightarrow} x_0(t)$ uniformly for $|t| \leq T$, for every sufficiently large k there exists a natural number m_k with

$$\delta_0 \leq m_k T_k \leq \delta_1, \quad x^{(k)}(T_k) = x^{(k)}(m_k T_k).$$

Then

$$x_0 = \lim_k x_k = \lim_k x^{(k)}(T_k) \notin U,$$

which is a contradiction. In this way we have established that $T_0 \neq 0$, and this completes the proof of the assertion. ♠

1.3. Wave front sets of distributions

In this section we collect some basic facts concerning wave fronts of distributions. For more details we refer the reader to the books of Hörmander [H1, H3].

Let $X \subset \mathbf{R}^n$ be an open set and let $\mathcal{D}'(X)$ be the *space of all distributions* on X. The *singular support* sing supp u of $u \in \mathcal{D}'(X)$ is a closed subset of X such that if $x_0 \notin$ sing supp u there exists an open neighbourhood U of x_0 in X and a smooth function $f \in C^\infty(U)$ such that

$$\langle u, \varphi \rangle = \int f(x)\varphi(x)\, \mathrm{d}x, \quad \varphi \in C_0^\infty(U).$$

To make a more precise analysis of sing supp u, it is useful to consider the directions $\xi \in \mathbf{R}^n \backslash \{0\}$ for which the *Fourier transform* $\widehat{\varphi u}(\xi)$ of the distribution $\varphi u \in \mathcal{E}'(X)$ is not rapidly decreasing, provided $\varphi \in C_0^\infty(X)$ and supp $\varphi \cap$ sing supp $u \neq \emptyset$.

Definition 1.3.1: Let $u \in \mathcal{D}'(X)$ and let $\mathcal{O}$ be the set of all $(x_0, \xi_0) \in X \times \mathbf{R}^n \backslash \{0\}$ for which there exists an open neighbourhood U of x_0 in X and an open conic neighbourhood V of ξ_0 in $\mathbf{R}^n$ so that for $\varphi \in C_0^\infty(U)$ and $\xi \in V$ we have

$$|\widehat{\varphi u}(\xi)| \le C_m(1 + |\xi|)^{-m}, \quad m \in \mathbf{N}.$$

The closed subset

$$WF(u) = (X \times (\mathbf{R}^n \backslash \{0\})) \backslash \mathcal{O}$$

of $X \times (\mathbf{R}^n \backslash \{0\})$ is called the *wave front set* of u.

It is easy to see that $WF(u)$ is a conic subset of $X \times (\mathbf{R}^n \backslash \{0\})$ having the property

$$\pi(WF(u)) = \text{sing supp } u,$$

$\pi : X \times \mathbf{R}^n \to X$ being the natural projection.

For our aims in Chapter 5 we shall describe the wave front sets of distributions given by oscillatory integrals. Such integrals have the form

$$\int e^{\mathrm{i}\varphi(x,\theta)} a(x,\theta)\, \mathrm{d}\theta. \tag{1.19}$$

Here the *phase* $\varphi(x, \theta)$ is a C^∞ real-valued function, defined for $(x, \theta) \in \Gamma \subset X \times (\mathbf{R}^N \backslash \{0\})$, and Γ is an open conic set, i.e. $(x, \theta) \in \Gamma$ implies $(x, t\theta) \in \Gamma$ for all $t > 0$. We assume that φ has the properties:

$$\varphi(x, t\theta) = t\varphi(x, \theta), \quad (x, \theta) \in \Gamma, \ t > 0,$$
$$\mathrm{d}_{x,\theta}\varphi(x, \theta) \neq 0, \quad (x, \theta) \in \Gamma.$$

The *amplitude* $a(x,\theta)$ belongs to the class of symbols $S^m(X \times \mathbf{R}^N)$, formed by C^∞ functions on $X \times \mathbf{R}^N$ such that for each compact $K \subset X$ and all multiindices α, β we have

$$\left| \partial_x^\alpha \partial_\theta^\beta a(x,\theta) \right| \le C_{\alpha,\beta,K}(1+|\theta|)^{m-|\beta|}, \quad x \in K, \ \theta \in \mathbf{R}^N. \qquad (1.20)$$

We endow $S^m(X \times \mathbf{R}^N)$ with the topology, defined by the seminorms

$$p_{\alpha,\beta,j}(a) = \sup_{x \in K_j, \theta \in \mathbf{R}^N} (1+|\theta|)^{-m+|\beta|} \left| \partial_x^\alpha \partial_\theta^\beta a(x,\theta) \right|,$$

where $\{K_j\}$ is an increasing sequence of compact sets with $\cup_{i=1}^\infty K_j = X$.

Let $F \subset \Gamma \cup (X \times \{0\})$ be a closed cone and let supp $a \subset F$. For $\psi \in C_0^\infty(X)$ we wish to define the integral

$$\iint e^{i\varphi(x,\theta)} a(x,\theta)\psi(x)\,\mathrm{d}x\mathrm{d}\theta$$

in order to obtain a distribution in $\mathcal{D}'(X)$. To do this, we need a regularization, since the integral in θ is not convergent for $m > -N$.

Choose a function $\chi \in C_0^\infty(\mathbf{R}^N)$ such that $\chi(\theta) = 1$ for $|\theta| \le 1$, $\chi(\theta) = 0$ for $|\theta| \ge 2$. For $0 < \epsilon \le 1$ the functions $\chi(\epsilon\theta)$ form a bounded subset of $S^0(X \times \mathbf{R}^N)$. Then the functions $a_\epsilon = a(x,\theta)\chi(\epsilon\theta)$ also form a bounded set in $S^0(X \times \mathbf{R}^N)$ and

$$a_\epsilon \to_{\epsilon \to 0} a \quad \text{in } S^{m'}(X \times \mathbf{R}^N)$$

for each $m' > m$.

Consider the operator

$$L = \sum_{j=1}^n a_j \frac{\partial}{\partial x_j} + \sum_{j=1}^N b_j \frac{\partial}{\partial \theta_j} + \chi$$

with

$$a_j = -i(1-\chi)\kappa^{-1}\varphi_{x_j}, \quad b_j = -i(1-\chi)\kappa^{-1}|\theta|^2 \varphi_{\theta_j},$$

and $\kappa = |\varphi_x|^2 + |\theta|^2 |\varphi_\theta|^2$. For each compact set $K \subset X$ we have

$$\kappa(x,\theta) \ge \delta_k |\theta|^2, \quad x \in K, \ (x,\theta) \in \Gamma,$$

where $\delta_k > 0$ depends only on K. Clearly,

$$L(e^{i\varphi}) = e^{i\varphi},$$

and the operator ^tL formally adjoint to L has the form

$$^tL = -\sum_{j=1}^{n} a_j \frac{\partial}{\partial x_j} - \sum_{j=1}^{N} b_j \frac{\partial}{\partial \theta_j} + c$$

with

$$a_j \in S^{-1}(X \times \mathbf{R}^N), \quad b_j \in S^{\circ}(X \times \mathbf{R}^N), \quad c \in S^{-1}(X \times \mathbf{R}^N).$$

The operator $^t(L)^k$ is a continuous map of S^m into S^{m-k}. Define the linear map $I_{\varphi,a} : C_0^{\infty}(X) \to \mathbf{R}$ by

$$I_{\varphi,a}(\psi) = \lim_{\epsilon \to 0} \iint e^{i\varphi(x,\theta)} a(x,\theta) \chi(\epsilon\theta) \psi(x) \, dx d\theta$$

$$= \lim_{\epsilon \to 0} \iint e^{i\varphi(x,\theta)} (^tL)^k [a(x,\theta) \chi(\epsilon\theta) \psi(x)] \, dx d\theta. \quad (1.21)$$

For $m - k < -N$ the integral on the right-hand side of (1.21) is absolutely convergent, and it is easy to see that $I_{\varphi,a}$ becomes a distribution in $\mathcal{D}'(X)$. Thus, we obtain the following.

Proposition 1.3.2: *Let $\varphi(x,\theta)$ and $a(x,\theta)$ be as above. Then the oscillatory integral (1.19) defines a distribution $I_{\varphi,a}$ given by (1.21).* ♠
  We are going to describe $WF(I_{\varphi,a})$.

Theorem 1.3.3: *We have*

$$WF(I_{\varphi,a}) \subset \{(x, \varphi_x(x,\theta)) : (x,\theta) \in F, \varphi_\theta(x,\theta) = 0\}. \quad (1.22)$$

Proof: Let $f \in C_0^{\infty}(X)$. Then the Fourier transform

$$\widehat{fI_{\varphi,a}}(\xi) = \iint e^{i(\varphi(x,\theta) - \langle x,\xi \rangle)} a(x,\theta) f(x) \, dx d\theta$$

is expressed by an oscillatory integral. Let $V \subset \mathbf{R}^N$ be a closed cone such that

$$V \cap \{\varphi_x(x,\theta) : (x,\theta) \in F, \ x \in \text{supp} f, \ \varphi_\theta(x,\theta) = 0\} = \emptyset.$$

By compactness, there exists $\delta > 0$ such that

$$\mu = |\xi - \varphi_x(x,\theta)|^2 + |\theta|^2 |\varphi_\theta(x,\theta)|^2 \geq \delta(|\theta| + |\xi|)^2 \quad (1.23)$$

for $(x,\theta) \in F, x \in \text{supp} f, \xi \in V$. To obtain (1.23), it suffices to observe that if the latter conditions are satisfied, then the left-hand side of (1.23) is positive and to use the homogeneity with respect to (θ, ξ). As above, consider the operator

$$\mathcal{L} = \sum_{j=1}^{n} a_j \frac{\partial}{\partial x_j} + \sum_{j=1}^{N} b_j \frac{\partial}{\partial \theta_j} + \chi$$

with

$$a_j = -\frac{\mathrm{i}(1-\chi)}{\mu}(\varphi_{x_j} - \xi_j), \quad b_j = -\frac{\mathrm{i}(1-\chi)}{\mu}|\theta|^2 \varphi_{\theta_j}.$$

Then

$$\widehat{fI_{\varphi,a}}(\epsilon) = \lim_{\epsilon \to 0} \iint e^{\mathrm{i}(\varphi(x,\theta) - \langle x, \xi \rangle)}({}^t\mathcal{L})^k[a(x,\theta)\chi(\epsilon\theta)f(x)]\,\mathrm{d}x\mathrm{d}\theta,$$

and applying (1.23) we conclude that

$$\left|\widehat{fI_{\varphi,a}}(\xi)\right| \le C_N(1+|\xi|)^{-N}, \quad \xi \in V.$$

This implies (1.22). ♠

For the asymptotics of oscillatory integrals depending on a parameter $\lambda \in \mathbf{R}$, we have the following.

Lemma 1.3.4: *Let* $u \in \mathcal{D}'(X)$, $f \in C_0^\infty(X)$ *and let* $\varphi \in C^\infty(X)$ *be a real-valued function. Assume*

$$WF(u) \cap \{(x, \varphi_x) : x \in \operatorname{supp} f\} = \emptyset.$$

Then for each $m \in \mathbf{N}$ *we have*

$$\left|\langle u, f(x)e^{-\mathrm{i}\lambda\varphi(x)}\rangle\right| \le C_m(1+|\lambda|)^{-m}, \quad \lambda \in \mathbf{R}.$$

Proof: Choosing a finite partition of unity, we can restrict our attention to the case $u \in \mathcal{E}'(X)$. Put

$$\Sigma_f = \{\xi \in \mathbf{R}^n \backslash \{0\} : \exists x \in \operatorname{supp} f \text{ with } (x, \xi) \in WF(u)\}.$$

Then

$$\langle u, f(x)e^{-\mathrm{i}\lambda\varphi(x)}\rangle = (2\pi)^{-n} \iint e^{\mathrm{i}(\langle x, \xi \rangle - \lambda\varphi(x))}f(x)\hat{u}(\xi)\,\mathrm{d}x\mathrm{d}\xi$$

$$= \int_X \int_W + \int_X \int_{\mathbf{R}^n \backslash W} = I_1(\lambda) + I_2(\lambda).$$

Here W is a closed conic set such that $\Sigma_f \subset W$,

$$W \cap \{\varphi_x(x) \,:\, x \in \operatorname{supp} f\} = \emptyset,$$

and $I_1(\lambda)$ is interpreted as an oscillatory integral. For $x \in \operatorname{supp} f$, $\xi \in W$ we have

$$|\xi - \lambda \varphi_x(x)| \geq \delta(|\xi| + |\lambda|), \quad \lambda \in \mathbf{R}$$

with $\delta > 0$. Using the same argument as in the proof of Theorem 1.3.3, we see that $I_1(\lambda) = O(|\lambda|^{-m})$ for all $m \in \mathbf{N}$. For $I_2(\lambda)$ we use the fact that if $\xi \in \mathbf{R}^n \backslash W$ and $\operatorname{supp} u \cap \operatorname{supp} f \neq \emptyset$, then $\hat{u}(\xi)$ is rapidly decreasing. This proves the assertion. ♠

Now let $\Gamma \subset X \times (\mathbf{R}^n \backslash \{0\})$ be a closed conic set. Introduce the space

$$\mathcal{D}'_\Gamma(X) = \{u \in \mathcal{D}'(X) \,:\, WF(u) \subset \Gamma\}.$$

Using an argument, similar to that in the proof of Lemma 1.3.4, it is easy to see that $u \in \mathcal{D}'_\Gamma(X)$ if and only if for each $\varphi \in C_0^\infty(X)$ and each closed cone $V \subset \mathbf{R}^n$ with

$$(\operatorname{supp} \varphi \times V) \cap \Gamma = \emptyset, \tag{1.24}$$

we have

$$\sup_{\xi \in V} |\xi|^m \, |\widehat{\varphi u}(\xi)| < \infty, \quad m \in \mathbf{N}.$$

This makes it possible to introduce the following.

Definition 1.3.5: Let $\{u_j\}_j \subset \mathcal{D}'_\Gamma(X)$ and $u \in \mathcal{D}'_\Gamma(X)$. We say that $\{u_j\}$ converges to u in $\mathcal{D}'_\Gamma(X)$ if:

(a) $u_j \to u$ weakly in $\mathcal{D}'(X)$,

(b) $\sup_{j \in \mathbf{N}} \sup_{\xi \in V} |\xi|^m \, |\widehat{\varphi u_j}(\xi)| < \infty$ for all $m \in \mathbf{N}$, each $\varphi \in C_0^\infty(X)$ and each closed cone V satisfying (1.24).

For every $u \in \mathcal{D}'_\Gamma(X)$ there exists a sequence $\{u_j\} \subset C_0^\infty(X)$ converging to u in $\mathcal{D}'_\Gamma(X)$. To prove this, consider two sequences $\chi_i, \varphi_j \in C_0^\infty(X)$ such that $\chi_j = 1$ on K_j, $\varphi_j \geq 0$, $\int \varphi_j(x)\,dx = 1$, $\operatorname{supp} \chi_j + \operatorname{supp} \varphi_j \subset X$. Then

$$u_j = \varphi_j * \chi_j u \in C_0^\infty(X)$$

and $u_j \to u$ in $\mathcal{D}'(X)$. Moreover, the condition (b) also holds, so $u_j \to u$ in $\mathcal{D}'_\Gamma(X)$.

For our aims in Chapter 5 we need to justify some operations on distributions, examined in more details in [H1]. For convenience of the reader we list these properties, including only the proof of the existence of the pull-back f^*. We use the notation from [H1].

Let $X \subset \mathbf{R}^n$ and $Y \subset \mathbf{R}^m$ be open sets and let $f : X \to Y$ be a smooth map. Consider a closed cone $\Gamma \subset Y \times \mathbf{R}^m \backslash \{0\}$ and set

$$N_f = \{(f(x), \eta) \in Y \times \mathbf{R}^n : {}^t f'(x) \eta = 0\},$$
$$f^* \Gamma = \{(x, {}^t f'(x) \eta) : (f(x), \eta) \in \Gamma\}.$$

For $u \in C_0^\infty(Y)$ consider the map

$$(f^* u)(x) = u(f(x)).$$

Theorem 1.3.6: *Let $N_f \cap \Gamma = \emptyset$. Then the map $f^* u$ can be extended uniquely on the space $\mathcal{D}'_\Gamma(Y)$ such that*

$$WF(f^* u) \subset f^* \Gamma. \tag{1.25}$$

Proof: Using a partition of unity, we may consider only the case when X and Y are small open neighbourhoods of $x_0 \in X$ and $y_0 = f(x_0) \in Y$, respectively. Set

$$\Gamma_y = \{\eta : (y, \eta) \in \Gamma\}.$$

Choose a small compact neighbourhood X_0 of x_0 and a closed conic neighbourhood V of Γ_{y_0} so that

$${}^t f'(x) \eta \neq 0 \quad \text{for } x \in X_0, \ \eta \in V.$$

Next, choose a small compact neighbourhood Y_0 of y_0 with $\Gamma_y \subset V$ for all $y \in Y_0$.

Now let $\chi \in C_0^\infty(X_0)$ and let $\{u_j\} \subset C_0^\infty(Y)$ be a sequence such that $u_j \to u$ in $\mathcal{D}'_\Gamma(Y)$. Choosing $\varphi \in C_0^\infty(Y_0)$ with $\varphi = 1$ on $f(X_0)$, we have

$$\langle f^* u_j, \chi \rangle = \langle f^*(\varphi u_j), \chi \rangle = (2\pi)^{-m} \int \widehat{\varphi u_j}(\eta) I_\chi \, d\eta$$

$$= \int_V \int_{\mathbf{R}^m \backslash V} = I_1 + I_2,$$

where

$$I_\chi(\eta) = \int e^{i\langle f(x), \eta \rangle} \chi(x) \, dx.$$

For $x \in \operatorname{supp} \chi$ and $\eta \in V$ we obtain

$$|\nabla_x \langle f(x), \eta \rangle| \geq \delta |\eta|, \quad \delta > 0.$$

Using the operator

$$L = \frac{-i}{|\nabla_x \langle f(x), \eta \rangle|^2} \sum_{j=1}^{n} \partial_{x_j}(\langle f(x), \eta \rangle) \frac{\partial}{\partial x_j},$$

we integrate by parts in $I_\chi(\eta)$ and get

$$\left| I_\chi(\eta) \right| \leq C_p (1 + |\eta|)^{-p}, \quad \eta \in V$$

for all $p \in \mathbf{N}$. On the other hand, there exists $M > 0$ such that

$$\left| \widehat{\varphi u_j}(\eta) \right| \leq C(1 + |\eta|)^M, \quad j \in \mathbf{N}.$$

Thus, I_1 is absolutely convergent, and we can consider the limit as $j \to \infty$. To deal with I_2, notice that $(\operatorname{supp} \varphi \backslash V) \cap \Gamma = \emptyset$. For $\eta \notin V$, (b) yields the estimates

$$\left| \widehat{\varphi u_j}(\eta) \right| \leq C'_p (1 + |\eta|)^{-p}, \quad p \in \mathbf{N} \tag{1.26}$$

uniformly with respect to j. Thus we can let $j \to \infty$ in I_2.

To establish (1.25), replace $\chi(x)$ by $\chi(x)e^{-i\langle x, \xi \rangle}$ and write

$$I_\chi(\eta, \epsilon) = (2\pi)^{-n} \int e^{i\langle f(x), \eta \rangle - i \langle x, \xi \rangle} \chi(x) \, dx.$$

Choose a small open conic neighbourhood W of the set

$$\{\xi = {}^t f'(x_0)\eta : (f(x_0), \eta) \in \Gamma\}$$

so that $x \in X_0$ and $\eta \in V$ imply ${}^t f'(x)\eta \in W$. As above, for $x \in X_0$, $\eta \in V$ and $\xi \notin W$ we deduce the estimate

$$\left| \xi - {}^t f'(x)\eta \right| \geq \delta(|\xi| + |\eta|), \quad \delta > 0.$$

For such ξ and η we integrate by parts in $I_\chi(\eta, \epsilon)$ and obtain

$$\left| I_\chi(\eta, \epsilon) \right| \leq C''_p (1 + |\xi| + |\eta|)^{-p}, \quad p \in \mathbf{N}.$$

For $\eta \notin V$, $\xi \notin W$ we choose a function $\psi(\xi) \in C_0^\infty(\mathbf{R})$ with $\psi(\xi) = 1$ for $|\xi| \leq 1$, and consider the operator

$$L = -i(1 - \psi(\xi)) |\xi|^{-2} \langle \xi, \frac{\partial}{\partial x} \rangle + \psi(\xi).$$

Then $L(e^{i\langle x,\xi\rangle}) = e^{i\langle x,\xi\rangle}$, and, as in the previous case, we get for $\eta \notin V$, $\xi \notin W$ the estimates

$$|I_\chi(\eta,\epsilon)| \leq C_p(1+|\eta|)^p(1+|\xi|)^{-p}, \quad p \in \mathbf{N}.$$

Combining these estimates with (1.26), we obtain

$$\left|\chi(\widehat{f^*u_j})(\xi)\right| \leq C_N(1+|\xi|)^{-N}$$

for $\xi \notin W$, where the constants C_N do not depend on j. Letting $j \to \infty$, we obtain (1.25). ♠

By an easy modification of the above argument one can prove the following modification of Theorem 1.3.6 for distributions depending on a parameter.

Corollary 1.3.7: *Let Z be a compact subset of $\mathbf{R}^p$ and let*

$$Z \ni z \mapsto u(.,z) \in \mathcal{D}'_\Gamma(Y)$$

be a continuous map. Then, under the assumptions of Theorem 1.3.6, the map

$$Z \ni z \mapsto f^*u(.,z) \in \mathcal{D}'_{f^*\Gamma}(X)$$

is continuous. ♠

Further, consider a linear continuous map

$$\mathcal{K} : C_0^\infty(Y) \to \mathcal{D}'(X).$$

By Schwartz's theorem (cf. Theorem 5.2.1 in [H1]) there exists a distribution $K \in \mathcal{D}'(X \times Y)$, called the *kernel* of $\mathcal{K}$ such that

$$\langle K, \varphi(x) \otimes \psi(y) \rangle = \langle (\mathcal{K}\psi)(x), \varphi(x) \rangle$$

for all $\varphi \in C_0^\infty(X)$, $\psi \in C_0^\infty(Y)$. By definition $WF(K)$ will be called *wave front set* of $\mathcal{K}$. We introduce the notation

$$WF'(K) = \{(x,y,\xi,\eta) : (x,y,\xi,-\eta) \in WF(K)\},$$
$$WF(K)_X = \{(x,\xi) : (x,y,\xi,0) \in WF(K) \quad \text{for some } y \in Y\},$$
$$WF'(K)_Y = \{(y,\eta) : (x,y,0,\eta) \in WF'(K) \quad \text{for some } x \in X\},$$

and consider the composition

$$WF'(K) \circ WF(u) =$$
$$\{(x,\xi) : \exists(y,\eta) \in WF(u) \text{ with } (x,y,\xi,\eta) \in WF'(K)\}.$$

The following two results will also be necessary for Chapter 5. Their proofs can be found in Section 8.2 of [H1].

Theorem 1.3.8: *For* $\psi \in C_0^\infty(Y)$ *we have*

$$WF(\mathcal{K}\psi) \subset \{(x,\xi) : (x,y,\xi,0) \in WF(K) \text{ for some } y \in \operatorname{supp}\psi\}.$$

Theorem 1.3.9: *There exists a unique extension of* $\mathcal{K}$ *on the set*

$$\{u \in \mathcal{E}'(Y) : WF(u) \cap WF'(K)_Y = \emptyset\}$$

such that for each compact $M \subset Y$ *and each closed conic set* Γ *with* $\Gamma \cap WF'(K)_Y = \emptyset$ *the map*

$$\mathcal{E}'(M) \cap \mathcal{D}'_\Gamma(Y) \ni u \mapsto \mathcal{K}u \in \mathcal{D}'(X)$$

is continuous. Moreover, the inclusion

$$WF(\mathcal{K}u) \subset WF(K)_X \cup WF'(K) \circ WF(u)$$

holds. ♠

The wave front of $u \in \mathcal{D}'(X)$ can be described by means of the characteristic sets of pseudo-differential operators in X. Denote by $L^m(X)$ the class of all pseudo-differential operators in X of order m. If $a(x,\xi) \in S^m(X \times \mathbf{R}^n)$ is the symbol of A, then the oscillatory integral

$$K_A(x,y) = (2\pi)^{-n} \int e^{i\langle x-y,\xi\rangle} a(x,\xi)\, d\xi$$

determines the kernel of A and $WF(A) = WF(K_A)$. The operator $A \in L^m(X)$ is called *properly supported* if for each compact $K \subset X$ there exists another compact $K' \subset X$ so that $\operatorname{supp} u \subset K$ implies $\operatorname{supp} Au \subset K'$, and if $u = 0$ on K', then $Au = 0$ on K. A point $(x_0,\xi_0) \in T^*X \setminus \{0\}$ is called *non-characteristic* for a properly supported pseudo-differential operator $A \in L^m(X)$ if there exists a properly supported pseudo-differential operator $B \in L^{-m}(X)$ so that

$$(x_0,\xi_0) \notin WF(AB - \mathrm{Id}) \cup WF(BA - \mathrm{Id}).$$

In this case A is called *elliptic* at (x_0,ξ_0).

Proposition 1.3.10: *If there exists a properly supported pseudo-differential operator* $A \in L^m(X)$, *elliptic at* (x_0,ξ_0), *such that* $Au \in C^\infty(X)$, *then* $(x_0,\xi_0) \notin WF(u)$. ♠

The reader may consult Section 18 in [H3] for the main properties of the pseudo-differential operators and for a proof of the above proposition.

Next, consider a closed domain $\Omega \subset \mathbf{R}^n$ with smooth boundary $\partial\Omega$ and set $Q = \Omega \times \mathbf{R}$. Let $u \in \mathcal{D}'(Q)$ be a solution of the problem

$$\begin{cases} (\partial_t^2 - \Delta_x)u = 0 & \text{in } Q^\circ \\ u_{|t>t_0, x\in\partial\Omega} \in C^\infty. \end{cases} \tag{1.27}$$

Here the trace $u_{|x\in\partial\Omega}$ exists, since the boundary $\partial\Omega \times \mathbf{R}$ is not characteristic for $\partial_t^2 - \Delta_x$. Consider the space $\tilde{T}^*(Q) = T^*(Q^\circ) \cup T^*\partial Q$ of equivalence classes in T^*Q with respect to the equivalence relation $\sim$, defined in the previous section. It will be called *compressed cotangent bundle*. For the solutions u of (1.27) we can define the *generalized wave front* $WF_b(u) \subset \tilde{T}^*(Q)\backslash\{0\}$ in such a way that

$$WF_b(u)_{|T^*(Q^\circ)} = WF(u_{|Q^\circ}),$$

and

$$WF_b(u)_{|T^*\partial Q} \subset \Sigma_b$$

(see Section 1.2 for the definition of Σ_b). To this end, as in Section 1.2, introduce local coordinates (x_1, x'), $x' = (x_2, \ldots, x_n, x_{n+1})$, $x_{n+1} = t$ in Q so that ∂Q is locally given by $x_1 = 0$. Let (ξ_1, ξ') be the coordinates dual to (x_1, x').

Now define $WF_b(u)$ as the subset of $T^*(\partial Q)\backslash\{0\}$ the complement of which consists of all $(x_0', \xi_0') \in T^*(\partial Q)\backslash\{0\}$ such that there exists a pseudo-differential operator $B(x, D')$, depending smoothly on x_1, elliptic at $(0, x_0', \xi_0')$, and such that $B(x, D_{x'})u \in C^\infty(Q)$. This definition does not depend on the choice of the local coordinates. The singularities of a solution u of (1.27) can be described by means of $WF_b(u)$. In fact, these singularities are propagating along the compressed generalized bicharacteristics of $\square$ in Ω. The following result was proved by Melrose and Sjöstrand [MS2] (see also Section 24 in [H3]).

Theorem 1.3.11: *Let $u \in \mathcal{D}'(Q)$ be a solution of the problem (1.27) and let*

$$\hat{z} \in WF_b(u) \cap \{(x, \xi) \in \tilde{T}^*(Q) : x_{n+1} = t > t_0\}.$$

Then there exists a maximal compressed generalized bicharacteristic $\tilde{\gamma}(\sigma) = (x(\sigma), \xi(\sigma))$ of $\square$, passing through $\hat{z}$ and staying in $WF_b(u)$ as long as $t(\sigma) = x_{n+1}(\sigma) > t_0$. ♠

1.4. Notes

The results in Section 1.1 are exposed with detailed proofs in [GG] and [Hir]. In Section 1.2 we follow [MS1], [MS2] and [H3]. Lemma 1.2.5 is proved in [MS1], while Lemmas 1.2.6, 1.2.7 and 1.2.10 are given in [H3]. The results in Section 1.3 concerning wave fronts of distributions and operators are due to Hörmander [H1], [H3]. The definition of the generalized wave front set $WF_b(u)$ has been introduced by Melrose and Sjöstrand [MS1], while Theorem 1.3.11 has been established in [MS1], [MS2]. The reader should consult section 24 in [H3] for more details concerning the generalized bicharacteristics and the propagation of singularities for the Dirichlet problem.

2 REFLECTING RAYS

In this chapter we begin with some elementary properties of the periodic reflecting rays in a domain Ω, including the Birkhoff theorem, relating these rays to the critical points of certain length functions. In a similar way we deal in Section 2.4 with the reflecting (ω, θ)-rays. For both kinds of rays the special case is considered when the complement of Ω is a finite disjoint union of strictly convex bounded domains. The results obtained in this case will be useful for our considerations in Chapter 10.

The linear Poincaré map P_γ of a periodic reflecting ray γ is one of the central notions in this book. It is defined in Section 2.3, where also a useful matrix representation of it is described. The analogue of a Poincaré map for a reflecting (ω, θ)-ray — the so-called differential cross section dJ_γ — is defined and studied in Section 2.4. This section contains also the main definitions concerning scattering rays, which will be used frequently in the next chapters.

We conclude this chapter by considering examples of rotative non-convex obstacles K in $\mathbf{R}^3$ and vectors $\omega, \theta \in S^2$ and describe the (ω, θ)-rays in the exterior of K.

2.1. Billiard ball map

Let Ω be a domain in $\mathbf{R}^n$, $n \geq 2$, with smooth boundary $X = \partial\Omega$. The *billiard flow* in Ω is the dynamical system generated by the motion of a material point in Ω. The point is moving with constant velocity in the interior of Ω making reflections at $\partial\Omega$ according to the usual law of geometrical optics 'the angle of incidence is equal to the angle of reflection'. The successive positions of the point at $\partial\Omega$ are described by the so-called *billiard ball map* which will be denoted hereafter by B. This map is defined on a subset M' of the set

$$M = \{(x, v) \in X \times S^{n-1} : \langle \nu(x), v \rangle > 0\},$$

$\nu(x)$ being the *unit normal* to $\partial\Omega$ at x pointing into the interior of Ω, as follows. Let $q = (x, v) \in M$ be such that the straightline ray γ starting at x with direction v has a common point with X and let y be the first such point, i.e. $y \in X$ and the open segment (x, y) does not contain points of X. The subset M' consists by definition of those q such that γ intersects transversally X at y. For such q define $w = v - 2\langle v, \nu(y) \rangle \nu(y)$ and set $B(q) = (y, w)$. Thus one obtains a map $B : M' \to M$. Clearly M' and M are open in $X \times S^{n-1}$ and they have natural structures of smooth manifolds. It is a standard exercise to show that B is a smooth map. From dynamical point of view it is more convenient to consider B as a map $B : M_0 \to M_0$, where

$$M_0 = \cap_{n=-\infty}^{\infty} B^n(M'),$$

$B^n = B \circ \cdots \circ B$ (n times). The points $q \in M_0$ with $B^k(q) = q$ for some integer $k > 0$ will be called *periodic points* of period k of B. Clearly B has no fixed points in M', i.e. points q with $B(q) = q$.

Remark: In general M' is a proper subset of M. For unbounded domains Ω this is clear, while for bounded but non-convex Ω this is due to the existence of rays tangent to $\partial\Omega$. In the latter case it is more convenient to deal with the generalized geodesic flow on $T^*\Omega$ introduced in Chapter 1. Using the standard identification of $T\Omega$ with $T^*\Omega$, one can consider the billiard flow as a subsystem of the generalized geodesic flow, and respectively the billiard ball map B as a map on some subset of the co-ball bundle $B^*\partial\Omega$.

Example 2.1.1: Let Ω be a strictly convex bounded domain in $\mathbf{R}^n$. Clearly, in this case $M_0 = M = M'$. Moreover, B can be naturally extended to a diffeomorphism $\bar{B} : \bar{M} \to \bar{M}$ by $B(q) = q$ for every $q \in \bar{M} \backslash M$, where

$$\bar{M} = \{(x, v) \in X \times S^{n-1} : \langle \nu(x), v \rangle \geq 0\}$$

is the closure of M in $X \times S^{n-1}$. In fact, $\bar{M}$ is a manifold with boundary $\partial\bar{M} = \bar{M} \backslash M$. The reader may consult [Ko] for the proof of the fact that B is smooth on $\bar{M}$.

It can be shown easily that for any integer $s \geq 2$ there exists a periodic point $q \in M$ of B of period s. Indeed, fix an arbitrary s and consider the function

$$F = F_s : \Omega^s \to R$$

given by

$$F(x_1, \ldots, x_s) = \sum_{i=1}^{s} \|x_i - x_{i+1}\|, \tag{2.1}$$

where $x_{s+1} = x_1$ by definition. Since Ω^s is compact and F is continuous, there exists $x = (x_1, \ldots, x_s) \in \Omega^s$ such that F has a maximum at x. A trivial argument shows that $x_i \in \partial\Omega$ and $x_i \neq x_{i+1}$ for every $i = 1, \ldots, s$. Then the restriction G of F on $(\partial\Omega)^s$ has a maximum at x. Since G is smooth on $(\partial\Omega)^{(s)}$, x is a critical point of G. It then follows that $x_1, \ldots, x_s$ are the successive reflection points of a periodic billiard trajectory (see Proposition 2.1.3 below for a rigorous proof), i.e. $B(x_i, v_i) = (x_{i+1}, v_{i+1})$, where $v_i = (x_{i+1} - x_i)/\|x_{i+1} - x_i\|$. Therefore $B^s(x_i, v_i) = (x_i, v_i)$, which shows that B has at least s distinct periodic points of period s.

In the case $n = 2$, i.e. $\Omega \subset \mathbf{R}^2$, one can modify the above argument to prove the existence of periodic points of B of arbitrary period s and rotation number $k \leq \frac{s}{2}$. Given an element $x = (x_1, \ldots, x_s) \in \Omega^s$, define the *rotation (winding) number* $r(x)$ as follows. Set $x_{s+1} = x_1$ and for any $i = 1, \ldots, s$ denote by l_i the length of the segment $[x_i, x_{i+1}]$ on $\partial\Omega$ with respect to the positive (counterclockwise) orientation of $\partial\Omega$ and by l'_i its length with respect to the negative orientation of $\partial\Omega$. Denote by L the *length* of $\partial\Omega$. The integer part of $\frac{1}{L}\sum_{i=1}^{s} l_i$ will be denoted by $r_+(x)$ and that of $\frac{1}{L}\sum_{i=1}^{s} l'_i$ by $r_-(x)$. Since $l_i + l'_i = L$, we have $r_+(x) + r_-(x) = s$, therefore

$$r(x) = \min\{r_+(x), r_-(x)\} \leq \frac{s}{2}.$$

The number $r(x)$ is called rotation number of $x = (x_1, \ldots, x_s)$.

Given two integers $s \geq 2$ and k, $1 \leq k \leq \frac{s}{2}$, applying the above argument to the function $F = F_s$ on the set of those $x \in \Omega^s$ with $r(x) = k$, one gets that there exists $x \in \Omega^s$ with $r(x) = k$ such that x is generated by a periodic point (x_1, v_1) of B of period s. ♠

For our needs in the following chapters it will be convenient to put the notion of periodic orbit of the billiard in a more general setting. This will allow us to consider such orbits in arbitrary domains Ω with smooth boundaries $X = \partial\Omega$.

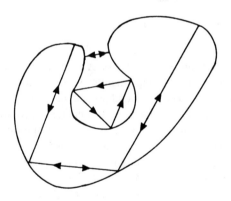

Figure* 2.1

Let X be a smooth $(n-1)$-dimensional submanifold of $\mathbf{R}^n$, $n \geq 2$. Given two linear segments l_1 and l_2 with a common end $x \in X$, we shall say that these segments *satisfy the law of reflection* at x with respect to X if l_1 and l_2 make

---

* Figures 2.1, 2.2, 2.3 and 2.4 are copied from [S1]. We are indebted to Cambridge University Press for permitting us to do this.

equal acute angles with one of the unit normals $\nu(x)$ to X at x, and l_1, l_2 and $\nu(x)$ lie in a common two-dimensional plane.

Definition 2.1.2: Let X be as above and γ be a curve in $\mathbf{R}^n$ of the form $\gamma = \cup_{i=1}^{k} l_i$, where $l_i = [x_i, x_{i+1}]$, $x_i \in X$ for each $i = 1, \ldots, k, k \geq 2$, and we set for convenience $x_{k+1} = x_1$ and $l_{k+1} = l_1$. We shall say that γ is a *periodic reflecting ray* for X if the following conditions are satisfied:

(a) for each $i = 1, \ldots, k$ the open segments l_i° do not intersect transversally X;

(b) for each $i = 1, \ldots, k$ the segments l_i and l_{i+1} satisfy the law of reflection at x_{i+1} with respect to X.

The points $x_1, \ldots, x_k$ will be called *reflection points* of γ.

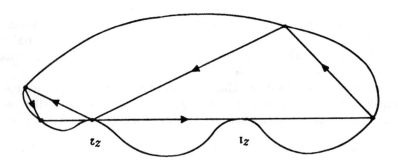

Figure 2.2

Note that if Ω is a domain with boundary X, then a periodic reflecting ray for X may lie in Ω as well as in the complement of Ω (see Figure 2.1). If γ contains a segment orthogonal to X at some of its end points, then γ will be called *symmetric*; otherwise it will be called *non-symmetric*. In general a periodic reflecting ray γ could contain a segment tangent to X at some of its interior points. If γ does not contain such segments, we will say that γ is *ordinary*. Let us mention that, with the exception of Section 2.2, points like z_1 on Figure 2.2 are not considered as reflection ones. In general a periodic reflecting ray γ can pass two or more times through some of its reflection points (Figure 2.3), and two different periodic reflecting rays could have a common reflection point (Figure 2.4). Given a periodic reflecting ray γ and an integer $k \geq 2$, one defines naturally the *k-multiple* δ of γ. In fact, as a subset of $\mathbf{R}^n$, δ coincides with γ, but the number of reflection points of δ is ks, where s is the number of reflection points of γ. We shall say that γ is *primitive* if it is not a multiple of a periodic reflecting ray.

We conclude this section by establishing an elementary fact, which however will be important for our next considerations. Namely, we shall show that there is a natural one-to-one correspondence between the periodic reflecting rays with s reflection points for a given submanifold X and some kind of critical points of the restriction of the map $F = F_s$, defined by (2.1), on X^s. Note that F is

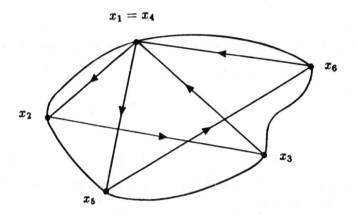

Figure 2.3

well-defined and continuous on $(\mathbf{R}^n)^s$, and F is smooth on the set U_s of those $y = (y_1, \ldots, y_s) \in (\mathbf{R}^n)^s$ such that $y_i \neq y_{i+1}$ for each $i = 1, \ldots, s$ (as before $y_{s+1} = y_1$ by definition).

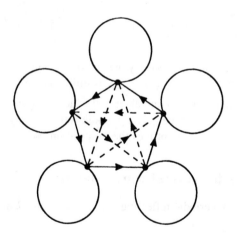

Figure 2.4

Proposition 2.1.3: *Let γ be a curve in $\mathbf{R}^n$ of the form $\gamma = \cup_{i=1}^s l_i$, where $l_i = [x_i, x_{i+1}]$, $x_i \in X$ for each $i = 1, \ldots, s, s \geq 2, x_{s+1} = x_1$, and such that the open segments l_i^0 do not transversally intersect X. Then γ is a periodic reflecting ray for X if and only if $x = (x_1, \ldots, x_s)$ is a critical point of the map $F_{|(X^s \cap U_s)}$.*

Proof: Take arbitrary smooth charts

$$\varphi_j : \mathbf{R}^{n-1} \rightarrow U_j \subset X, \quad j = 1, \ldots, s,$$

such that $\varphi_j(0) = x_j$. Then $\{\frac{\partial \varphi_j}{\partial u_j^{(t)}}(0)\}_{t=1}^{n-1}$ is a basis of the tangential space to X at x_j. Here we use the notation $u_j = (u_j^{(1)}, \ldots, u_j^{(n-1)}) \in \mathbf{R}^{n-1}$. Consider the function

$$G : (\mathbf{R}^{n-1})^s \rightarrow \mathbf{R}$$

defined by

$$G(u_1, \ldots, u_s) = F(\varphi_1(u_1), \ldots, \varphi_s(u_s)).$$

Let $\varphi_j = (\varphi_j^{(1)}, \ldots, \varphi_j^{(n)})$. Clearly for any $u \in (\mathbf{R}^{n-1})^s$ sufficiently close to 0, G is differentiable at u and

$$\frac{\partial G}{\partial u_j^{(t)}}(u) = \left\langle \frac{\varphi_j(u_j) - \varphi_{j-1}(u_{j-1})}{\|\varphi_j(u_j) - \varphi_{j-1}(u_{j-1})\|} + \frac{\varphi_j(u_j) - \varphi_{j+1}(u_{j+1})}{\|\varphi_j(u_j) - \varphi_{j+1}(u_{j+1})\|}, \frac{\partial \varphi_j}{\partial u_j^{(t)}}(u_j) \right\rangle.$$

Setting $v_{i,j} = (x_i - x_j)/\|x_i - x_j\|$, we get

$$\frac{\partial G}{\partial u_j^{(t)}}(0) = \left\langle v_{j,j-1} + v_{j,j+1}, \frac{\partial \varphi_j}{\partial u_j^{(t)}}(0) \right\rangle.$$

Note that the segments l_j and l_{j+1} satisfy the law of reflection at x_{j+1} with respect to X if and only if the vector $v_{i,j-1} + v_{i,j+1}$ is orthogonal to X at x_{j+1}. According to the above argument, this is equivalent to the fact that $(\partial G)/(\partial u_j^{(t)})(0) = 0$ for each $t = 1, \ldots, n-1$. Hence γ is a periodic reflecting ray for X if and only if 0 is a critical point of G. This proves the proposition. ♠

2.2. Periodic rays for several convex bodies

In this section we study periodic reflecting rays in a domain Ω in $\mathbf{R}^n$ such that the complement $K = \mathbf{R}^n \backslash \Omega$ has the form

$$K = K_1 \cup \ldots \cup K_s, \tag{2.2}$$

$s \geq 3$. Here each K_i is a compact convex domain in $\mathbf{R}^n$ with C^2-smooth boundary $\Gamma_i = \partial K_i$, and $K_i \cap K_j = \emptyset$ whenever $i \neq j$.

Our aim in what follows is to provide a coding for the periodic reflecting rays in Ω. Namely, we associate with any periodic reflecting ray γ a finite sequence

$$\alpha_\gamma = (i_1, \ldots, i_k) \in \{1, \ldots, s\}^k,$$

where k is the number of reflections of γ, such that the jth successive reflection point of γ belongs to K_{i_j} for any $j = 1, \ldots, k$. Clearly, for such a sequence we have $i_j \neq i_{j+1}$ for $j = 1, \ldots, k-1$ and $i_k \neq i_1$. Every

$$\alpha = (i_1, \ldots, i_k) \in \{1, \ldots s\}^k \tag{2.3}$$

having the latter property will be called a *configuration* of length $|\alpha| = k$. Denote by $\mathcal{A}_k$ the *set of all configurations of length* k. We shall establish that if all K_i are strictly convex, then the correspondence $\gamma \to \alpha_\gamma \in \mathcal{A}_k$ is invertible, and under some additional assumption, it is moreover bijective. It is easy to construct examples showing that in general this map is not surjective.

Clearly,

$$\Gamma = \partial\Omega = \Gamma_1 \cup \ldots \cup \Gamma_s.$$

For $q \in \Gamma$ the *unit normal vector* to Γ at q, pointing into the interior of Ω will be denoted by $\nu(q)$. The second fundamental form of Γ is non-positively definite at any $q \in \Gamma$ with respect to this choice of the normal field.

Definition 2.2.1:   We will say that Ω (resp. K) satisfies the **condition (H)** if for any $i \neq j$ the convex hull of $K_i \cup K_j$ has no common points with K_r for each $r \in \{1, \ldots, s\} \setminus \{i, j\}$.

In what follows the periodic reflecting rays for Γ contained in $\bar{\Omega}$ will be called briefly *reflecting rays in* $\bar{\Omega}$.

Proposition 2.2.2:   *Let Ω satisfy the condition (H). Then for each integer $K \geq 2$ and each $\alpha \in \mathcal{A}_k$ there exists a periodic reflecting ray γ in $\bar{\Omega}$ such that $\alpha_\gamma = \alpha$.*

Proof: Fix an arbitrary α of the form (2.3) and consider the function

$$F = F_\alpha : K_\alpha = K_{i_1} \times \cdots \times K_{i_k} \to \mathbf{R}, \tag{2.4}$$

defined by

$$F(q_1, \ldots, q_k) = \sum_{j=1}^{k} \| q_j - q_{j+1} \|, \tag{2.5}$$

where we use the notation $q_{k+1} = q_1$. Since F is continuous and K_α is compact, there exists $q = (q_1, \ldots, q_k) \in K_\alpha$ such that F has a total minimum at q. A simple geometrical argument shows that

$$q_j \in \Gamma_{i_j}$$

for any $j = 1, \ldots, k$. It follows by the condition (H) that each of the open linear segments (q_j, q_{j+1}) has no common points with K. Now applying Proposition 2.1.3, one gets that $q_1, \ldots, q_k$ are the successive reflection points of a periodic reflecting ray γ for Γ. Clearly, $\gamma \subset \bar{\Omega}$ and $\alpha_\gamma = \alpha$. This proves the assertion. ◆

In what follows up to the end of this section we deal with the general case, i.e. we do not assume the condition (H) to be satisfied. Now it is more convenient to consider the points of tangency of a periodic reflecting ray γ to Γ as reflection points of γ (we shall do the same in Section 2.4). To this end we need an extension of the billiard ball map B similar to that in Example 2.1.1.

Let $L_q\Gamma$ be the tangent hyperplane to Γ at $q \in \Gamma$. An element $x = (q,v) \in \Gamma \times S^{n-1}$ will be called *regular* if either $\langle \nu(q), v \rangle > 0$ or $\langle \nu(q), v \rangle = 0$ and there exists a neighbourhood U of q in Γ such that $U \cap L_q\Gamma = \{q\}$. If all K_i are strictly convex, then each point x with $\langle \nu(q), v \rangle \geq 0$ is regular, but in the general case there could exist non-regular points.

Denote by M'' the set of those regular elements x such that the straightline ray δ, starting at q with direction v, has a common point with Γ, and if p is the first common point (i.e. the open segment (q,p) has no common points with Γ), then $y = (p, \omega)$ is a regular element of $\Gamma \times S^{n-1}$, where $\omega = v - 2\langle \nu(p), v \rangle \nu(p)$. We set $B(x) = y$, thus extending B to a map $B : M'' \to \Gamma \times S^{n-1}$. We shall be interested in the restriction

$$B : M_1 \to M_1$$

of B, where $M_1 = \cap_{m=0}^{\infty} B^{-m}(M'')$. More specifically, we study the periodic points $x \in M_1$ of B, i.e. the points for which there exist $k \in \mathbf{N}$ with $B^k(x) = x$.

Let $\pi : \Gamma \times S^{n-1} \to \Gamma$ be the natural *projection* and let $\alpha \in \mathcal{A}_k$. A point $x = (q,v) \in M_1$ will be called a *periodic point of type* α for B if $B^k(x) = x$ and

$$q_j = \pi \circ B^{j-1}(x) \in \Gamma_{i_j} \qquad (2.6)$$

for each $j = 1, \ldots, k$. If the segment $[q_j, q_{j+1}]$ is tangent to Γ at q_j, we will say that q_j is a *tangential reflection point* of the corresponding periodic billiard trajectory $\gamma(x)$; otherwise q_j will be called a *proper reflection point*.

In general there could exist distinct periodic points of B having one and the same type α. One can construct examples considering periodic reflecting rays with parallel corresponding segments, arranging suitably obstacles with flat pieces on their boundaries (see Figure 2.5). As we see from the next theorem, these are in fact the only possibilities to construct such examples.

Theorem 2.2.3: *Let* $\alpha \in \mathcal{A}_k, k \in \mathbf{N}$, *and suppose there exist two distinct periodic points* (q,v), (p,w) *of type* α *for* B *and let* $q_j = \pi \circ B^{j-1}(q,v)$, $p_j = \pi \circ B^{j-1}(p,w)$, $j = 1, 2, \ldots$ *Then* $v = w$ *and for any* $j \geq 1$ *the segments* $[q_j, q_{j+1}]$ *and* $[p_j, p_{j+1}]$ *are parallel. If* q_j *is a proper reflection point, then*

$$tq_j + (1-t)p_j \in \Gamma_{i_j} \qquad (2.7)$$

for all $t \in [0,1]$. *If all* q_j *are proper reflection points, then for each* $t \in (0,1)$ *sufficiently close to 1 the points* $(tq + (1-t)p, v)$ *are periodic points of type* α *for* B, *generating periodic billiard trajectories in* $\bar{\Omega}$ *with equal lengths and parallel corresponding segments.*

In other words, for any $\alpha \in \mathcal{A}_k$ there are three possibilities: (a) there are no periodic points of type α for B; (b) there is exactly one periodic point of type α;

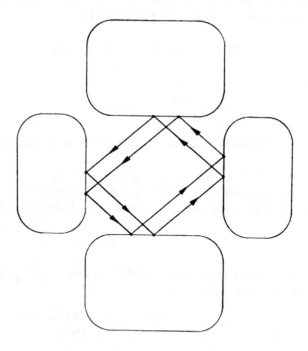

Figure 2.5

(c) the periodic points of type α generate a family (which might be discrete) of periodic billiard trajectories in $\bar{\Omega}$ of equal lengths and having parallel corresponding segments.

Before proceeding with the proof of this theorem, we consider some direct consequences of it.

Corollary 2.2.4: *Let α have the form (2.3) and let Γ_{i_j} be strictly convex for each $j = 1, \ldots, k$. Then there exists at most one periodic point of type α for B.* ♠

For $k \geq 2$ set $a_k = \sharp \mathcal{A}_k$. Clearly

$$a_2 = s(s-1), \quad a_3 = s(s-1)(s-2).$$

On the other hand, one gets easily that for $k \geq 4$ we have

$$a_k = (s-2)a_{k-1} + (s-1)a_{k-2}.$$

Consequently,

$$a_k = (s-1)^k + (-1)^k(s-1)$$

for any $k \geq 2$. Combining the latter equality with Corollary 2.2.4 and Proposition 2.2.2, we deduce the following.

Corollary 2.2.5:   *Let K_i be strictly convex for each $i = 1, \ldots, s$ and let P_k be the number of the periodic points of B of period k. Then*

$$P_k \leq a_k = (s-1)^k + (-1)^k(s-1), \tag{2.8}$$

and therefore

$$\limsup_{k \to \infty} \frac{\log P_k}{k} \leq \log(s-1). \tag{2.9}$$

If in addition Ω satisfies the condition (H), then there are equalities in both (2.8) and (2.9). ♠

Here $\log = \log_c$, with $c > 1$ an arbitrary fixed number.

The rest of this section is devoted to the proof of Theorem 2.2.3.

Fix α of the form (2.3) and consider the function (2.4) defined by (2.5). Set

$$\Gamma_\alpha = \Gamma_{i_1} \times \cdots \times \Gamma_{i_k}.$$

Clearly, K_α is a compact convex subset of $(\mathbf{R}^n)^k$, however Γ_α is not its boundary. In fact, Γ_α is a 'very thin' subset of ∂K_α.

Lemma 2.2.6:   *Let $x = (q, v)$ be a periodic point of type α of B and let the points q_j be defined by (2.6) for each $j = 1, \ldots, k$. Then:*

(a) The map $F : K_\alpha \to \mathbf{R}$ has a local minimum at $\tilde{q} = (q_1, \ldots, q_k)$;

(b) If there exists at least one j such that Γ is strictly convex at q_j and q_{j+1} is a proper reflection point, then F has a strict local minimum at $\tilde{q}$.

Proof: Since the case $k = 2$ is trivial, we assume $k \geq 3$.

There exist C^2-smooth charts

$$\varphi_j : \mathbf{R}^{n-1} \to U_j \subset \Gamma_{i_j}$$

such that $\varphi_j(0) = q_j$. Consider the function

$$G : (\mathbf{R}^{n-1})^k \to \mathbf{R},$$

defined by

$$G(u_1, \ldots, u_k) = F(\varphi_1(u_1), \ldots, \varphi_k(u_k)).$$

First, we show that G has a local minimum at 0; this will imply that $F_{|\Gamma_\alpha}$ has a local minimum at $\tilde{q}$.

Let $\varphi_j = (\varphi_j^{(1)}, \ldots, \varphi_j^{(n)})$ and $u = (u_1, \ldots, u_k) \in (\mathbf{R}^{n-1})^k$. In what follows we use also the following notation: $I_j = \{j-1, j+1\}$, $u_j = (u_j^{(1)}, \ldots, u_j^{(n-1)}) \in \mathbf{R}^{n-1}$, $a_{ij} = 1/\|q_i - q_j\|$, $v_{ij} = a_{ij}(q_i - q_j)$. Clearly, $a_{ij} = a_{ji} > 0$ and $v_{ji} = -v_{ij} \in S^{n-1}$.

For $j = 1, \ldots, k$, $t = 1, \ldots, n-1$ and u sufficiently close to 0 we have

$$\frac{\partial G}{\partial u_j^{(t)}}(u) = \sum_{i \in I_j} \left\langle \frac{\varphi_j(u_j) - \varphi_i(u_i)}{\|\varphi_j(u_j) - \varphi_i(u_i)\|}, \frac{\partial \varphi_j}{\partial u_j^{(t)}}(u_j) \right\rangle. \tag{2.10}$$

By Proposition 2.1.3, 0 is a critical point of G. We shall prove that the second fundamental form of G at 0 is non-negatively definite. To this end we need the derivatives

$$\frac{\partial^2 G}{\partial u_j^{(t)} \partial u_i^{(m)}}(0) \tag{2.11}$$

for $i, j = 1, \ldots, k$ and $t, m = 1, \ldots, n-1$. Having fixed j, there are three possibilities for i.

Case 1. $i \notin I_j \cup \{j\}$. Then clearly the derivative (2.11) is 0.

Case 2. $i \in I_j$. In this case (2.10) implies

$$\frac{\partial^2 G}{\partial u_j^{(t)} \partial u_i^{(m)}}(0) = -a_{ij} \left\langle \frac{\partial \varphi_i}{\partial u_j^{(t)}}(0), \frac{\partial \varphi_i}{\partial u_i^{(m)}}(0) \right\rangle$$

$$+ a_{ji} \left\langle \frac{\partial \varphi_j}{\partial u_j^{(t)}}(0), v_{ji} \right\rangle \left\langle \frac{\partial \varphi_i}{\partial u_i^{(m)}}(0), v_{ji} \right\rangle.$$

Case 3. $i = j$. Then

$$\frac{\partial^2 G}{\partial u_j^{(t)} \partial u_j^{(m)}}(0) = \sum_{i \in I_j} \left\langle v_{ji}, \frac{\partial^2 \varphi_j}{\partial u_j^{(t)} \partial u_j^{(m)}}(0) \right\rangle + \sum_{i \in I_j} a_{ji} \left\langle \frac{\partial \varphi_j}{\partial u_j^{(t)}}(0), \frac{\partial \varphi_j}{\partial u_j^{(m)}}(0) \right\rangle$$

$$- \sum_{i \in I_j} a_{ji} \left\langle \frac{\partial \varphi_j}{\partial u_j^{(t)}}(0), v_{ji} \right\rangle \left\langle \frac{\partial \varphi_j}{\partial u_j^{(m)}}(0), v_{ji} \right\rangle.$$

Fix an arbitrary vector $\xi = (\xi_j^{(t)})_{1 \leq j \leq k, 1 \leq t \leq n-1} \in (\mathbf{R}^{n-1})^k$. We have to establish that

$$\sigma = \sum_{j,i=1}^{k} \sum_{t,m=1}^{n-1} \frac{\partial^2 G}{\partial u_j^{(t)} \partial u_i^{(m)}}(0) \xi_j^{(t)} \xi_i^{(m)} \geq 0. $$

Set $\xi_j = (\xi_j^{(1)}, \ldots, \xi_j^{(n-1)}) \in \mathbf{R}^{n-1}$ and $z_j = \sum_{t=1}^{n-1} \xi_j^{(t)} \frac{\partial \varphi_j}{\partial u_j^{(t)}}(0)$. Note that for $\nu_j = \nu(q_j)$ there exists $\lambda_j > 0$ with $v_{jj-1} + v_{jj+1} = -\lambda_j \nu_j$.

Since the hypersurface $U_j = \varphi_j(\mathbf{R}^{n-1}) \subset \Gamma$ is convex at g_j, the choice of the normal field ν shows that the second fundamental form B_j of U_j at q_i is non-positively definite. That is

$$B_j(\xi_j, \xi_j) = \sum_{t,m=1}^{n-1} \left\langle \nu_j, \frac{\partial^2 \varphi_j}{\partial u_j^{(t)} \partial u_j^{(m)}}(0) \right\rangle \xi_j^{(t)} \xi_j^{(m)} \le 0$$

for any $\xi_j \in \mathbf{R}^{n-1}$.

Using the expressions for the second derivatives of G at 0, we find

$$
\begin{aligned}
\sigma &= \sum_{j=1}^{k} \sum_{t,m=1}^{n-1} \frac{\partial^2 G}{\partial u_j^{(t)} \partial u_j^{(m)}}(0) \xi_j^{(t)} \xi_j^{(m)} \\
&\quad + \sum_{j=1}^{k} \sum_{i \in I_j} \sum_{t,m=1}^{n-1} \frac{\partial^2 G}{\partial u_j^{(t)} \partial u_i^{(m)}}(0) \xi_j^{(t)} \xi_i^{(m)} \\
&= \left(-\sum_{j=1}^{k} \lambda_j \sum_{t,m=1}^{n-1} \left\langle \nu_j, \frac{\partial^2 \varphi_j}{\partial u_j^{(t)} \partial u_i^{(m)}}(0) \right\rangle \xi_j^{(t)} \xi_i^{(m)} \right. \\
&\quad + \sum_{j=1}^{k} \sum_{i \in I_j} \sum_{t,m=1}^{n-1} a_{ji} \left\langle \frac{\partial \varphi_j}{\partial u_j^{(t)}}(0), \frac{\partial \varphi_j}{\partial u_j^{(m)}}(0) \right\rangle \xi_j^{(t)} \xi_j^{(m)} \\
&\quad \left. - \sum_{j=1}^{k} \sum_{i \in I_j} \sum_{t,m=1}^{n-1} a_{ji} \left\langle \frac{\partial \varphi_j}{\partial u_j^{(t)}}(0), v_{ji} \right\rangle \left\langle \frac{\partial \varphi_j}{\partial u_j^{(m)}}(0), v_{ji} \right\rangle \xi_j^{(t)} \xi_j^{(m)} \right) \\
&\quad + \left(-\sum_{j=1}^{k} \sum_{i \in I_j} \sum_{t,m=1}^{n-1} a_{ji} \left\langle \frac{\partial \varphi_j}{\partial u_j^{(t)}}(0), \frac{\partial \varphi_i}{\partial u_i^{(m)}}(0) \right\rangle \xi_j^{(t)} \xi_i^{(m)} \right. \\
&\quad \left. + \sum_{j=1}^{k} \sum_{i \in I_j} \sum_{t,m=1}^{n-1} a_{ji} \left\langle \frac{\partial \varphi_j}{\partial u_i^{(t)}}(0), v_{ji} \right\rangle \left\langle \frac{\partial \varphi_i}{\partial u_i^{(m)}}(0), v_{ji} \right\rangle \xi_j^{(t)} \xi_i^{(m)} \right) \\
&= -\sum_{j=1}^{k} \lambda_j B_j(\xi_j, \xi_j) + \sum_{j=1}^{k} \sum_{i \in I_j} a_{ji} \langle z_j, z_j \rangle - \sum_{j=1}^{k} \sum_{i \in I_j} a_{ji} \langle z_j, v_{ji} \rangle^2 \\
&\quad - \sum_{j=1}^{k} \sum_{i \in I_j} a_{ji} \langle z_j, z_i \rangle + \sum_{j=1}^{k} \sum_{i \in I_j} a_{ji} \langle z_j, v_{ji} \rangle \langle z_i, v_{ji} \rangle.
\end{aligned}
$$

Since $i \in I_j$ is equivalent to $j \in I_i$, according to $a_{ji} = a_{ij}$ and $v_{ji} = -v_{ij}$, we can rewrite the last expression for σ as follows:

$$
\begin{aligned}
\sigma &= -\sum_{j=1}^{k} \lambda_j B_j(\xi_j, \xi_j) + \sum_{j=1}^{k} a_{jj+1} \left(\|z_j\|^2 - \langle z_j, v_{jj+1} \rangle^2 \right. \\
&\quad \left. - \langle z_j, z_{j+1} \rangle + \langle z_j, v_{jj+1} \rangle \langle z_{j+1}, v_{jj+1} \rangle + \|z_{j+1}\|^2 \right.
\end{aligned}
$$

$$- \langle z_{j+1}, v_{j+1j} \rangle^2 - \langle z_{j+1}, z_j \rangle + \langle z_{j+1}, v_{j+1j} \rangle \langle z_j, v_{j+1j} \rangle \Big)$$

$$= - \sum_{j=1}^{k} \lambda_j B_j(\xi_j, \xi_j) + \sum_{j=1}^{k} a_{jj+1} \left(\|z_j - z_{j+1}\|^2 - \langle z_j - z_{j+1}, v_{jj+1} \rangle^2 \right).$$

Since $\|v_{jj+1}\| = 1$, we have $\langle z_j - z_{j+1}, v_{jj+1} \rangle^2 \leq \|z_j - z_{j+1}\|^2$, which yields $\sigma \geq 0$.

Next, assume that $\xi \neq 0$ and $\sigma = 0$, and let there exist j such that Γ is strictly convex at q_j and q_{j+1} is a proper reflection point. By $\sigma = 0$ it follows that $B_j(\xi_j, \xi_j) = 0$ and the vector $z_j - z_{j+1}$ is collinear with v_{jj+1}, i.e. with the segment $[q_j, q_{j+1}]$. Since B_j is definite, one gets $\xi_j = 0$, i.e. $z_j = 0$. On the other hand, the vector z_{j+1} is collinear with the tangent hyperplane to Γ at q_{j+1}, therefore $[q_j, q_{j+1}]$ is tangent to Γ at q_{j+1} and we obtain a contradiction. This is why the assumption in (b) implies $\sigma > 0$ for any $\xi \neq 0$.

In this way we have established that G has a local minimum at 0, hence $F_{|\Gamma_\alpha}$ has a local minimum at $\tilde{q}$. Moreover, if the assumption in (b) is satisfied, then $F_{|\Gamma_\alpha}$ has a strict local minimum at $\tilde{q}$.

Further, for every j fix a neighbourhood V_j of q_j in K_{i_j} such that $F(\tilde{q}) \leq F(\tilde{p})$ whenever $\tilde{p} \in V \cap \Gamma_\alpha$, where $V = V_1 \times \cdots \times V_k$. Since the points $B^{j-1}(q, v)$ are regular, we can assume that the neighbourhoods V_j are chosen in such a way that for each $\tilde{p} \in V$ and each $j = 1, \ldots, k$ the straightline determined by the segment $[p_j, p_{j+1}]$ intersects Γ_{i_j} and $\Gamma_{i_{j+1}}$ at points of V_j and V_{j+1}, respectively. Indeed, if q_j is a tangential reflection point, we may define V_j by

$$V_j = \{ p_j \in K_{i_j} : \langle p_j - q_j, \nu(q_j) \rangle > -\epsilon_j \}$$

for some sufficiently small $\epsilon_j > 0$. If q_j is a proper reflection point, consider an open ball D_j in $\mathbf{R}^n$ centred at q_j and having sufficiently small radius $\epsilon_j > 0$, and set $V_j = K_{i_j} \cap D_j$.

Let $\tilde{p} = (p_1, \ldots, p_k) \in V$. Denote by p_1' the point of intersection of Γ_{i_1} with the segment $[p_1, p_2]$. Then $p_1' \in V_1$, and the triangle inequality implies

$$F(p_1, \ldots, p_k) \geq F(p_1', p_2, \ldots p_k).$$

Next, for the intersection point p_2' of Γ_{i_2} and the segment $[p_1', p_2]$ we get

$$F(p_1', p_2, p_3, \ldots, p_k) \geq F(p_1', p_2', p_3, \ldots, p_k),$$

etc. Thus, for any j we find a point $p_j' \in \Gamma_{i_j} \cap V_j$ such that $F(\tilde{p}) \geq F(\tilde{p}')$ holds for $\tilde{p}' = (p_1', \ldots, p_k') \in \Gamma_\alpha \cap V$. It then follows by the choice of V that $F(\tilde{p}') \geq F(\tilde{q})$, therefore $F(\tilde{p}) \geq F(\tilde{q})$. This concludes the proof of (a).

The proof of (b) follows easily from the above arguments. We leave the details to the reader. ♠

Proof of Theorem 2.2.3:   Fix α of the form (2.3) and let

$$F : K_\alpha \to \mathbf{R}$$

be defined as above. Clearly, F is a convex function, i.e.

$$F(\tilde{q} + (1-t)\tilde{p}) \leq tF(\tilde{q}) + (1-t)F(\tilde{p})$$

for each $t \in [0,1]$ and all $\tilde{q}, \tilde{p} \in K_\alpha$.

Assume that there exist two distinct periodic points (q, v) and (p, w) of type α for B. Set

$$\tilde{q} = (q_1, \ldots, q_k), \quad \tilde{p} = (p_1, \ldots, p_k).$$

Then $\tilde{q}, \tilde{p} = K_\alpha$, and Lemma 2.2.5 implies that F has local minima at both these points. For $t \in [0,1]$ set

$$q_j^{(t)} = tq_j + (1-t)p_j, \quad \tilde{q}^{(t)} = (q_1^{(t)}, \ldots, q_k^{(t)}).$$

Clearly, $\tilde{q}^{(t)} = t\tilde{q} + (1-t)\tilde{p} \in K_\alpha$.

We shall show that $F(\tilde{q}) = F(\tilde{p})$. Suppose that $F(\tilde{q}) > F(\tilde{p})$. Then for each $t \in (0,1)$ we have

$$F(\tilde{q}^{(t)}) = F(t\tilde{q} + (1-t)\tilde{p}) \leq tF(\tilde{q}) + (1-t)F(\tilde{p}) < F(\tilde{q}). \quad (2.12)$$

Since $\tilde{q}^{(t)} \to \tilde{q}$ as $t \to 1$, we get a contradiction with the fact that F has a local minimum at $\tilde{q}$. Therefore $F(\tilde{q}) \leq F(\tilde{p})$. In the same way one obtains $F(\tilde{q}) \geq F(\tilde{p})$, which implies $F(\tilde{q}) = F(\tilde{p})$.

Combining the latter with (2.12), we get something more, namely that

$$F(\tilde{q}^{(t)}) = F(\tilde{q}) = F(\tilde{p}). \quad (2.13)$$

holds for all t sufficiently close to 0 or 1. Now the convexity of F yields that (2.13) holds for each $t \in [0,1]$.

Let us recall that for $p \neq p'$, $q \neq q'$ and $t \in (0,1)$ the equality

$$\|(tq + (1-t)p) - (tq' + (1-t)p')\| = t\|q - q'\| + (1-t)\|p - p'\|$$

holds if and only if the segments $[p, p']$ and $[q, q']$ are collinear. In our situation this yields that the segments $[q_j, q_{j+1}]$ and $[p_j, p_{j+1}]$ are collinear for every j. In particular, $v = w$.

Choose the neighbourhoods V_j of q_j in K_{i_j} as in the proof of Lemma 2.2.5. There exists $t_0 \in (0,1)$ such that $q_j^{(t)} \in V_j$ for any $t \in (t_0, 1]$. Clearly, F has a minimum at $q_j^{(t)}$ in $V = V_1 \times \cdots \times V_k$ for every $t \in (t_0 1]$. Let q_j be a proper reflection point for some $j \leq k$, and assume that $q_j^{(t)} \notin \Gamma_{i_j}$, for some $t \in (t_0, 1)$. Set

$$\tilde{r} = (q_1^{(t)}, \ldots, q_{j-1}^{(t)}, q_j', q_{j+1}^{(t)}, \ldots, q_k^{(t)}),$$

where q'_j is the intersection point of Γ_{i_j} with $[q_j^{(t)}, q_{j+1}^{(t)}]$. Since q_j is a proper reflection point, it follows by the above remark that

$$\left\| q_{j-1}^{(t)} - q_j^{(t)} \right\| + \left\| q_j^{(t)} - q_{j+1}^{(t)} \right\| > \left\| q_j^{(t)} - q'_j \right\| + \left\| q'_j - q_{j+1}^{(t)} \right\|,$$

therefore $F(\tilde{q}^{(t)}) > F(\tilde{r})$. This is a contradiction with the minimality of $F(\tilde{q}^{(t)})$. Hence $q_j^{(t)} \in \Gamma_{i_j}$, for every $t \in (t_0, 1]$ which is sufficiently close to 1.

Finally, if all q_j are proper reflection points, the last argument shows that for $t \in (0, 1)$ sufficiently close to 1, the points $(t_q + (1-t)p, v)$ are periodic points of type α of B. Clearly, they generate periodic billiard trajectories in Ω with lengths $F(\tilde{q}) = F(\tilde{p})$ and parallel corresponding segments. ♠

2.3. Poincaré map

Throughout this section Ω will be a closed domain in $\mathbf{R}^n$ with smooth boundary $X = \partial\Omega$ and γ will be an ordinary periodic reflecting ray in Ω with successive reflection points $q_1, \ldots, q_m, q_{m+1} = q_1$ and period (length) $T > 0$. Here we define the linear Poincaré map P_γ of γ and find a useful representation of it. There are different ways to define this map, but all of them are equivalent in the sense that the *spectrum* spec P_γ is the same.

Let $\tilde{\gamma}$ be a generalized bicharacteristic of $\square$ in $T^*(\mathbf{R} \times \Omega)$ such that $\pi_x(\tilde{\gamma}) = \gamma$, where

$$\pi_x : T^*(\mathbf{R} \times \Omega) \to \Omega$$

is the superposition of the natural projections

$$T^*(\mathbf{R} \times \Omega) \to \mathbf{R} \times \Omega \to \Omega.$$

Given $\rho \in \tilde{\gamma}$ with $\pi_x(\rho) \neq q_i, i = 1, \ldots, m$, there exists a small conic neighbourhood $V \subset T^*(\Omega^\circ)$ of ρ such that for each $(y, \eta) \in V$ the generalized bicharacteristic of $\square$, parametrized by the time and issued from (y, η) has just m reflections on $\partial\Omega$ for $0 \leq t \leq T$. The Hamiltonian flow Φ^T, introduced in Section 1.2, maps V into a conic neighbourhood W of ρ and we can define the map

$$(d\Phi^T)(\rho) : T_\rho(T^*(\Omega)) \to T_\rho(T^*(\Omega)).$$

Clearly, the tangent vector e to $\tilde{\gamma}$ at ρ and the direction of the cone axis at ρ are invariant with respect to $(d\Phi^T)(\rho)$. Let E_ρ be the two-dimensional space generated by e and f, and let

$$\Sigma_\rho = T_\rho(T^*(\Omega))/E_\rho$$

be the corresponding quotient space. The linear map

$$P_\gamma(\rho) = d\Phi^T(\rho)_{|\Sigma_\rho}$$

will be called the *(linear) Poincaré map* of γ at ρ. Clearly, P_γ preserves the natural symplectic structure of Σ_ρ (cf. [AbM]).

If ρ and μ are two different points on $\tilde{\gamma}$ such that $\pi_x(\rho) \notin \partial\Omega$, $\pi_x(\mu) \notin \partial\Omega$, then for some $\tau \in \mathbf{R}$ we have $\Phi^\tau(\rho) = \mu$, and therefore

$$\Phi^\tau \circ \Phi^T(\sigma) = \Phi^T \circ \Phi^\tau(\sigma), \quad \sigma \in V.$$

Thus,

$$d\Phi^\tau(\rho) \circ d\Phi^T(\rho) = d\Phi^T(\mu) \circ d\Phi^\tau(\rho),$$

and the Poincaré map $P_\gamma(\rho)$ is conjugated to the Poincaré map $P_\gamma(\mu)$. Consequently, the *spectrum* spec P_γ of $P_\gamma(\rho)$ is independent of the choice of ρ. We shall say that γ is *non-degenerate* if $1 \notin \operatorname{spec} P_\gamma$.

Denote by Π_i the hyperplane in $\mathbf{R}^n$ passing through the point q_i and orthogonal to the line $q_i q_{i+1}$, and by ω_i the unit vector collinear with $\overrightarrow{q_i q_{i+1}}$. Hereafter we assume for convenience that for $j \equiv i \pmod m$, $\Pi_j = \Pi_i$, $q_j = q_i$, etc. We assume also that for each i the hyperplane Π_i is endowed with a linear basis such that $q_i = 0$.

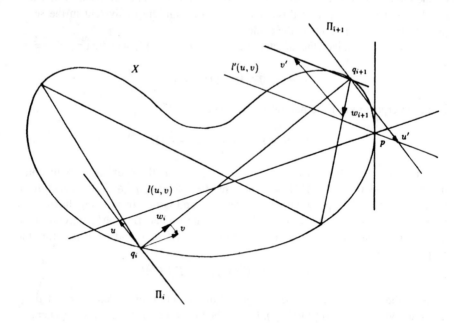

Figure 2.6

For a pair $(u, v) \in \Pi_i \times \Pi_i$ sufficiently close to $(0, 0)$, let $l(u, v)$ be the oriented line passing through u and having direction $\omega_i + v$. Here we identify

the point v with the vector v. If (u, v) is sufficiently close to $(0, 0)$, then $l(u, v)$
intersects transversally X at some point $p = p(u, v)$ close to q_{i+1}. Let $l'(u, v)$ be
the oriented line, symmetric to $l(u, v)$ with respect to the tangential hyperplane to
X at p, and denote by u' the intersection point of $l'(u, v)$ with Π_{i+1} (such a point
clearly exists for (u, v) close to $(0, 0)$). There is $v' \in \Pi_{i+1}$ such that $\omega_{i+1} + v'$
is collinear with the direction of $l'(u, v)$ (cf. Figure 2.6). Thus we obtain a map

$$\Phi_{i+1} : \Pi_i \times \Pi_i \ni (u, v) \to (u', v') \in \Pi_{i+1} \times \Pi_{i+1},$$

defined for (u, v) in a small neighbourhood of $(0, 0)$. The smoothness of this map
follows from the smoothness of the billiard ball map. Consider the composition

$$\mathcal{P}_\gamma = \Phi_m \circ \ldots \circ \Phi_1 : \Pi_m \times \Pi_m \to \Pi_m \times \Pi_m$$

and the linear map

$$d\mathcal{P}_\gamma(0,0) : \Pi_m \times \Pi_m \to \Pi_m \times \Pi_m,$$

Let $\rho \in \tilde{\gamma}$ be a fixed point such that $z = \pi_x(\rho)$ lies on the open segment
(q_m, q_1). Consider the hyperplane Π_z passing through z and orthogonal to $\overrightarrow{q_m q_1}$.
Then we can identify $\Pi_z \times \Pi_z$ with the space $T_\rho(T^*(\Omega))/E_\rho$. Given $(u, v) \in$
$\Pi_m \times \Pi_m$, sufficiently close to $(0, 0)$, consider the oriented line $l(u, v)$, passing
through u with direction $\omega_m + v$. Let $l(u, v)$ intersect Π_z at $\tilde{u}$. Write the unit
vector $\omega(u, v)$ collinear with $l(u, v)$ in the form $\omega(u, v) = w_m + \tilde{v}$, where
$\tilde{v} \in \Pi_z$. Thus we obtain a map

$$\Phi_z : \Pi_m \times \Pi_m \ni (u, v) \to (\tilde{u}, \tilde{v}) \in \Pi_z \times \Pi_z,$$

defined for (u, v) sufficiently close to $(0, 0)$, such that $\Phi_z(0,0) = (0,0)$.
For $(\tilde{u}, \tilde{v}) \in \Pi_z \times \Pi_z$ let $t(\tilde{u}, \tilde{v})$ be the minimal positive number such that

$$\Phi^{t(\tilde{u},\tilde{v})}(\tilde{u}, w_m + \tilde{v}) = (p, w_m + q)$$

with $(p, q) \in \Pi_z \times \Pi_z$. Thus, we obtain a map Q_z, defined in a small neighbourhood of $(0, 0)$ in $\Pi_z \times \Pi_z$ by

$$Q_z(\tilde{u}, \tilde{v}) = (p, q) \in \Pi_z \times \Pi_z.$$

Clearly, $t(0,0) = T, Q_z(0,0) = (0,0)$, and for (u, v) close to $(0, 0)$ we have

$$(\Phi_z \circ \mathcal{P}_\gamma)(u, v) = (Q_z \circ \Phi_z)(u, v).$$

By using the local smoothness of $\Phi^t(\sigma)$ with respect to t and σ, we get

$$(d\Phi_z)(0,0) \circ (d\mathcal{P}_\gamma)(0,0) = (d\Phi^T)_{|\Pi_z \times \Pi_z}(0,0) \circ (d\Phi_z)(0,0).$$

Therefore, $(d\mathcal{P}_\gamma)(0,0)$ is conjugated to the Poincaré map P_γ or γ.

Very often we shall use the notation P_γ for $(d\mathcal{P}_\gamma)(0,0)$ and shall call it *Poincaré map* of γ.

Next, we proceed to describe a representation of $P_\gamma = (d\mathcal{P}_\gamma)(0,0)$. To this end we need some additional notation. Set

$$\lambda_i = \|q_{i-1} - q_i\|, \tag{2.14}$$

and denote by α_i the *tangential hyperplane* to X at q_i, by σ_i the *symmetry* with respect to α_i, and by Π_i' the hyperplane, passing through q_i and orthogonal to ω_{i-1} (i.e. Π_i' is parallel to Π_{i-1}; see Figure 2.7). Clearly,

$$\sigma_i(\omega_{i-1}) = \omega_i, \quad \sigma_i(\Pi_i') = \Pi_i.$$

Choose a continuous *normal field* $\nu_i(q)$ to X for $q \in X$ close to q_i such that

$$\langle \nu_i(q_i), \omega_i \rangle > 0.$$

For $u \in \Pi_i'$ close to 0 (i.e. to q_i) denote by $l(u,0)$ the line through u, orthogonal to Π_i'. The intersection points of $l(u,0)$ with X and α_i will be denoted by $f_i(u)$ and $\pi_i u$, respectively (we choose $f_i(u)$ close to q_i). Then

$$\pi_i : \Pi_i' \rightarrow \alpha_i$$

is the *natural projection* along the vector ω_{i-1}, while f_i is a *local diffeomorphism*

$$f_i : (\Pi_i', q_i) \rightarrow (X, q_i),$$

which can be considered as a parameterization of X around q_i. The *second fundamental form* $S(\xi, \eta)$ of X at q_i is defined for $\xi, \eta \in \alpha_i$ by

$$S(\xi, \eta) = \langle G_i(\xi), \eta \rangle,$$

where

$$G_i = d\nu_i(q_i) : \alpha_i \rightarrow \alpha_i.$$

Since S is symmetric and billinear (cf. [GKM] for example), there exists a unique symmetric linear map

$$\tilde{\psi}_i : \Pi_i \rightarrow \Pi_i \tag{2.15}$$

such that

$$\langle \tilde{\psi}_i \sigma_i(\xi), \sigma_i(\eta) \rangle = -2\langle \omega_{i-1}, \nu_i(q_i) \rangle \langle G_i(\pi_i \xi), \pi_i \eta \rangle \tag{2.16}$$

for all $\xi, \eta \in \Pi_i'$.

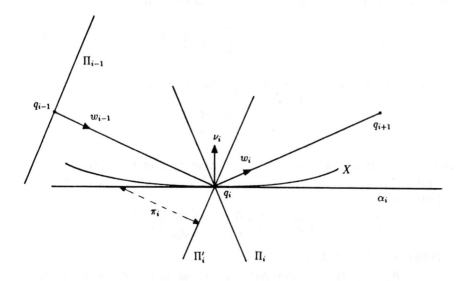

Figure 2.7

Since $\sigma_i(\Pi'_i) = \Pi_i$, with respect to the linear basis fixed in Φ_i and that in Π'_i, obtained by identifying the latter hyperplane with Π_{i-1}, using the translation along the line $q_{i-1}q_i$, we may regard $\sigma_{i|\Pi'_i}$ as a real symmetric $(n-1) \times (n-1)$ matrix. For the sake of brevity we shall denote this matrix again by σ_i. Next, we write the linear maps between the products of the type $\Pi_j \times \Pi_j$ as $2(n-1) \times 2(n-1)$ block-matrices

$$\begin{pmatrix} A & B \\ C & D \end{pmatrix},$$

where A, B, C, D are real $(n-1) \times (n-1)$ matrices. By I we denote the *identity matrix* on Π_j.

After these preparations, fix an arbitrary $i = 1, \dots, m$, and write the map Φ_i in the form $\Phi_i = \Phi_i^{(r)} \circ \Phi_i^{(t)}$, where

$$\Phi_i^{(t)} : \Pi_{i-1} \times \Pi_{i-1} \rightarrow \Pi'_i \times \Pi'_i,$$
$$\Phi_i^{(r)} : \Pi'_i \times \Pi'_i \rightarrow \Pi_i \times \Pi_i$$

are defined in small neighbourhoods of $(0, 0)$ as follows. For $u, v \in \Pi_{i-1}$ let $l(u, v)$ be the oriented line, defined as above. Denote by u' the intersection point of $l(u, v)$ with Π'_i and set $\Phi_i^{(t)}(u, v) = (u', v')$ where $v' \in \Pi'_i$ is such that the vector $v' + w_{i-1}$ is collinear with the direction (orientation) of $l(u, v)$. Finally, set

$\Phi_i^{(r)} = \Phi_i \circ (\Phi_i^{(t)})^{-1}$. Clearly, $\Phi_i^{(t)}$ is a linear map and

$$\Phi_i^{(t)} = \begin{pmatrix} I & \lambda_i I \\ 0 & I \end{pmatrix}. \tag{2.17}$$

Further, write the linear map $R_i = \mathrm{d}\Phi_i^{(r)}(0,0)$ in the form

$$R_i = \begin{pmatrix} A & B \\ C & D \end{pmatrix}.$$

We are going to find the matrices A, B, C, D in terms of λ_i, σ_i and $\tilde{\psi}_i$.

Take $u = 0$ and $v \in \Pi_i'$ close to 0. Then clearly $\Phi_i^{(r)}(0, v) = (0, \sigma_i(v))$, which yields

$$\begin{pmatrix} A & B \\ C & D \end{pmatrix} \begin{pmatrix} 0 \\ v \end{pmatrix} = \begin{pmatrix} 0 \\ \sigma_i v \end{pmatrix},$$

and therefore $B = 0$, $D = \sigma_i$.

Now take $u \in \Pi_i'$ close to 0 and $v = 0$, and set $(u', v') = \Phi_i^{(r)}(u, 0) \in \Pi_i \times \Pi_i$. It follows by the definition of $\Phi_i^{(r)}$ that

$$u' = f_i(u) + t\omega', v' = \omega' - \langle \omega', w_i \rangle w_i \tag{2.18}$$

for some $t \in \mathbf{R}$, where ω' is the vector symmetric to ω_{i-1} with respect to the tangential hyperplane to X at $f_i(u)$, i.e.

$$\omega' = \omega_{i-1} - 2\langle \omega_{i-1}, \nu_i(f_i(u)) \rangle \nu_i(f_i(u)).$$

Setting $\nu_i = \nu_i(q_i)$, we have

$$\nu_i(f_i(u)) = \nu_i + G_i(\pi_i u) + \mathrm{O}(\|u\|^2),$$

and, taking into account that

$$\sigma_i(\omega_{i-1}) = \omega_i = \omega_{i-1} - 2\langle \omega_{i-1}, \nu_i \rangle \nu_i,$$

we find

$$\omega' = \omega_i - 2\langle \omega_{i-1}, G_i(\pi_i u) \rangle \nu_i - 2\langle \omega_{i-1}, \nu_i \rangle G_i(\pi_i u) + \mathrm{O}(\|u\|^2). \tag{2.19}$$

Since $\langle \omega_{i-1}, \nu_i \rangle = \langle \omega_i, \nu_i \rangle$ and $G_i(\pi_i u) \in \alpha_i$, (2.19) implies

$$\begin{aligned} \langle \omega', \omega_i \rangle &= 1 - 2\langle \omega_{i-1}, G_i(\pi_i u) \rangle \langle \nu_i, \omega_i \rangle \\ &\quad - 2\langle \omega_{i-1}, \nu_i \rangle \langle G_i(\pi_i u), \omega_i \rangle + \mathrm{O}(\|u\|^2) \\ &= 1 + \mathrm{O}(\|u\|^2). \end{aligned}$$

Therefore, it follows by (2.18) and (2.19) that

$$v' = -2\langle \omega_i, G_i(\pi_i u)\rangle \nu_i + 2\langle \omega_i, \nu_i\rangle G_i(\pi_i u) + \mathrm{O}(\|u\|^2). \qquad (2.20)$$

To compute u', first combine (2.18), (2.19) and (2.20) to get

$$u' = f_i(u) + t(\omega_i - v') + \mathrm{O}(\|u\|^2) = \pi_i u + t(\omega_i - v') + \mathrm{O}(\|u\|^2). \quad (2.21)$$

Since $\langle u', \omega_i\rangle = \langle v', \omega_i\rangle = 0$, (2.21) yields $t = -\langle \pi_i u, \omega_i\rangle + \mathrm{O}(\|u\|^2)$, therefore

$$u' = \pi_i u - \langle \pi_i u, \omega_i\rangle \omega_i + \mathrm{O}(\|u\|^2) = \sigma_i(\pi_i u - \langle \pi_i u, \omega_i\rangle \omega_{i-1}) + \mathrm{O}(\|u\|^2).$$

On the other hand, it is easily seen that $u = \pi_i u - \langle \pi_i u, \omega_i\rangle \omega_{i-1}$. Combining this with the latter expression for u', we find

$$u' = \sigma_i(u) + \mathrm{O}(\|u\|^2). \qquad (2.22)$$

Now from (2.22) and (2.20) it follows that the components A and C of $R = \Phi_i^{(r)}(0,0)$ have the form $A = \sigma_i$ and

$$Cu = -2\langle \omega_i, G_i(\pi_i u)\rangle \nu_i + 2\langle \omega_i, \nu_i\rangle G_i(\pi_i u). \qquad (2.23)$$

In fact, it is easy to see that $C = \tilde{\psi}_i \sigma_i$. Indeed, since $\sigma_i(v) = \pi_i v - \langle \pi_i v, \omega_i\rangle \omega_i$ for $v \in \Pi_i'$ and $\langle Cu, \omega_i\rangle = 0$, it follows from (2.23) that

$$\langle Cu, \sigma_i(v)\rangle = \langle Cu, \pi_i v\rangle = 2\langle \omega_i, \nu_i\rangle\langle G_i(\pi_i u), \pi_i v\rangle$$
$$= -2\langle \omega_{i-1}, \nu_i\rangle\langle G_i(\pi_i, u), \pi_i v\rangle.$$

Combining the latter with (2.16) one gets

$$\langle (C - \tilde{\psi}_i \sigma_i)(u), \sigma_i(v)\rangle = 0$$

for all $u, v \in \pi_i'$. Therefore $C = \tilde{\psi}_i \sigma_i$, which shows that

$$\Phi_i^{(r)}(0,0) = \begin{pmatrix} \sigma_i & 0 \\ \tilde{\psi}_i \sigma_i & \sigma_i \end{pmatrix}.$$

Finally, the latter and (2.17) imply

$$d\Phi_i(0,0) = \begin{pmatrix} \sigma_i & 0 \\ \tilde{\psi}_i \sigma_i & \sigma_i \end{pmatrix} \begin{pmatrix} I & \lambda_i I \\ 0 & I \end{pmatrix} = \begin{pmatrix} \sigma_i & \lambda_i \sigma_i \\ \tilde{\psi}_i \sigma_i & (I + \lambda_i \tilde{\psi}_i)\sigma_i \end{pmatrix}$$
$$= \begin{pmatrix} I & \lambda_i I \\ \tilde{\psi}_i & I + \lambda_i \tilde{\psi}_i \end{pmatrix} \begin{pmatrix} \sigma_i & 0 \\ 0 & \sigma_i \end{pmatrix}.$$

Next, we identify the hyperplanes Π_{i-1} and Π_i' using the translation along the line $q_{i-1}q_i$ (which is orthogonal to both hyperplanes). Then we can write $\sigma_i(\Pi_{i-1}) = \Pi_i$, and for

$$s_i = \sigma_i \circ \sigma_{i-1} \circ \ldots \circ \sigma_1 \qquad (2.24)$$

one has $s_i(\Pi_m) = \Pi_i$. Consider the symmetric linear map

$$\psi_i = s_i^{-1} \tilde{\psi}_i s_i : \Pi_m \to \Pi_m. \qquad (2.25)$$

Now we can view the matrix

$$\begin{pmatrix} I & \lambda_i I \\ \psi_i & I + \lambda_i \psi_i \end{pmatrix} = \begin{pmatrix} s_i^{-1} & 0 \\ 0 & s_i^{-1} \end{pmatrix} \begin{pmatrix} I & \lambda_i I \\ \tilde{\psi}_i & I + \lambda_i \tilde{\psi}_i \end{pmatrix} \begin{pmatrix} s_i & 0 \\ 0 & s_i \end{pmatrix}$$

as a linear map $\Pi_m \times \Pi_m \to \Pi_m \times \Pi_m$. Combining this with the representation of $d\Phi_i(0,0)$, we obtain the following

Theorem 2.3.1:   *Under the assumptions and conventions above, the Pioncaré map*

$$P_\gamma : \Pi_m \times \Pi_m \to \Pi_m \times \Pi_m$$

is a linear symplectic map which has the following matrix representation

$$P_\gamma = \begin{pmatrix} s_m & 0 \\ 0 & s_m \end{pmatrix} \begin{pmatrix} I & \lambda_m I \\ \psi_m & I + \lambda_m \psi_m \end{pmatrix} \cdots \begin{pmatrix} I & \lambda_1 I \\ \psi_1 & I + \lambda_1 \psi_1 \end{pmatrix}, \qquad (2.26)$$

where s_m, ψ_j, λ_j *are given by* (2.24), (2.25), (2.16), (2.14). ♠.

Let us recall that $X = \partial\Omega$ is called *strictly convex* (convex) at q_i with respect to the normal field $\nu_i(q)$ provided the linear operator $G_i(q_i)$, defined above, is positively definite (non-negatively semi-definite). It follows by (2.16) that the latter is equivalent to the fact that the linear symmetric map $\tilde{\psi}_i$ is positively definite (non-negatively semi-definite).

Proposition 2.3.2:   *Let Ω and γ be as at the beginning of this section, and assume that $X = \partial\Omega$ is strictly convex at q_i with respect to the normal field $\nu_i(q)$ for every $i = 1, \ldots m$. Then the Poincaré map P_γ of γ is hyperbolic, i.e. spec P_γ has no common points with the unit circle.*

Proof: For any k denote by $\mathcal{M}_k$ the space of all linear symmetric maps M : $\Pi_k \to \Pi_k$. Recall that $\Pi_0 = \Pi_m$, according to our notation introduced above.

Let $M_0 \in \mathcal{M}_0$ be non-negatively semi-definite (we shall denote this by $M_0 \geq 0$). Consider the linear subspace

$$L_0 = \{(u, M_0 u) : u \in \Pi_0\}$$

of $\Pi_0 \times \Pi_0$. In fact, L_0 is a Lagrangian subspace of $\Pi_0 \times \Pi_0$ with respect to the natural symplectic structure on $\Pi_0 \times \Pi_0$. Then $d\Phi_1(0,0)(L_0)$ coincides with the linear subspace

$$L_1 = \{\sigma_1(I + \lambda_1 M_0)u, \sigma_1((I + \lambda_1 \psi_1)M_0 + \psi_1)u : u \in \Pi_0\}$$

of $\Pi_1 \times \Pi_1$. It is convenient to introduce the operators

$$\mathcal{F}_i : \mathcal{M}_{i-1} \to \mathcal{M}_i \qquad (2.27)$$

defined by

$$\mathcal{F}_i(M) = \sigma_i M(I + \lambda_i M)^{-1} + \tilde{\psi}_i \qquad (2.28)$$

Then for $M_1 = \mathcal{F}_1(M_0)$ we have

$$L_1 = \{(v, M_1 v : v \in \Pi_1\}.$$

Define inductively

$$M_k = \mathcal{F}_k(M_{k-1}), \quad L_k = \{(u, M_k u) : u \in \Pi_k\}.$$

Then $M_k \in \mathcal{M}_k$ and L_k is a linear subspace of $\Pi_k \times \Pi_k$.

Choose an arbitrary $\epsilon > 0$ such that $\tilde{\psi}_i \geq \epsilon I$ for every $i = 1, \ldots, m$, and denote by $\mathcal{M}_k(\epsilon)$ the subspace of $\mathcal{M}_k$ consisting of all $M \in \mathcal{M}_k$ such that $M \geq \epsilon I$. Note that $\mathcal{F}_k(\mathcal{M}_{k-1}(\epsilon)) \subset \mathcal{M}_k(\epsilon)$ and for any $A, B \in \mathcal{M}_{k-1}(\epsilon)$ we have

$$\mathcal{F}_k(A) - \mathcal{F}_k(B) = \sigma_k((I + \lambda_k A)^{-1}(A - B)(I + \lambda_k B)^{-1})\sigma_k.$$

Therefore

$$\|\mathcal{F}_k(A) - \mathcal{F}_k(B)\| \geq (1 + \epsilon\lambda_k)^{-2}\|A - B\| \geq \frac{\|A - B\|}{(1 + \epsilon\lambda)^2},$$

where $\lambda = \min \lambda_k$. This shows that for any k the map $\mathcal{F}_k$ is a contraction from $\mathcal{M}_{k-1}(\epsilon)$ into $\mathcal{M}_k(\epsilon)$. Then the map

$$\mathcal{F} = \mathcal{F}_m \circ \ldots \mathcal{F}_2 \circ \mathcal{F}_1$$

is a contraction from $\mathcal{M}_0(\epsilon)$ into $\mathcal{M}_0(\epsilon)$. Consequently, there exists a (unique) fixed point $M_0 \in \mathcal{M}_0(\epsilon)$ of $\mathcal{F}$. Now taking into account (2.26) and (2.27), we see that for any $u \in \Pi_m = \Pi_0$

$$P_\gamma \begin{pmatrix} u \\ M_0 u \end{pmatrix} = \begin{pmatrix} Su \\ M_0 S_u \end{pmatrix},$$

where the linear map $S : \Pi_m \rightarrow \Pi_m$ is defined by

$$
\begin{aligned}
S = \sigma_m(I + \lambda_m \mathcal{F}'_{m-1}(M_0)) \circ \\
\sigma_{m-1}(I + \lambda_{m-1} \mathcal{F}'_{m-2}(M_0)) \circ \dots \circ \\
\sigma_2(I + \lambda_2 \mathcal{F}'_1(M_0)) \circ \sigma_1(I + \lambda_1 M_0),
\end{aligned}
$$

and $\mathcal{F}'_k = \mathcal{F}_k \circ \dots \circ \mathcal{F}_2 \circ \mathcal{F}_1$. Moreover, it follows by the expression for S that

$$
\|Sx\| \geq \prod_{i=1}^{m}(1 + \lambda_i \epsilon) \|x\|
$$

for any $x \in \Pi_m$. Consequently,

$$
\operatorname{spec} S \subset \{z \in \mathbf{C} : |z| > 1\}.
$$

The eigenvalues of S are clearly eigenvalues of P_γ. Hence P_γ has $n-1$ eigenvalues z_j with $|z_j| > 1$. Since P_γ is symplectic, $1/z_j$ are eigenvalues, too. This proves the proposition. ♠

Corollary 2.3.3: *Let* $\Omega = \mathbf{R}^n \backslash (K_1 \cup K_2)$, *where* K_1 *and* K_2 *are compact strictly convex disjoint domains with smooth boundaries. Let* γ *be the unique periodic reflecting ray in* Ω *with two reflection points* $q_1 \in K_1, q_2 \in K_2$. *Then*

$$
\operatorname{spec} P_\gamma \subset (0,1) \cup (1,\infty).
$$

Proof: We apply the argument from the proof of the previous proposition. In the special case under consideration we have $\lambda_1 = \lambda_2 = \lambda$, moreover, Π_1 and Π_2 can be identified. Thus

$$
S = \sigma_2(I + \lambda \mathcal{F}'_1(M_0)) \circ \sigma_1(I + \lambda M_0),
$$

and setting $M_2 = M_0(I + \lambda M_0)^{-1} + \psi_1$, we obtain

$$
(I + \lambda M_2)^{-1/2} S (I + \lambda M_2)^{1/2} = (I + \lambda M_2)^{-1/2}(I + \lambda M_0)(I + \lambda M_2)^{1/2}.
$$

Therefore the eigenvalues of S are real and greater than 1, which proves the assertion. ♠

2.4. Scattering rays

Let Ω be a closed connected domain in $\mathbf{R}^n$, $n \geq 2$, with smooth boundary $X = \partial\Omega$ and bounded complement. Set

$$
K = \overline{\mathbf{R}^n \backslash \Omega}, \tag{2.29}
$$

then clearly K is a compact domain with smooth boundary $\partial K = X$ and the complement of Ω coincides with $K \backslash X$. Let ω and θ be two fixed unit vectors in $\mathbf{R}^n$.

Definition 2.4.1: Let γ be a curve in Ω of the form

$$\gamma = \cup_{i=0}^{k} l_i,$$

$k \geq 1$, where $l_i = [x_i, x_{i+1}]$ are finite segments for $i = 1, \ldots, k-1, x_i \in X$ for all i, and l_0 (resp. l_k) is the infinite segment starting at x_1 (resp. at x_k) and having direction $-\omega$ (resp. θ). The curve γ is called a *reflecting* (ω, θ)-*ray* in Ω if for every $i = 0, 1, \ldots, k-1$ the segments l_i and l_{i+1} satisfy the law of reflection at x_{i+1} with respect to X. For such γ the points $x_1, \ldots, x_k$ will be called *reflection points* of γ.

Clearly, for every generalized (ω, θ)-geodesic $c : \mathbf{R} \to \Omega$ in Ω, which has no gliding segments on $\partial \Omega$ and only finitely many reflection points, the curve $\gamma = \operatorname{Im} c$ is a reflecting (ω, θ)-ray in Ω (cf. Section 1.2). It is easy to construct examples showing that the converse is not true.

By a *scattering ray* we mean a reflecting (ω, θ)-ray for some unit vectors ω and θ. Such a ray γ will be called *symmetric* if some segment of γ is orthogonal to X at some of its end points; otherwise γ will be called *non-symmetric*. Clearly, a symmetric reflecting $(\omega, \theta)-$ ray may exists only if $\theta = -\omega$, and for such a ray γ we have $k = 2m + 1$ and $l_{m-i} = l_{m+i-1}$ for all $i = 0, 1, \ldots, m$. A scattering ray without segments tangent to X will be called *ordinary*.

Next, we define two important notions related to a scattering ray. To this end fix an arbitrary open ball U_0 with radius $a > 0$ containing K. For any $\xi \in S^{n-1}$ denote by Z_ξ the hyperplane, orthogonal to ξ and tangent to U_0 such that ξ is pointing into the interior of the open halfspace H_ξ, having boundary Z_ξ and containing U_0. Let

$$\pi_\xi : \mathbf{R}^n \to Z_\xi$$

be the orthogonal projection. For a reflecting (ω, θ)-ray γ in Ω with successive reflection points $x_1, \ldots, x_k$, the *sojourn time* T_γ of γ is defined by

$$T_\gamma = \|\pi_\omega(x_1) - x_1\| + \sum_{i=1}^{k-1} \|x_i - x_{i+1}\| + \|x_k - \pi_{-\theta}(x_k)\| - 2a. \quad (2.30)$$

Clearly, $T_\gamma + 2a$ coincides with the length of this part of γ which lies in $H_\omega \cap H_{-\theta}$ (see Figure 2.8). In fact, T_γ does not depend on the choice of the ball U_0. Indeed, it is easily seen that

$$\|\pi_\omega(x_1) - x_1\| = a + \langle \omega, x_1 \rangle, \quad \|x_k - \pi_{-\theta}(x_k)\| = a - \langle \theta, x_k \rangle.$$

Then (2.30) implies

$$T_\gamma = \langle \omega, x_1 \rangle + \sum_{i=1}^{k-1} \| x_i - x_{i+1} \| - \langle \theta, x_k \rangle, \qquad (2.31)$$

which shows in particular that T_γ does not depend on the choice of U_0.

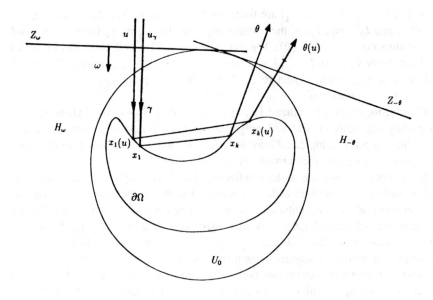

Figure 2.8

Let γ be a reflecting (ω, θ)-ray as above. Set $u_\gamma = \pi_\omega(x_1)$ and assume that γ is ordinary, i.e. it does not contain segments tangent to $X = \partial\Omega$. Then there exists a neighbourhood W_γ of u_γ in Z_ω such that for every $u \in W_\gamma$ there are unique $\theta(u) \in S^{n-1}$ and points $x_1(u), \ldots, x_k(u) \in X$ which are the successive reflection points of a reflecting $(\omega, \theta(u))$-ray in Ω with $\pi_\omega(x_1(u)) = u$. We set $J_\gamma(u) = \theta(u)$, thus obtaining a map

$$J_\gamma : W_\gamma \to S^{n-1}.$$

It follows immediately from the smoothness of the billiard ball map, related to an appropriate domain $\Omega' \subset \Omega \cap H_\omega \cap H_{-\theta}$, that the map J_γ is smooth, too. It is an easy exercise to check the latter fact directly.

Next, applying Theorem 2.3.1, we shall obtain a matrix representation of $dJ_\gamma(u_\gamma)$. To this end set $m = k+2, q_i = x_i$ for $i = 1, \ldots, k, q_0 = \pi_\omega(q_1)$, $q_{k+1} = \pi_{-\theta}(q_k), \lambda_i = \| q_{i-1} q_i \|, \Pi_0 = Z_\omega, \Pi_{k+1} = Z_{-\theta}$. For $i = 1, \ldots, k$

introduce Π_i, σ_i, $\tilde{\psi}_i$ as in Section 2.3. We assume again that in every Π_i a linear basis is fixed with $q_i = 0$. Define the maps

$$\Phi_{i+1} : \Pi_i \times \Pi_i \to \Pi_{i+1} \times \Pi_{i+1},$$

$i = 0, 1, \ldots, k$, as in Section 2.3. Then by the same argument one gets

$$d\Phi_i(0,0) = \begin{pmatrix} \sigma_i & \lambda_i \sigma_i \\ \tilde{\psi}_i \sigma_i & \sigma_i + \lambda_i \tilde{\psi}_i \sigma_i \end{pmatrix}, \quad i = 0, 1, \ldots, k+1.$$

On the other hand,

$$dJ_\gamma(q_0)u = pr_2(d\Phi_{k+1}(0,0) \circ \cdots \circ d\Phi_1(0,0) \begin{pmatrix} u \\ 0 \end{pmatrix})$$

for $u \in \Pi_0$, where $pr_2 \begin{pmatrix} u \\ v \end{pmatrix} = v$. Therefore

$$dJ_\gamma(q_0)u = pr_2 \begin{pmatrix} \sigma_k & \lambda_k \sigma_k \\ \sigma_k \tilde{\psi}_k & \sigma_k + \lambda_k \sigma_k \tilde{\psi}_k \end{pmatrix} \cdots$$
$$\times \begin{pmatrix} \sigma_1 & \lambda_1 \sigma_1 \\ \tilde{\psi}_1 \sigma_1 & \sigma_1 + \lambda_1 \tilde{\psi}_1 \sigma_1 \end{pmatrix} \begin{pmatrix} u \\ 0 \end{pmatrix}. \tag{2.32}$$

The next proposition treats a special case of scattering rays.

Proposition 2.4.2: *Under the notation and conventions above, assume in addition that for every $i = 1, \ldots, k$, $X = \partial K$ is (strictly) convex at $q_i = x_i$ with respect to the normal field ν_i pointing to the interior of Ω. Then for every $u \in \Pi_0 = Z_\omega$*

$$dJ_\gamma(q_0)u = M_k \sigma_k (I + \lambda_k M_{k-1}) \sigma_{k-1} (I + \lambda_{k-1} M_{k-2}) \cdots$$
$$\times \sigma_2 (I + \lambda_2 M_1) \sigma_1 u, \tag{2.33}$$

holds, where

$$M_i : \Pi_i \to \Pi_i, \quad i = 1, \ldots, k, \tag{2.34}$$

are non-negatively semi-definite (positively definite) symmetric linear maps defined inductively by

$$M_1 = \tilde{\psi}_1, \quad M_i = \sigma_i M_{i-1}(I + \lambda_i M_{i-1})^{-1} \sigma_i + \tilde{\psi}_i, \tag{2.35}$$

$i = 2, 3, \ldots, k+1$. *In particular, $\det dJ_\gamma(q_0) \neq 0$.*

Proof: Since ∂K is (strictly) convex at q_i, it follows by the definitions of $\tilde{\psi}_i$ (cf. Section 2.3) that they are non-negatively semi-definite (positively definite) symmetric linear maps. Now we can use (2.34) to define the maps M_i inductively; the definition is correct and $M_i \geq 0$ (resp. $M_i > 0$) for any $i = 1, \ldots, k$.

Set

$$\begin{pmatrix} u_i \\ v_i \end{pmatrix} = d\Phi_i(0,0) \circ \cdots \circ d\Phi_1(0,0) \begin{pmatrix} u \\ 0 \end{pmatrix}.$$

Clearly, $u_1 = \sigma_1 u$, $v_1 = \tilde{\psi}_1 \sigma_1 u$. We shall prove by induction that

$$u_i = \sigma_i(I + \lambda_i M_{i-1})u_{i-1}, \quad v_i = M_i u_i \qquad (2.36)$$

for every $i = 2, 3, \ldots, k+1$. From this the equality (2.33) follows immediately. Assume that $v_{i-1} = M_{i-1}u_{i-1}$ for some $i > 1$. Then

$$\begin{pmatrix} u_i \\ v_i \end{pmatrix} = d\Phi_i(0,0) \begin{pmatrix} u_{i-1} \\ v_{i-1} \end{pmatrix} = \begin{pmatrix} I & \lambda_i I \\ \tilde{\psi}_i & I + \lambda_i \tilde{\psi}_i \end{pmatrix} \begin{pmatrix} \sigma_i u_{i-1} \\ \sigma_i M_{i-1} u_{i-1} \end{pmatrix}$$

$$= \begin{pmatrix} \sigma_i(I + \lambda_i M_{i-1})u_{i-1} \\ (\tilde{\psi}_i \sigma_i + \sigma_i M_{i-1} + \lambda_i \tilde{\psi}_i \sigma_i M_{i-1})u_{i-1} \end{pmatrix}.$$

Thus, $u_i = \sigma_i(I + \lambda_i M_{i-1})u_{i-1}$ and

$$\begin{aligned} v_i &= (\sigma_i M_{i-1} + \tilde{\psi}_i \sigma_i(I + \lambda_i M_{i-1}))u_{i-1} \\ &= (\sigma_i M_{i-1}(I + \lambda_i M_{i-1})^{-1}\sigma_i + \tilde{\psi}_i)\sigma_i(I + \lambda_i M_{i-1})u_{i-1} \\ &= M_i u_i. \end{aligned}$$

Therefore (2.36) holds, which proves the assertion. ♠

From now on till the end of this Section we assume that K has the form (2.2), where $s \geq 2$ and K_i are disjoint strictly convex compact domains in $\mathbf{R}^n$ with smooth *boundaries* $\partial K_i = \Gamma_i$. As before ω and θ will be two fixed unit vectors and U_0 an open ball containing K. We shall use also the notation Z_ξ, H_ξ and π_ξ introduced above.

For any $u \in Z_\omega$ consider the *billiard semi-trajectory*

$$\gamma(u) = \{S_t(u) : t \geq 0\},$$

such that $S_0(u) = u$ and $N_0(u) = \omega$, where $N_t(u)$ is the *velocity vector* of the trajectory $S_t(u)$ at time t (i.e. the unit vector determining the direction of the trajectory). For $y = S_t(u) \in X$ we set

$$N_{-t}(u) = \lim_{\epsilon \to 0} N_{t-\epsilon}, \quad \epsilon > 0,$$

and $N_{+t}(u) = \sigma_y(N_{-t}(u))$, σ_y being the *symmetry with respect to the tangent hyperplane* to X at y. By $x_1(u)$, $x_2(u)$, ... we denote the successive reflection points of $\gamma(u)$, and by $t_1(u)$, $t_2(u)$, ... the corresponding times (moments) of reflection. Notice that the points $x_j(u)$ include not only the *proper (transversal) reflection points* of $\gamma(u)$ but its *tangent (reflection) points* as well. For convenience set

$$x_0(u) = u, \quad t_0(u) = 0,$$

and denote by $r(u)$ the *number of reflections* of $\gamma(u)$. Thus $r(u)$ is a non-negative integer or ∞.

Denote by $\mathcal{A}'_k$ the set of all symbols of the form (2.3) such that $i_j \neq i_{j+1}$ for each $j = 1, \ldots, k-1$. Here we do not assume that $i_1 \neq i_k$, so for $k > 1$ and $s > 1$ the set $\mathcal{A}_k$, introduced in Section 2.2, is a proper subset of $\mathcal{A}'_k$. The elements of the latter set will be called again *configurations* of length k.

Let $\alpha \in \mathcal{A}'_k$ be a fixed configuration of the form (2.3). Set

$$F_\alpha = \{u \in Z_\omega : r(u) \geq k, x_j(u) \in K_{i_j} \quad \text{for } 1 \leq j \leq k\}, \quad (2.37)$$

and denote by U_α the set of those $u \in F_\alpha$ such that $x_j(u)$ is a proper reflection point of $\gamma(u)$ for each $j = 1, \ldots, k$. Clearly, F_α is a bounded subset of Z_ω (it is contained in $\pi_\omega(K_{i_1})$) consisting of these $u \in Z_\omega$ such that the trajectory $\gamma(u)$ has at least k reflection points and the first k reflections 'follow' the configuration α. In general F_α is not a closed subset of Z_ω. However, it is easy to see that U_α is open in Z_ω, $U_\alpha \subset F_\alpha$.

For every $u \in F_\alpha$ set

$$J_\alpha(u) = N_{t+}(u) \in S^{n-1}, \quad t = t_k(u).$$

Thus we obtain a continuous map

$$J_\alpha : F_\alpha \to S^{n-1}. \quad (2.38)$$

It is clear that J_α is smooth on U_α.

The rest of this section is devoted to the study of the map J_α. First we consider some immediate consequences from the definition of this map.

Lemma 2.4.3: *(a) For every $u \in \bar{F}_\alpha$ there exists $\beta = (i'_1, \ldots i'_p) \in \mathcal{A}'_p$, $p \geq k$, such that $u \in F_\beta$ and there is a sequence $p_1 < p_2 < \cdots < p_k = p$ with $i_{p_j} = i_j$ for any $j = 1, \ldots, k$. Moreover, if $x_r(u)$ is a proper reflection point of $\gamma(u)$ for some $r = 1, \ldots, k$, then $i_r = i_{p_j}$ for some $j = 1, \ldots, k$;*
(b) J_α can be extended to a continuous map

$$J_\alpha : \bar{F}_\alpha \to S^{n-1}. \spadesuit$$

Before going on, let us introduce some additional notation. Set

$$L_\alpha = \{u \in \bar{F}_\alpha : N_t(u) = \omega, \ t \geq 0\}, \quad (2.39)$$
$$M_\alpha = \bar{F}_\alpha \backslash L_\alpha, \quad E_\alpha = J_\alpha(\bar{F}_\alpha). \quad (2.40)$$

Note that L_α is a compact subset of Z_ω contained in the boundary (in Z_ω) of the convex subset $\pi_\omega(K_{i_1})$. Hence L_α has Lebesgue measure zero and empty interior in Z_ω. In fact, it is a smooth compact $(n-2)$-dimensional submanifold. It is clear from the definition that L_α is non-empty only for special K and special configurations α. Since $\bar{F}_\alpha$ is compact, E_α is a compact subset of S^{n-1}. Finally, since $J_\alpha(u) = \omega$ for any $u \in L_\alpha$, the set $J_\alpha(M_\alpha)$ coincides either with E_α (in most of the cases) or with $E_\alpha \setminus \{\omega\}$.

To prove the next property of J_α we make use of the main result of Section 2.2.

Proposition 2.4.4: *For every configuration α, the map*

$$J_\alpha : M_\alpha \rightarrow J_\alpha(M_\alpha) \tag{2.41}$$

is a homeomorphism.

Proof: It is sufficient to show that if $u \in M_\alpha$, $v \in \bar{F}_\alpha$ and $J_\alpha(u) = J_\alpha(v)$, then $u = v$. Indeed, assume this assertion is true. Then (2.41) is a continuous bijection, and we have to verify that its inverse is continuous. Let $\{u_k\} \subset M_\alpha$ and $u \in M_\alpha$ be such that $\lim_{k \to \infty} J_\alpha(u_k) = J_\alpha(u)$ and let v be an arbitrary cluster point of $\{u_k\}$. Then $v \in \bar{F}_\alpha$ and clearly $J_\alpha(u) = J_\alpha(v)$. Therefore $u = v$, and according to the compactness of $\bar{F}_\alpha$ we conclude that $\lim_{k \to \infty}(u_k) = u$. Consequently, the map (2.41) is a homeomorphism.

Let α have the form (2.3) and let $u \in M_\alpha$, $v \in \bar{F}_\alpha$, $v \neq u$. It follows by the definition of M_α that $\gamma(u)$ has at least one proper reflection point. Let $j_1 < \cdots < j_m (m \leq k)$ be all natural numbers, not greater than k, such that $y_l = x_{i_{j_l}}(u)$ are all proper reflection points of $\gamma(u)$. It follows by Lemma 2.4.2(a) that there exists a configuration β, with the properties listed in the lemma, such that $v \in F_\beta$. Moreover, there exist reflection points $z_l \in K_{i_{j_l}} (l = 1, \ldots, m)$ of $\gamma(v)$ such that the successive proper reflection points of $\gamma(v)$ form a subsequence of $z_1, \ldots, z_m$.

Assume that $J_\alpha(u) = J_\alpha(v) = \eta \in S^{n-1}$. Consider arbitrary convex domains K_0 and K_{s+1} with smooth boundaries in $\mathbf{R}^n$ such that $K_0 \subset \mathbf{R}^n \setminus H_\omega$, $K_{s+1} \subset H_\eta$ and $\pi_\omega(K) \subset \partial K_0$, $\pi_{-\eta}(K) \subset \partial K_{s+1}$.

Now we are in a position to apply Lemma 2.2.5 to

$$K' = K_0 \cup K_1 \cup \ldots \cup K_s \cup K_{s+1}$$

and the configuration

$$\lambda = (0, i_{j_1}, \ldots, i_{j_m}, s+1, i_{j_m}, \ldots, i_{j_1}, 0).$$

Consider the convex domain

$$K_\lambda = K_0 \times K_{i_{j_l}} \times \ldots \times K_{i_{j_m}} \times K_{s+1} \times K_{i_{j_m}} \times \ldots \times K_{i_{j_l}} \times K_0$$

in $(\mathbf{R}^n)^{2m+3}$ and the corresponding length function

$$F = F_\lambda : K_\lambda \rightarrow \mathbf{R}$$

(cf. (2.4) and (2.5)). Then by Lemma 2.2.6(b), F has a strict local minimum at the

point
$$\tilde{q} = (q_0, q_1, \ldots, q_m, q_{m+1}, q_m, \ldots, q_1, q_0),$$
where $q_i = x_i(u)$ for $i = 0, 1, \ldots, m$ and $q_{m+1} = \pi_\eta(q_m)$. It then follows by the convexity of F that it has no other local minima in K_λ. On the other hand, by Lemma 2.2.6(a), F has also a local minimum at the point
$$\tilde{p} = (p_0, p_1, \ldots, p_m, p_{m+1}, p_m, \ldots, p_1, p_0),$$
where $p_i = x_i(v)$ for $i = 0, 1, \ldots, m$ and $p_{m+1} = \pi_\eta(p_m)$. Therefore $\tilde{q} = \tilde{p}$ which implies $u = q_0 = p_0 = v$ in contradiction with $u \neq v$.

We have shown in this way that $J_\alpha(u) \neq J_\alpha(v)$ which proves the proposition. ♠

Combining Propositions 2.4.2 and 2.4.4 we get the following.

Corollary 2.4.5: *For every configuration α the map*
$$J_\alpha : U_\alpha \rightarrow J_\alpha(U_\alpha)$$
is a diffeomorphism. ♠

Let α be a configuration of the form (2.3) and γ be a reflecting (ω, θ)-ray in Ω with successive reflection points $x_1, \ldots, x_k$. We shall say that γ *is of type α* if $x_j \in K_{i_j}$ for all $j = 1, \ldots, k$.

The following assertion is another direct consequence of Proposition 2.4.4.

Corollary 2.4.6: *If $\omega \neq \theta$, then for every configuration α there exists at most one reflecting (ω, θ)-ray of type α in Ω.* ♠

Under some conditions on α (resp. K), ω and θ, it is shown in Section 10.3 that there exists a (unique) reflecting (ω, θ)-ray of type α in Ω.

2.5. Examples

In this section we describe the reflecting (ω, θ)-rays for some special connected domains Ω in $\mathbf{R}^3$ such that the corresponding obstacles K, given by (2.29), are non-convex. The choice of the vectors ω and θ is also special.

The domains K will be obtained by rotation with respect to the axis $x = 0$ of domains D in $\mathbf{R}^2 = \{(x, y)\}$ with smooth boundaries $\Gamma = \partial D$, symmetric with respect to $x = 0$.

Example 2.5.1: Let $\Gamma = \Gamma_1 \cup \Gamma_2$, where Γ_1 is the graph of an even continuous function $f : [-m_1, m] \rightarrow [0, \infty)$, which is C^∞ in $(-m, m)$, while Γ_2 is the graph of another even continuous function
$$\tilde{f} : [-m, m] \rightarrow (-\infty, 0].$$

We will assume that $\tilde{f}$ is chosen in such a way that Γ_2 is strictly convex, and the compact domain D in $\mathbf{R}^2$ with $\partial D = \Gamma$ is star-shaped. Namely, we may choose

$\tilde{f}$ so that for every $z \in D$ the segment connecting z and $(0, \tilde{f}(0))$ lies entirely in D.

For f we assume that there exist real numbers b, c with $0 < c < b < a = b + c < m$ and such that:

$$f'(x) > 0 \quad \forall\, x \in (0, a), \quad f'(x) < 0 \quad \forall\, x \in (a, m), \tag{2.42}$$
$$f'(0) = f'(a) = 0, f'(c) = 1, \text{ and}$$
$$f''(x) > 0 \quad \forall\, x \in [0, c), \quad f''(x) < 0 \quad \forall\, x \in (c, m). \tag{2.43}$$

Let $\epsilon \in (0, \frac{1}{2})$ be such that

$$2\epsilon < \frac{f(a) - f(b)}{a + b}. \tag{2.44}$$

We shall assume also that f' satisfies the inequalities

$$0 < 1 - |f'(x)| < \epsilon, \quad x \in [-b, -c]. \tag{2.45}$$

Define the vectors ω and θ by

$$\omega = (0, -1, 0), \quad \theta = (\cos\theta_0, \sin\theta_0, 0), \tag{2.46}$$

where θ_0 is chosen so that

$$\frac{f(a) - f(b)}{2b} < \tan\theta_0 \leq \tan\frac{\pi}{8} \leq \frac{f(b) - f(c)}{b + 2c}. \tag{2.47}$$

We use also the notation $\omega = (0, -1)$, $\theta = (\cos\theta_0, \sin\theta_0)$.

Finally, we assume that f is strictly concave for $x \in [a, m)$ and $f(m) = 0$. Now we choose an even continuous function $\tilde{f}$ which is smooth and convex in $(-m, m)$ and $\tilde{f}(m) = 0$. Moreover, we may choose f and $\tilde{f}$ such that the curve Γ is smooth at $(m, 0)$, and D is a star-shaped domain (see Figure 2.9).

Clearly, there exists a family of quadruples $(f, \tilde{f}, \epsilon, \theta_0)$, satisfying all the conditions above. Next, we consider a fixed quadruple of this type, and define K and Ω as at the beginning of this section.

We are going to establish that *the set* $L_{\omega,\theta}(\Omega)$ *of all generalized* (ω, θ)*- geodesics in* Ω *has the form*

$$L_{\omega,\theta}(\Omega) = \{\gamma_0, \gamma_1, \gamma_2\}, \tag{2.48}$$

where $\gamma_0, \gamma_1, \gamma_2$ *are ordinary reflecting* (ω, θ)*-rays in* Ω, γ_0 *has two reflection points while each of* $\gamma_i, i = 1, 2,$ *has one reflection point. Moreover, for the corresponding sojourn times* $T_i = T_{\gamma_i}$ *and differential cross-sections* $\mathrm{d}J_i = \mathrm{d}J_{\gamma_i}$ *we have* $T_0 > T_1 > T_2$ *and* $\det \mathrm{d}J_i \neq 0$ *for* $i = 0, 1, 2$.

To prove this assertion, we first consider reflecting (ω, θ)-rays for Γ in $\mathbf{R}^2$.

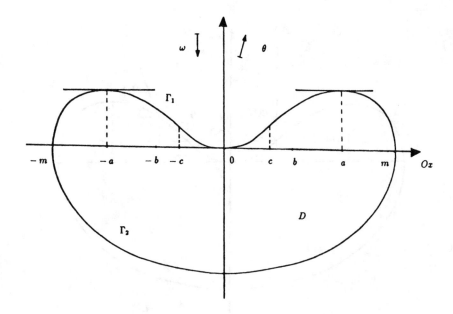

Figure 2.9

Let γ be a reflecting (ω, θ)-ray for Γ with N reflection points. Clearly, the first reflection point has the form $q_1 = (x, f(x))$, $x \in (-a, a)$. Denote by l_x the ray starting at $(x, f(x))$ and having direction, symmetric to ω with respect to the unit normal $\nu(x)$ (see Figure 2.10), and by $\varphi(x)$ the measure of the oriented angle between the axis Ox and the ray l_x.

We begin with the case $x \in (-a, 0)$. Then

$$\varphi(x) = \frac{\pi}{2} - 2\alpha(x) = \frac{\pi}{2} + 2\tan^{-1} f'(x), \tag{2.49}$$

where $\tan \alpha(x) = |f'(x)| \le 1$. Since $\alpha(x) \in \left[0, \frac{\pi}{4}\right]$, we have $\varphi(x) \in \left[0, \frac{\pi}{4}\right]$. Moreover, it follows from

$$\varphi'(x) = \frac{2f''(x)}{(1 + f'(x))^2}$$

that φ is decreasing in $[-a, -c]$ and increasing in $[-c, 0]$.

Assume that the ray l_x has a common point $q_2 = (y, f(y))$ with Γ (i.e. we assume that $N \ge 2$). Then clearly $y \in [0, a]$. In general l_x could be tangent to Γ at q_2. By l_y we denote the ray starting at q_2 with direction, symmetric to the direction of l_x with respect to the normal $\nu(y)$, and by $\psi(x)$ the measure of the

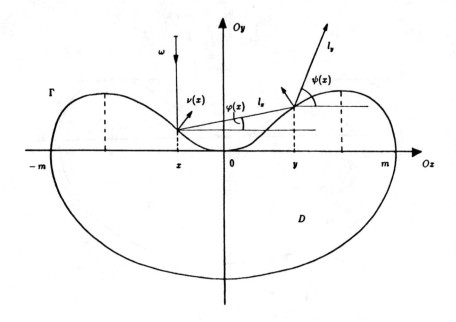

Figure 2.10

oriented angle between the axis Ox and l_y (see Figure 2.10). Then

$$\psi(x) = 2\beta(y) - \varphi(x),$$

$\beta(y)$ being the measure of the angle between Ox and the tangent line to Γ at q_2. It is easily seen that $\beta(y) \geq \varphi(x)$. Since $\beta(y) \leq \dfrac{\pi}{4}$, we get

$$0 \leq \psi(x) \leq 2\beta(y) \leq \frac{\pi}{2}. \tag{2.50}$$

Before going on, let us mention that the case $N \geq 2, x \geq 0$ is impossible. This follows easily, applying an argument, similar to the above one. Therefore, if $N \geq 2$, then the first two reflection points of γ have the form $q_1 = (x, f(x)), x \in (-a, 0), q_2 = (y, f(y)), y \in (0, a)$. Next, we show that

$$x \in (-a, -c), \quad y \in (c, a]. \tag{2.51}$$

Assume $x \in [-c, 0)$. It follows by $|f'(y)| \leq 1$ that

$$\frac{\pi}{2} - 2\alpha(x) = \varphi(x) \leq \frac{\pi}{4},$$

hence $\alpha(x) \geq \frac{\pi}{4}$. This is why $x \in (-c, -d)$ for $d \in (0, c)$, determined by $f'(d) = \tan \frac{\pi}{8}$. Then clearly $y > |x| > d$. Applying a simple geometrical argument, one gets

$$\theta_0 \geq \psi(x) = 2\beta(y) - 2\varphi(x) \geq \beta(y).$$

If $y \leq b$, then $y > b$ implies $\frac{\pi}{8} < \beta(y)$, i.e. $\theta_0 > \frac{\pi}{8}$ in contradiction with (2.47). Let $y > b$. Then $N = 2$ and $\theta_0 \geq \psi(x) > \varphi(x)$. On the other hand,

$$\tan \varphi(x) = \frac{f(y) - f(x)}{y - x} > \frac{f(b) - f(c)}{b + 2c},$$

therefore

$$\theta_0 > \tan^{-1} \frac{f(b) - f(c)}{b + 2c} \geq \frac{\pi}{8},$$

which is a contradiction with (2.47) again.

In this way we have established that (2.51) holds. Since f is strictly concave for $y \in (c, a)$, we see that $N = 2$.

Therefore every reflecting (ω, θ)-ray for Γ has at most two reflection points. Now we are going to show that there exists exactly one such ray with two reflection points.

There exists $\delta \in (-a, -c)$ with $\varphi(\delta) = \frac{\pi}{4}$. Then $\tan \varphi(x)$ is strictly decreasing for $x \in [\delta, -c]$. Since

$$f' : [c, a] \to [0, 1]$$

is strictly decreasing, there exists a continuous function

$$g : [\delta, -c] \to [c, a]$$

such that $g(x) = (f')^{-1}(\tan \varphi(x))$, i.e. $f'(g(x)) = \tan \varphi(x)$. The latter means that the ray l_x is collinear with the tangent line to Γ at $(g(x), f(g(x)))$. For $x \in (-\delta, -c)$ we have

$$f''(g(x)) \cdot g'(x) = \frac{\varphi'(x)}{\cos^2 \varphi(x)} < 0,$$

therefore $g'(x) > 0$. Consequently, g is strictly increasing in $[\delta, -c]$. Set

$$h(x) = f(g(x)) - f(x) - (g(x) - x) \tan \varphi(x)$$

for $x \in [\delta, -c]$. One checks by a simple calculation that $h'(x) > 0$ for any $x \in (\delta, -c)$. Moreover,

$$h(\delta) = f(c) - f(\delta) - c + \delta \leq \delta - f(\delta) < 0, \quad h(-c) = f(a) - f(c) > 0.$$

Therefore, there exists unique $\xi \in (\delta, -c)$ with $h(\xi) = 0$, $h < 0$ in $[\delta, \xi)$ and $h > 0$ in $(\xi, -c]$. Geometrically, this means that l_x is tangent to Γ at

$(g(\xi), f(g(\xi)))$, l_x intersects transversally Γ for $x \in (\xi, -c]$ and l_x has no common points with Γ for $x \in [-a, \xi)$.

Further, we show that l_{-b} intersects transversally Γ. This will imply $\xi < -b$. Let l_{-b} intersects the line $x = a$ at the point (a, λ). It is sufficient to show that $\lambda < f(a)$. Set $\alpha = \alpha(-b)$. Then $\tan \alpha = |f'(b)|$ and by (2.49), $\varphi(-b) = \frac{\pi}{2} - 2\alpha$. Now (2.45) implies

$$\tan \varphi(-b) = \cot 2\alpha = \frac{1 - \tan^2 \alpha}{2 \tan \alpha} = \frac{1 - |f'(-b)|^2}{2|f'(-b)|}$$

$$= \epsilon(1 + |f'(-b)|) \leq 2\epsilon,$$

since $2|f'(-b)| > 1$. Therefore, according to the choice of ϵ, we get

$$\lambda - f(b) = (a + b) \tan \varphi(-b) < 2\epsilon(a + b) < f(a) - f(b).$$

Consequently, $\lambda < f(a)$, which means that l_{-b} intersects transversally Γ. Thus $\xi < -b$.

Set $y = f(\xi)$. As we have mentioned above, l_ξ is tangent to Γ at $(y, f(y))$. Then

$$\psi(\xi) = \varphi(\xi) = \tan^{-1} \frac{f(y) - f(\xi)}{y - \xi} < \tan^{-1} \frac{f(a) - f(b)}{2b} < \theta_0.$$

Next, for $x \in [\xi, -b)$ define $y(x) \in (0, a]$ to be the first intersection point of l_x with Γ. Then

$$f(y(x)) - f(x) = (y(x) - x) \tan \varphi(x), \quad x \in [\xi, -c)$$

and $c \leq y(x) \leq g(x) \leq a$. Moreover, $y(x) < g(x)$ for $x \in (\xi, -b]$. Consider the function

$$F(x, y) = f(y) - f(x) - (y - x)f'(g(x))$$

for $y < g(x)$, $x \in (\xi, -c]$. Since

$$\frac{\partial F}{\partial y} = f'(y) - f'(g(x)) > 0,$$

by the implicit function theorem we obtain that $y(x)$ is differentiable for $x \in (\xi, -c]$ and

$$y'(x) = \frac{f'(x) - f'(g(x)) + (y - x)f''(g(x))g'(x)}{f'(y(x)) - f'(g(x))} > 0.$$

It is easily seen that $y(x)$ is continuous at $x = \xi$. Then $\psi(x) = 2\beta(y(x)) - \varphi(x)$ is continuous for $x \in [\xi, -c]$. Moreover, this function is differentiable in $(\xi, -c)$ and

$$\psi'(x) = 2\beta'(y(x))y'(x) - \varphi'(x) > 0,$$

which shows that ψ is strictly increasing in $[\xi, -c]$. Since $\psi(\xi) = \beta(\xi) = \varphi(\xi) < \theta_0 < \frac{\pi}{2}$ and $\psi(-c) = 2\beta(c) - \varphi(-c) = \frac{\pi}{2}$, there exists unique $x_0 \in (\xi, -c)$ with $\psi(x_0) = \theta_0$. In this way we have proved that there exists a unique reflecting (ω, θ)-ray γ_0 with two reflection points, and the reflection points of γ_0 are $q_1 = (x_0, f(x_0))$ and $q_2 = (y(x_0), f(y(x_0)))$.

Finally, we have to find all reflecting (ω, θ)-rays for Γ with one reflection point. This means to find all $x \in (-a, m)$ with $\varphi(x) = \theta_0$. A simple argument shows that there exist unique $x_1 \in (-a, \xi)$ and $x_2 \in (a, m)$ with $\varphi(x_i) = \theta_0$, $i = 1, 2$. Consequently, there exist exactly two reflecting (ω, θ)-rays γ_1 and γ_2 each of them having one reflection point, resp. $p_i = (x_i, f(x_i)), i = 1, 2$.

Note that there exits (unique) $z \in [\xi, 0]$ with $\varphi(z) = \theta_0$, but there is no reflecting (ω, θ)-ray with one reflection point $(z, f(z))$, because the ray l_z intersects Γ transversally. Indeed let (a, λ) be the intersection point of l_z with the line $x = a$. If $\lambda \leq f(a)$, then

$$\tan \theta_0 = \frac{\lambda - f(z)}{a - z} > \frac{f(a) - f(c)}{b + 2c} > \frac{f(b) - f(c)}{b + 2c} \geq \tan \frac{\pi}{8},$$

which is a contradiction with (2.47). Hence γ_1 and γ_2 are the only reflecting (ω, θ)-rays for Γ with one reflection point.

The above considerations show also that any generalized (ω, θ)-geodesic for Γ is in fact a reflecting (ω, θ)-ray for Γ, i.e. it coincides with one of the rays $\gamma_0, \gamma_1, \gamma_2$.

Let T_i be the sojourn times of $\gamma_i, i = 0, 1, 2$. Then $T_i = \langle p_i, \omega - \theta \rangle$ for $i = 1, 2$ and

$$T_0 = \langle q_1, \omega \rangle - \langle q_2, \theta \rangle + \|q_1 q_2\|.$$

Consider the function $H(p) = \langle p, \theta - \omega \rangle$, $p \in \Gamma$. Clearly, H has a global maximum at p_2, therefore $T_1 = -H(p_1) > -H(p_2) = T_2$. Since

$$\Gamma' = \{(x, f(x)) : x \in [-a, -c]\}$$

is strictly convex, the restriction of H on Γ' has a total maximum at $p_1 \in \Gamma'$. Hence

$$\begin{aligned} T_0 &= \langle q_1, \omega - \theta \rangle + \langle q_1 - q_2, \theta \rangle + \|q_1 q_2\| \\ &> \langle q_1, \omega - \theta \rangle = -H(q_1) \geq -H(p_1) = T_1. \end{aligned}$$

Thus $T_0 > T_1 > T_2$.

Now we consider the bounded domain K in $\mathbf{R}^3$ with boundary ∂K, obtained by rotating Γ around the axis Oy. First, note that any reflecting (ω, θ)-ray in the unbounded domain Ω, determined by (2.29), lies in the plane Oxy in $\mathbf{R}^3$. This follows from the fact that the normal $\nu(z)$ to ∂K for every $z \in \partial K$ is parallel to the plane A_z containing z and the axis Oy. Therefore if γ is a reflecting (ω, θ)-ray in Ω with first reflection point z, then $\gamma \subset A_z$. In particular, θ is parallel to A_z which means that z belongs to the plane, determined by the axis Oy and θ, i.e. $z \in Oxy$. Clearly, the latter implies $\gamma \subset Oxy$. In this way we have shown that the only (ω, θ)-rays in Ω are $\gamma_i, i = 0, 1, 2$. A similar argument shows that any generalized (ω, θ)-geodesic in Ω is contained in Oxy, therefore (2.48) holds.

Fix an arbitrary $y_0 > \max f$ and consider the plane

$$Z = \{(x, y_0, z) : x, z \in \mathbf{R}\}.$$

Denote by u_i the first intersection point of γ_i with Z and by $\tilde{J}_i$ the restriction of the map

$$J_i = J_{\gamma_i} : Z \to S^2$$

to the line $Z \cap Oxy$. Clearly, for $i = 1, 2$ we have

$$\tilde{J}_i(x, y_0) = (\cos \alpha(x), \sin \alpha(x)),$$

which shows that $\det d\tilde{J}_i(x_i, y_0) \neq 0$ for $i = 1, 2$. On the other hand

$$\tilde{J}_0(x, y_0) = (\cos \psi(x), \sin \psi(x)),$$

therefore $\det d\tilde{J}_0(x_0, y_0) \neq 0$. Next, using the fact that ∂K is obtained by rotating Γ around the axis Oy, we deduce

$$\left| \det J_i(u_i) \right| = \left| \det \tilde{J}_i(x_i, y_0) \right| \neq 0$$

for any $i = 0, 1, 2$. ♠

Next, we consider an example of a domain Ω in $\mathbf{R}^3$ and vector ω and θ such that there are infinitely many reflecting (ω, θ)-rays in Ω. In fact, Ω is the exterior of a two-dimensional tours in $\mathbf{R}^3$.

Example 2.5.2:   Let $\Gamma = \Gamma_1 \cup \Gamma_2$, where Γ_1 and Γ_2 are the circles in the plane Oxy with radii $r < a$ and centres $(-a, 0)$ and $(a, 0)$, respectively. Define ω and θ as in the previous example with $0 < \theta_0 \leq \frac{\pi}{2}$. We assume that

$$\theta_0 > \frac{r}{a - r}.$$

Then it is easy to see (cf. Section 10.3, where a more general fact is proved) that for any $m \in \mathbf{N}, i = 1, 2$ there exists a reflecting (ω, θ)-ray $\gamma_m^{(i)}$ for Γ with exactly m reflection points the first of which belongs to Γ_i. It follows by Corollary 2.4.6 that such a ray is unique.

Denote by ∂K the two-dimensional torus in $\mathbf{R}^3$, obtained by rotating Γ around the axis Oy, and by Ω the unbounded domain in $\mathbf{R}^3$ with $\partial \Omega = \partial K$. Applying arguments similar to those at the end of the previous example, we see that

$$\mathcal{L}_{\omega, \theta}(\Omega) = \left\{ \gamma_m^{(i)} : m \in \mathbf{N}, \quad i = 1, 2 \right\}.$$

Moreover, all elements of $\mathcal{L}_{\omega, \theta}(\Omega)$ are ordinary and the corresponding maps dJ_γ are non-degenerate. ♠

2.6. Notes

The material discussed in Section 2.1 was already known to Birkhoff [Bir]. The reader may consult [KozT] for more details on this and related subjects.

The results in Section 2.2 are taken from [S2], where the more general case of semi-dispersing billiards is considered. In the particular case when condition (H) is satisfied, the assertion of Corollary 2.2.4 was proved by Ikawa [I4] using different technique.

The representation of the linear Poincaré map P_γ, described in Theorem 2.3.1, was obtained in [PV]. Proposition 2.3.2 generalizes a result from [BGR] concerning two disjoint strictly convex domains. Most of the material in Section 2.4 is taken from [PS5], the proof of Proposition 2.4.4 being different from that in [PS5]. Example 2.5.1 is contained in [PS6].

A general definition of a billiard on a Riemannian manifold with boundary can be found in [CFS], see also [Sin2]. A more general type of dynamical systems is studied in [KS]. Ergodic properties of billiards, connected with certain problems in statistical mechanics, have been studied very intensively during the last 20 years by Sinai, Bunimovich, Chernov and others, see [Sin1], [BunS], [BCS], [Cher1], [KSS] for more information in this direction. It follows by the results in [BCS] that for dispersing billiards in the plane there exists an exponential estimate from below of the number P_k of periodic points with period k (cf. also Corollary 2.2.5), and the set of periodic points is dense in the phase space of such billiards. Further results concerning the periodic points of the billiard are contained in [Cher2]. As it is mentioned in Section 3.4 of [Cher2], according to some arguments from [BCS], the estimate (2.8) in Corollary 2.2.5 does not hold for dispersing billiards on the flat torus; for the latter it may even happen that $P_k = \infty$ for some k. It follows from [Cher2] that for dispersing billiards

$$\liminf \frac{\log P_k}{k}$$

can be estimated from below by the metric entropy of the billiard ball map (see [CFS] or [Wa] for the definition of entropy). Let us mention that for the geodesic flow φ_t on a compact manifold with constant negative curvature Margulis [Marg] proved the formula

$$\lim_{T \to +\infty} \frac{P_T}{T} = h(\varphi_1),$$

where P_T is the number of closed geodesics with lengths not greater than T and $h(f)$ denotes the topological entropy of the map f. The reader may consult [Kat2] for more information in this direction. It is an open question whether a similar assertion is true for dispersing billiards. In the special case of a billiard in the exterior of a finite number of disjoint strictly convex domains in the plane, satisfying the condition (H) of Ikawa, an affirmative answer of this question is given by Morita [Mor].

3 GENERIC PROPERTIES OF REFLECTING RAYS

In this chapter we establish several generic properties of periodic reflecting rays in bounded domains and scattering rays in domains with bounded complements. These properties will be used in Chapters 7 and 9 to investigate certain inverse spectral problems for generic domains. In Section 3.1 we prove a general theorem which provides existence of residual sets of smooth embeddings of a given submanifold of $\mathbf{R}^n$ satisfying certain conditions. As a consequence of it some elementary generic properties are derived of the two kinds of reflecting rays under consideration. In particular, we show that the Herman Weyl conjecture is true for generic bounded domains. Not only the theorem in Section 3.1 but also the scheme of its proof will be used several times in this book. Following this scheme, we prove in the present chapter that for generic domains the reflecting rays under consideration are ordinary and non-degenerate.

3.1. Generic properties of smooth embeddings

Let X be a smooth $(n-1)$-dimensional submanifold of $\mathbf{R}^n$, $n \geq 2$. In this Section we prove a general theorem establishing the existence of a residual set of a special type of smooth embeddings f of X into $\mathbf{R}^n$. This theorem will be applied several times in the present chapter.

Theorem 3.1.1: *Let $n \geq 2$, $s \geq 2$, p and q be natural numbers, U an open subset of $(\mathbf{R}^n)^{(s)}$ and X be a smooth $(n-1)$-dimensional submanifold of $\mathbf{R}^n$. Let*

$$H = (H_1, \ldots, H_p) : U \to \mathbf{R}^p$$

be a smooth map such that for every $y \in U$ and every $i = 1, \ldots, s$ there is r_i, $1 \leq r_i \leq p$, with

$$\operatorname{grad}_{y_i} H_{r_i}(y) \neq 0, \tag{3.1}$$

$y = (y_1, \ldots, y_s)$.
    (A) Let $p = 1$ and let T_1 be the set of those $f \in \mathbf{C}(X)$ for which the critical points x of $H \circ f^s$ with $f^s(x) \in U$ form a discrete subset of $X^{(s)}$. Then T_1

contains a residual subset of $\mathbf{C}(X)$;
   (B) Let p be arbitrary and let

$$L = (L_1, \ldots, L_q) : U \to \mathbf{R}^q$$

be a smooth map such that $dL(y) \neq 0$ *for any* $y \in U$ *with* $L(y) = 0$. *Let* T_2 *be the set of all* $f \in \mathbf{C}(X)$ *such that if* x *is a critical point of* $H \circ f^s$ *with* $f^s(x) \in U$, *then* $L(f^s(x)) \neq 0$. *Then* T_2 *contains a residual subset of* $\mathbf{C}(X)$.
   Here we use the notation

$$\operatorname{grad}_{y_i} H_{r_i}(y) = \left(\frac{\partial H_{r_i}}{\partial y_i^{(j)}}(y) \right)_{j=1}^n \in \mathbf{R}^n.$$

For the proof of the theorem we need some preparation.
   First, let us note that in (B) it is sufficient to consider only the case $q = 1$. Indeed, assume the assertion in (B) is true for $q = 1$. For $m = 1, \ldots, q$ set

$$U_m = \{y \in U : dL_m(y) \neq 0\}.$$

Then $U_1, \ldots, U_q$ are open subsets of U and $U = \cup_{k=1}^q U_k$. Moreover,

$$\cap_{k=1}^q T_2^{(k)} \subset T_2,$$

where $T_2^{(k)}$ is the set of those $f \in \mathbf{C}(X)$ such that if x is a critical point of $H \circ f^s$ with $f^s(x) \in U_k$, then $L_k(f^s(x)) \neq 0$. It follows by our assumption that $T_2^{(k)}$ contains a residual subset of $\mathbf{C}(X)$ for every $k = 1, \ldots, q$. Hence T_2 has the same property.
   We may assume in addition that for every $i = 1, \ldots, s$ there exists $r_i = 1, \ldots, p$ such that (3.1) holds whenever $y \in U$. Indeed, for any $\mathbf{r} = (r_1, \ldots, r_s)$ let

$$U_{\mathbf{r}} = \{y \in U : (3.1) \quad \text{holds for } i = 1, \ldots, s\}.$$

Then $U = \cup U_{\mathbf{r}}$, where $\mathbf{r}$ runs over all s-tuples of the considered type (there are finitely many of them), and therefore

$$\cap_{\mathbf{r}} T_2^{(\mathbf{r})} \subset T_2,$$

where $T_2^{(\mathbf{r})}$ is the set of all $f \in \mathbf{C}(X)$ such that if x is a critical point of $H \circ f^s$ with $f^s(x) \in U_{\mathbf{r}}$, then $L(f^s(x)) \neq 0$. Hence if any $T_2^{(\mathbf{r})}$ contains a residual subset of $f \in \mathbf{C}(X)$, then T_2 has the same property.
   From now on we assume that $q = 1$. We consider simultaneously (A) and (B).
   Let $J_s^1(X, \mathbf{R}^n)$ be the s-fold bundle of 1-jets, and let α and β be the corresponding source and target maps (cf. Section 1.1). Set

$$M = (\alpha^s)^{-1}(X^{(s)}) \cap (\beta^s)^{-1}(U). \tag{3.2}$$

Clearly, M is an open subset (and therefore a submanifold) of $J_s^1(X, \mathbf{R}^n)$. Let Σ_1 be the set of those

$$\sigma = (j^1 f_1(x_1), \ldots, j^1 f_s(x_s)) \in M, \tag{3.3}$$

such that $x = (x_1, \ldots, x_s)$ is a critical point of the map $H \circ (f_1 \times \ldots \times f_s)$. For a given $f \in C(X)$ set

$$A_f = \{x \in X^{(s)} : j_s^1 f(x) \in \Sigma_1\}.$$

It then follows that

$$T_1 = \{f \in C(X) : A_f \text{ is a discrete subset of } X^{(s)}\}. \tag{3.4}$$

To describe T_2 in a similar way, we introduce the set Σ_2 of those $\sigma \in \Sigma_1$ of the form (3.3) such that $L \circ (f_1 \times \ldots \times f_s) = 0$. Then clearly

$$T_1 = \{f \in C(X) : j_s^1 f(X^{(s)}) \cap \Sigma_2 = \emptyset\}. \tag{3.5}$$

The central point in the proof of Theorem 3.1.1 is the study of the singular sets Σ_1 and Σ_2. In fact, we do not need precise information on the structure of these sets; for our aim it is enough to show that each of them can be covered by a countable family of smooth submanifolds of M of sufficiently big codimension.

Lemma 3.1.2: *For $p = 1, \Sigma_1$ is a smooth submanifold of M with*

$$\operatorname{codim} \Sigma_1 = (n-1)s. \tag{3.6}$$

For every $p \in \mathbf{N}$ there exists a finite or countable family $\{W_m\}$ of smooth submanifold of M with

$$\operatorname{codim} W_m = (n-1)s+1, \tag{3.7}$$

such that

$$\Sigma_2 \subset \cup_m W_m. \tag{3.8}$$

Proof of Lemma 3.1.2: For an arbitrary $\sigma_0 \in M$ we shall construct a chart for M, defined in a neighbourhood of σ_0. There exist coordinate neighbourhoods $V_1, \ldots, V_s$ of elements of X such that $V_i \cap V_j = \emptyset$ for $i \neq j$ and

$$\sigma_0 \in \prod_{i=1}^s J^1(V_i, \mathbf{R}^n) \subset J_s^1(X, \mathbf{R}^n).$$

Set

$$D = M \cap \left(\prod_{i=1}^s J^1(V_i, \mathbf{R}^n) \right). \tag{3.9}$$

Clearly, D is an open neighbourhood of σ_0 in M (cf. Section 1.1). Consider arbitrary charts $\varphi_i : V_i \to \mathbf{R}^{n-1}$ and define the chart

$$\varphi : D \to (\mathbf{R}^{n-1})^{(s)} \times (\mathbf{R}^n)^{(s)} \times \mathbf{R}^{n(n-1)s} \qquad (3.10)$$

by

$$\varphi(\sigma) = (u; v; a), \qquad (3.11)$$

where σ has the form (3.3) and

$$u = (u_1, \ldots, u_s), \; v = (v_1, \ldots, v_s), \; a = \left(a_{ij}^{(t)}\right)_{1 \le i \le s, 1 \le j \le n-1, 1 \le t \le n}, \qquad (3.12)$$

$$u_i = \varphi_i(x_i), \quad v_i = f_i(x_i), \qquad (3.13)$$

$$a_{ij}^{(t)} = \frac{\partial(f_i^{(t)} \circ \varphi_i^{-1})}{\partial u_i^{(j)}}(u_i) \qquad (3.14)$$

for $i = 1, \ldots, s$, $j = 1, \ldots, n-1$, $t = 1, \ldots, n$. Here we use the notation

$$f_i = \left(f_i^{(1)}, \ldots, f_i^{(n)}\right), \quad u_i = \left(u_i^{(1)}, \ldots, u_i^{(n-1)}\right) \in \mathbf{R}^{n-1},$$

$$v_i = \left(v_i^{(1)}, \ldots, v_i^{(n)}\right) \in \mathbf{R}^n.$$

Let us mention that if $F : U \to \mathbf{R}$ is a smooth function, then

$$\frac{\partial(F \circ ((f_1 \circ \varphi_1^{-1}) \times \ldots \times (f_s \circ \varphi_s^{-1})))}{\partial u_i^{(j)}}(u) = \sum_{t=1}^{n} \frac{\partial F}{\partial y_i^{(t)}}(v) a_{ij}^{(t)}, \qquad (3.15)$$

where $y_i = (y_i^{(1)}, \ldots, y_i^{(n)})$.

Since each of the sets Σ_1 and Σ_2 can be covered with a countable number of coordinate neighbourhoods, it is sufficient to prove that if D is an arbitrary coordinate neighbourhood of the above form, then for $p = 1$, $\varphi(D \cap \Sigma_1)$ is a smooth submanifold of $\varphi(D)$ of codimension $(n-1)s$, and for arbitrary p, $\varphi(D \cap \Sigma_2)$ is contained in a smooth submanifold of $\varphi(D)$ of codimension $(n-1)s + 1$.

We shall write the elements ξ of $\varphi(D)$ of the form

$$\xi = (u; v; a) \in (\mathbf{R}^{n-1})^{(s)} \times (\mathbf{R}^n)^{(s)} \times \mathbf{R}^{n(n-1)s},$$

where u, v and a are determined by (3.12), (3.13) and (3.14). It follows by our assumptions that for every $i = 1, \ldots, s$ there exists $r_i = 1, \ldots, p$ such that (3.1) holds for all $y \in U$. For $i = 1, \ldots, s$ and $j = 1, \ldots, n-1$ set

$$b_{ij}(\xi) = \sum_{t=1}^{n} \frac{\partial H_{r_i}}{\partial y_i^{(t)}}(v) a_{ij}^{(t)}. \qquad (3.16)$$

Define the maps

$$R_1 : \varphi(D) \rightarrow \mathbf{R}^{(n-1)s}, \quad R_2 : \varphi(D) \rightarrow \mathbf{R}^{(n-1)s} \times \mathbf{R}$$

by

$$R_1(\xi) = (b_{ij}(\xi))_{1 \leq i \leq s,\, 1 \leq j \leq n-1}, \quad R_2(\xi) = (R_1(\xi);\ \tilde{L}(\xi)),$$

where $\tilde{L}(\xi) = L(v)$ by definition.

Note that for $p = 1$ we have $r_i = 1$ for any i, therefore

$$\varphi(D \cap \Sigma_1) = R_1^{-1}(0). \tag{3.17}$$

For p arbitrary, the definitions of Σ_2 and R_2 yield

$$\varphi(D \cap \Sigma_2) \subset R_2^{-1}(0). \tag{3.18}$$

We are going to show that R_2 is a submersion on $R_2^{-1}(0)$. Let $\xi = (u; v; a) \in R_2^{-1}(0)$. Assume that

$$\sum_{i=1}^{s} \sum_{i=1}^{n-1} B_{ij} \operatorname{grad} b_{ij}(\xi) + C \operatorname{grad} \tilde{L}(\xi) = 0 \tag{3.19}$$

for some constants C and B_{ij}, $i = 1, \ldots, s$, $j = 1, \ldots, n-1$. Here we consider $\operatorname{grad} b_{ij}(\xi)$ and $\operatorname{grad} \tilde{L}(\xi)$ as vectors in $(\mathbf{R}^{n-1})^{(s)} \times (\mathbf{R}^n)^{(s)} \times \mathbf{R}^{n(n-1)s}$. It follows from (3.11)–(3.14), $D \subset M$ and $\xi \in \varphi(D)$ that $v \in U$. Fix arbitrary $i = 1, \ldots, s$ and $j = 1, \ldots, n-1$. There exists $t = 1, \ldots, n$ such that

$$\frac{\partial H_{r_i}}{\partial y_i^{(t)}}(v) \neq 0.$$

Then, according to (3.16), we get

$$\frac{\partial b_{ij}}{\partial a_{ij}^{(t)}}(\xi) = \frac{\partial H_{r_i}}{\partial y_i^{(t)}}(v) \neq 0, \quad \frac{\partial b_{i'j'}}{\partial a_{ij}^{(t)}}(\xi) = 0 \quad \text{for } i' \neq i,\ j' \neq j.$$

Moreover, $\frac{\partial \tilde{L}}{\partial a_{ij}^{(t)}}(\xi) = 0$. Considering in (3.19) the derivatives with respect to $a_{ij}^{(t)}$, we deduce $B_{ij} \frac{\partial H_{r_i}}{\partial y_i^{(t)}}(v) = 0$, hence $B_{ij} = 0$. The latter holds for any choice of $i = 1, \ldots, s$ and $j = 1, \ldots, n-1$. Now (3.19) becomes $C \cdot \operatorname{grad} \tilde{L}(\xi) = 0$. Since $\xi \in R_2^{-1}(0)$, we have $R_2(\xi) = 0$. In particular, $L(v) = \tilde{L}(\xi) = 0$. The assumption on L and $v \in U$ now imply that $dL(v) \neq 0$. Therefore $C = 0$ which shows that R_2 is a submersion at ξ.

In this way we have proved that R_2 is a submersion at any point of $R_2^{-1}(0)$.

Hence (cf. Section 1.1) $R_2^{-1}(0)$ is a smooth submanifold of $\varphi(D)$ with

$$\operatorname{codim} R_2^{-1}(0) = \dim(\mathbf{R}^{(n-1)s} \times \mathbf{R}) = (n-1)s + 1.$$

Applying the above argument for $p = 1$, we see that R_1 is a submersion at any point of $R_1^{-1}(0)$. Therefore $R_1^{-1}(0)$ is a smooth submanifold of $\varphi(D)$ with

$$\operatorname{codim} R_1^{-1}(0) = \dim \mathbf{R}^{(n-1)s} = (n-1)s.$$

Now the assertion follows from (3.17) and (3.18). ♠.

Proof of Theorem 3.1.1: (A) Let $p = 1$. According to the above lemma, Σ_1 is a smooth submanifold of M with (3.6). Since M is open in $J_s^1(X, \mathbf{R}^n)$, Σ_1 is a smooth submanifold of $J_s^1(X, \mathbf{R}^n)$ of the same codimension. Applying the multijet transversality theorem (cf. Section 1.1), we get that the set

$$T_1' = \{f \in C^\infty(X, \mathbf{R}^n) : j_s^1 f \pitchfork \Sigma_1\}$$

is residual in $C^\infty(X, \mathbf{R}^n)$. Since $\mathbf{C}(X)$ is open in $C^\infty(X, \mathbf{R}^n)$, $T_1'' = T_1' \cap \mathbf{C}(X)$ is residual in $\mathbf{C}(X)$. To prove (A) it is sufficient to establish the inclusion $T_1'' \subset T_1$. Let $f \in T_1''$, then $j_s^1 f \pitchfork \Sigma_1$. Since

$$j_s^1 f : X^{(s)} \to J_s^1(X, \mathbf{R}^n)$$

and $\dim X^{(s)} = (n-1)s = \operatorname{codim} \Sigma_1$, we deduce that A_f is a discrete subset of Σ_1 (cf. Section 1.1), i.e. $f \in T_1$. Hence $T_1'' \subset T_1$.

(B) It follows by Lemma 3.1.2 that there exists a finite or countable family $\{W_m\}$ of smooth submanifolds of M (and therefore of $J_s^1(X, \mathbf{R}^n)$) satisfying (3.7) and (3.8). Applying the multijet transversality theorem again, we obtain that for any m the set

$$S_m = \{f \in \mathbf{C}(X) : j_s^1 f \pitchfork W_m\}$$

is residual in $\mathbf{C}(X)$. It remains to check the inclusion

$$\cup_m S_m \subset T_2. \tag{3.20}$$

Let f belong to the left-hand side of (3.20). Then for every m, $f \in S_m$ implies $j_s^1 f \pitchfork W_m$. Since

$$j_s^1 f : X^{(s)} \to J_s^1(X, \mathbf{R}^n)$$

and by (3.16), $\dim X^{(s)} = (n-1)s < \operatorname{codim} W_m$, $j_s^1 f \pitchfork W_m$ is equivalent to $j_s^1 f(X^{(s)}) \cap W_m = \emptyset$ (cf. Section 1.1). The latter holds for any m, so by (3.8) we get $j_s^1 f(X^{(s)}) \cap \Sigma_2 = \emptyset$. This and (3.5) imply $f \in T_2$ which proves the inclusion (3.20). Therefore T_2 contains a residual subset of $\mathbf{C}(X)$. ♠

3.2. Elementary generic properties

In this section we apply Theorem 3.1.1 to obtain some properties of the generic domains in $\mathbf{R}^n$ concerning the behaviour of periodic reflecting rays and scattering rays in Ω. In particular we establish that for generic Ω the following properties are satisfied:

(a) The lengths of any two distinct primitive periodic reflecting rays are independent over the rationals;

(b) for every $x \in \partial\Omega$ there exists at most one direction $\xi \in S^{n-1}$ (up to the symmetry with respect to the normal $\nu(x)$ to $\partial\Omega$ at x) such that (x, ξ) generates a periodic reflecting ray in Ω.

Similar properties are considered for scattering rays.

Clearly, in the general case neither (a) nor (b) are satisfied. A simple example is a disc Ω in the plane. Let us mention that the property (b) is equivalent to the following: any two distinct primitive periodic reflecting rays in Ω have no common reflection point, and any primitive periodic reflecting ray passes only once through each of its reflection points. In fact, precisely speaking the latter is true only for non-symmetric rays (cf. Section 2.1); for symmetric ones this property can be formulated appropriately in a similar way.

We begin with a simple combinatorial classification of the periodic reflecting rays.

Let $k \geq s \geq 2$ be natural numbers and let

$$\alpha : \{1, \ldots, k\} \rightarrow \{1, \ldots, s\} \tag{3.21}$$

be a map with

$$\alpha(i) \neq \alpha(i+1) \tag{3.22}$$

for $i = 1, \ldots, k$. Here we set for convenience $\alpha(m) = \alpha(i)$ for $m = i + pk$, $1 \leq i \leq k$, p being an integer. If

$$\{\alpha(i), \ \alpha(i+1)\} \neq \{\alpha(j), \alpha(j+1)\} \tag{3.23}$$

holds for all $1 \leq i < j \leq k$, we shall say that α is an *ns-map* (non-symmetric map). If $k = 2m$ and there is i_0 with $1 \leq i_0 \leq k$ such that (3.23) holds for $i_0 \leq i < j \leq i_0 + m$ and

$$\alpha(i_0 + m + j) = \alpha(i_0 + m - j), \quad j = 1, \ldots, m-1, \tag{3.24}$$

we shall say that α is an *s-map* (symmetric map). By an *admissible map* we mean a map (3.21) which is either ns- or s-map.

Next, we assume that an admissible map of the form (3.21) is fixed. Introduce the sets

$$I_i = I_i(\alpha) = \{j : \exists t = 1, \ldots, k \text{ with } \{i, j\} = \{\alpha(t), \alpha(t+1)\}\} \tag{3.25}$$

for $i = 1, \ldots, s$. Denote by U_α the set of all $y = (y_1, \ldots, y_s) \in (\mathbf{R}^n)^{(s)}$ such that

$$y_i \notin \operatorname{conv} \{y_j : j \in I_i\}, \quad i = 1, \ldots, s,$$

and for any $i = 1, \ldots, s$, if q, j, r, t are different elements of I_i, then at least one of the triples y_i, y_q, y_j, and y_i, y_r, y_t consists of non-collinear points. It is easily seen that U_α is open in $(\mathbf{R}^n)^{(s)}$, and the function

$$F = F_\alpha : U_\alpha \to \mathbf{R}, \qquad (3.26)$$

given by

$$F(y) = \sum_{i=1}^{k} \left\| y_{\alpha(i)} - y_{\alpha(i+1)} \right\|, \qquad (3.27)$$

is smooth.

Notice that if γ is a periodic reflecting ray and $y_1, \ldots, y_s$ are all the reflection points of γ (their ordering does not matter in this case), there exist $k \geq s$ and an admissible map (3.21) such that

$$y_{\alpha(1)}, \ldots, y_{\alpha(k)} \qquad (3.28)$$

are the successive reflection points of γ. In such a case we shall say that γ is of type α. Then we have $y = (y_1, \ldots, y_s) \in U_\alpha$ and $F(y)$ is equal to the length T_γ of γ. Moreover, for any $i = 1, \ldots, s$ the relation $j \in I_i$ is equivalent to the fact that there exists a segment of γ connecting y_i and y_j. Clearly, the type α of a periodic reflecting ray is not uniquely determined.

In what follows X will be an arbitrary fixed smooth $(n-1)$-dimensional sub-manifold of $\mathbf{R}^n$.

The following proposition is a consequence of Proposition 2.1.3.

Proposition 3.2.1: *Let α be an admissible map of the form (3.21) and let $x_1, \ldots, x_s$ be all distinct reflection points of a periodic reflecting ray γ of type α for X, i.e. $x_{\alpha(1)}, \ldots, x_{\alpha(k)}$ are the successive reflection points of γ. Then $x = (x_1, \ldots, x_s)$ is a critical point of the map $F_{\alpha | X^{(s)}}$.*

Proof: Define the maps

$$g : U_\alpha \to (\mathbf{R}^n)^k, G : (\mathbf{R}^n)^k \to \mathbf{R}$$

by

$$g(y) = (y_{\alpha(1)}, \ldots, y_{\alpha(k)})$$

and

$$G(z_1, \ldots, z_k) = \sum_{i=1}^{k} \left\| z_i - z_{i+1} \right\|.$$

Then $F_\alpha = G \circ g$. Moreover, $g(X^{(s)}) \subset X^k$. It follows by our assumption and Proposition 2.1.3 that $g(x)$ is a critical point of $G_{|X^k}$. Therefore x is a critical point of $F_{\alpha|X^{(s)}}$. ♠

In order to apply the above proposition we need another property of the map $F = F_\alpha$.

Lemma 3.2.2:   *Let α be an admissible map of the form (3.21). For every $i = 1, \ldots, s$ and every $y \in U_\alpha$ there exists $j = 1, \ldots, n$ such that*

$$\frac{\partial F}{\partial y_i^{(j)}}(y) \neq 0. \tag{3.29}$$

Proof: Fix arbitrary $i = 1, \ldots, s$ and $y \in U_\alpha$. For any $j = 1, \ldots, n$ we have

$$\frac{\partial F}{\partial y_i^{(j)}}(y) = a \sum_{t \in I_i} \frac{y_i^{(j)} - y_t^{(j)}}{\|y_i - y_t\|}, \tag{3.30}$$

where $a = 1$ for an ns-map α, and $a = 2$ if α is an s-map.

Assume that the derivative (3.30) is 0 for any j. Then

$$\sum_{t \in I_i} \frac{y_i^{(j)} - y_t^{(j)}}{\|y_i - y_t\|} = 0,$$

which can be rewritten as

$$y_i = \sum_{t \in I_i} a_t y_t,$$

where

$$a_t = \frac{1}{\|y_i - y_t\| \, \Sigma_{j \in I_i} \dfrac{1}{\|y_i - y_j\|}}.$$

Since $\Sigma_{t \in I_i} a_t = 1$, we deduce

$$y_i \in \text{conv} \, \{y_t : t \in I_i\},$$

which is a contradiction with $y \in U_\alpha$. This proves the assertion. ♠

Denote by $\mathcal{R}$ the set of those $f \in \mathbf{C}(X)$ such that every two primitive periodic reflecting rays for $f(X)$ have rationally independent lengths. Consider also the set $\mathcal{A}$ of those $f \in \mathbf{C}(X)$ such that for every $y \in f(X)$ there exists at most one direction $\xi \in S^{n-1}$ (up to the symmetry with respect to the normal to $f(X)$ at y) for which (y, ξ) generates a periodic reflecting ray for $f(X)$.

Theorem 3.2.3:   *Each of the sets $\mathcal{R}$ and $\mathcal{A}$ contains a residual subset of $\mathbf{C}(X)$.*

Proof: Fix an arbitrary surjective ns-map (3.21) and suppose that $k > s$. Without loss of generality we may assume that

$$\alpha(1) = 1, \quad |\alpha^{-1}(1)| > 1.$$

Denote by $\mathcal{A}_\alpha$ the set of those $f \in C(X)$ such that there does not exist a periodic reflecting ray of type α for $f(X)$. We shall show that $\mathcal{A}_\alpha$ contains a residual subset of $C(X)$. To this end we are going to apply Theorem 3.1.1 (B) with $U = U_\alpha$, $p = 1$, $H = F_\alpha$.

Choose arbitrary distinct elements j_1, j_2 of $\alpha^{-1}(1)$. Then

$$q = \alpha(j_1 - 1), \quad j = \alpha(j_1 + 1), \quad r = \alpha(j_2 - 1), \quad t = \alpha(j_2 + 1)$$

are distinct elements of $I_1 = I_1(\alpha)$. Set

$$L_u(y) = \left\langle \frac{y_q - y_1}{\|y_q - y_1\|} + (-1)^u \frac{y_j - y_1}{\|y_j - y_1\|}, \frac{y_r - y_1}{\|y_r - y_1\|} - (-1)^u \frac{y_t - y_1}{\|y_t - y_1\|} \right\rangle$$

for $u = 1, 2$, $y \in U = U_\alpha$, and define

$$L : U_\alpha \to \mathbf{R}^2$$

by $L(y) = (L_1(y), L_2(y))$. We have to check that if $L(y) = 0$ for some $y \in U$, then $\mathrm{d}L(y) \neq 0$. Let $y \in U$ be such that $L(y) = 0$. If

$$\frac{\partial L_1}{\partial y_m^{(l)}}(y) = 0, \quad l = 1, \ldots, n,$$

a simple calculation implies that $y_q - y_1$ is collinear with the vector

$$v = \frac{y_r - y_1}{\|y_r - y_1\|} + \frac{y_t - y_1}{\|y_t - y_1\|}.$$

Note that $y \in U$ implies $v \neq 0$. Since $L_1(y) = 0$, and $(y_q - y_1)/\|y_q - y_1\|$ and $(y_j - y_1)/\|y_j - y_1\|$ are unit vectors, we deduce that $y_j - y_1$ is collinear with v, too. Therefore y_1, y_q, y_j are collinear. Next, assume that

$$\frac{\partial L_2}{\partial y_r^{(l)}}(y) = 0, \quad l = 1, \ldots, n.$$

In the same way, one obtains that y_1, y_r, y_t are collinear, which is a contradiction with $y \in U = U_\alpha$ and the definition of U_α. Hence $\mathrm{d}L(y) \neq 0$.

Finally, note that if $y_1, \ldots, y_s$ are the reflection points of a periodic reflecting ray of type α, then for $y = (y_1, \ldots, y_s)$ we have $y \in U$ and $L(y) = 0$. Now, applying Theorem 3.1.1 (B), we obtain that $\mathcal{A}_\alpha$ contains a residual subset of $C(X)$.

Next, we consider the case when α is a surjective s-map of the form (3.21). Let $k = 2m$ and let $y_1, \ldots, y_s$ be all different reflection points of a periodic reflecting ray γ for X such that (3.28) are the successive reflection points of γ (i.e. γ is of type α). Then y is a critical point of the map $F_{|X^{(s)}}$. Moreover,

$$F(y) = 2 \sum_{i=1}^{m} \|y_{\alpha(i)} - y_{\alpha(i+1)}\| .$$

Assume that $k > 2s - 2$ and define $\mathcal{A}_\alpha$ as in the non-symmetric case. Using a slight modification of the above argument, replacing F by G and applying Theorem 3.1.1 (B) again, we deduce that $\mathcal{A}_\alpha$ contains a residual subset of $\mathbf{C}(X)$.

Consequently, the set

$$\mathcal{A}_1 = \cap_\alpha \mathcal{A}_\alpha,$$

where α runs over the surjective admissible maps (3.21) with $k > s$ for ns-maps α and $k > 2s - 2$ for s-maps α, contains a residual subset of $\mathbf{C}(X)$.

Let $Z = f(X)$, where f is an arbitrary element of $\mathcal{A}_1$. Suppose that there exist two different primitive periodic reflecting rays γ and δ for Z which have a common reflection point. We may assume that $z_1, \ldots, z_s$ are the successive reflection points of γ, $u_1, \ldots, u_t$ those of δ, and $z_1 = u_1$. For $k \in \mathbf{N}$ set

$$\alpha_k = \mathrm{id} : \{1, \ldots, k\} \rightarrow \{1, \ldots, k\}$$

and

$$F_k = F_{\alpha_k} : U_k = U_{\alpha_k} \rightarrow \mathbf{R}.$$

Let $z_i = f(x_i)$, $u_j = f(y_j)$ for $i = 1, \ldots, s$ and $j = 1, \ldots, t$. Then for $x = (x_1, \ldots, x_s)$ and $y = (y_1, \ldots, y_t)$ we have $z = f^s(x) \in U_s$, $u = f^t(y) \in U_t$, x is a critical point of $F_s \circ f^s$ and y is a critical point of $F_t \circ f^t$. Hence $(x, y) \in U_s \times U_t$ and (x, y) is a critical point of the map $H \circ f^{s+t}$, where

$$H : U_s \times U_t \rightarrow \mathbf{R}^2 \qquad\qquad (3.31)$$

is defined by

$$H(z, u) = (F_s(z), F_t(u)). \qquad\qquad (3.32)$$

The fact that $z_1 = u_1$ can be expressed by

$$L(f^s(x), f^t(y)) = 0,$$

where

$$L : U_s \times U_t \rightarrow \mathbf{R}^n$$

is given by $L(z, u) = z_1 - u_1$.

It is easily seen that $U = U_s \times U_t$, H, L satisfy the assumptions of Theorem 3.1.1 (B). Consequently, there exists a residual subset $\mathcal{A}_2(s, t)$ of $\mathbf{C}(X)$ such that if $f \in \mathcal{A}_2(s,t)$, then $L(f^s(x), f^t(y)) \neq 0$ holds for any critical point (x,y) of H with $(f^s(x), f^t(y)) \in U_s \times U_t$. Then

$$\mathcal{A}_2 = \cap\{\mathcal{A}_2(s,t) : s,t \geq 2\}$$

is also a residual subset of $\mathbf{C}(X)$.

It follows from the definitions of $\mathcal{A}_1$ and $\mathcal{A}_2$ that if $f \in \mathcal{A}_1 \cap \mathcal{A}_2$, then any two different primitive periodic reflecting rays for $f(X)$ have no common reflection point. Therefore $\mathcal{A}_1 \cap \mathcal{A}_2 \subset \mathcal{A}$ which shows that $\mathcal{A}$ contains a residual subset of $\mathbf{C}(X)$.

Next, we consider the subset $\mathcal{R}$ of $\mathbf{C}(X)$. Let $Z = f(X), f \in \mathcal{A}$. Suppose that there exist two different primitive periodic reflecting rays γ and δ for Z such that $pT_\gamma = qT_\delta$ for some $p, q \in \mathbf{N}$. Let $z_1, \ldots, z_s$ and $u_1, \ldots, u_t$ be the successive reflection points of γ and δ, respectively. Set $x_i = f^{-1}(z_i), y_j = f^{-1}(u_j)$, $x = (x_1, \ldots, x_s), y = (y_1, \ldots, y_t)$. Consider the map H, given by (3.31) and (3.32). One gets as above that (x,y) is a critical point of $H \circ f^{s+t}$. Moreover, for the map

$$K : U_s \times U_t \to \mathbf{R}^n,$$

defined by $K(z,u) = pF_s(z) - qF_t(u)$, we have $K \circ f^{s+t}(x,y) = 0$. It follows by Lemma 3.2.2 that $dK(z,u) \neq 0$ for all $(z,u) \in U_s \times U_t$. Now applying Theorem 3.1.1 (B), we deduce that there exists a residual subset $\mathcal{R}(p,q,s,t)$ of $\mathbf{C}(X)$ such that if $f \in \mathcal{R}(p,q,s,t)$, $(x,y) \in X^s \times X^t$, $(f^s(x), f^t(y)) \in U_s \times U_t$ and (x,y) is a critical point of $H \circ f^{s+t}$, then $K(f^s(x), f^t(y)) \neq 0$. This yields the inclusion

$$\cap\{\mathcal{R}(p,q,s,t) : p,q,s,t \in \mathbf{N}, s,t \geq 2\} \cap \mathcal{A} \subset \mathcal{R},$$

which shows that $\mathcal{R}$ contains a residual subset of $\mathbf{C}(X)$. This completes the proof of the theorem. ♠

Let us mention that the property (b) from the beginning of the present section implies that the measure of the set of periodic points in the phase space of the billiard system, related to Ω, is zero. More precisely, let Σ be the set of those $(x, \xi) \in X \times S^{n-1}$ such that there exists a periodic reflecting ray for X passing through x with direction ξ. If μ denotes the standard Lebesgue measure on $X \times S^{n-1}$, then for generic X we have $\mu(\Sigma) = 0$. In Section 3.4 we shall establish a stronger result; namely we shall show that for generic X in $\mathbf{R}^n$ there exist at most countably many periodic reflecting rays for X. However, the weaker property proved in this section, is already sufficient to make an application.

As a direct consequence of the result of Ivrii [Iv1], concerning the asymptotic (0.4) for the counting function $N(\lambda)$, related to the point spectrum of the Laplacian, and Theorem 3.2.3 we obtain the following.

Corollary 3.2.4: *For every bounded domain Ω in $\mathbf{R}^n$, $n \geq 2$, with smooth boundary $X = \partial\Omega$ there exists a residual subset T of $\mathbf{C}(X)$ such that for every $f \in T$ the asymptotic (0.4) holds replacing Ω by Ω_f.* ♠

In other words, the Herman Weyl conjecture (cf. the Introduction) is true for generic domains in $\mathbf{R}^n$.

The second part of this section is devoted to some properties of the generic compact domains K in $\mathbf{R}^n$ with smooth boundaries $X = \partial K$, concerning the reflecting (ω, θ)-rays in $\Omega = \overline{\mathbf{R}^n \backslash K}$. These properties are analogues of the properties of the periodic reflecting rays, considered in the first part of this section, and we use the same technique for their proofs. This is why we omit some of the details of the considerations below.

Hereafter X will be a fixed compact smooth $(n-1)$-dimensional submanifold of $\mathbf{R}^n$, $n \geq 2$, and ω and θ will be fixed unit vectors in $\mathbf{R}^n$.

It is convenient to introduce a notion slightly more general than the notion of a reflecting (ω, θ)-ray. Let γ be a curve of the form

$$\gamma = \cup_{i=1}^{k} l_i \subset \mathbf{R}^n,$$

such that $l_i = [x_i, x_{i+1}]$ are finite linear segments for $i = 1, \ldots, k-1$, $x_i \in X$ for all i, and l_0 (resp. l_k) is the infinite linear segment starting at x_1 (resp. x_k) and having direction $-\omega$ (resp. θ). Then γ will be called an (ω, θ)-*trajectory* for X if for any $i = 0, 1, \ldots, k-1$ the segments l_i and l_{i+1} satisfy the law of reflection at x_{i+1} with respect to X. The points $x_1, \ldots, x_k$ will be called again reflection points of γ, and the sojourn time T_γ of γ is defined by (2.31). As for reflecting (ω, θ)-rays (cf. Section 2.4), we distinguish two types of (ω, θ)-trajectories — symmetric and non-symmetric ones. The definitions are the same.

We should note that in general the segments of an (ω, θ)-trajectory for X may intersect (transversally) X. According to Definition 2.4.1, we see that every reflecting (ω, θ)-ray for X is an (ω, θ)-trajectory for X but the converse is not true in general.

Define the subsets $\mathcal{B}, \mathcal{P}, \mathcal{S}$ of $\mathbf{C}(X)$ as the sets of those $f \in \mathbf{C}(X)$ such that:

(a) $\mathcal{B}$: for every $x \in f(X)$ there exists at most one direction $\xi \in S^{n-1}$ (up to the symmetry with respect to the normal $\nu(x)$ to $f(X)$) so that there exists an (ω, θ)-trajectory for $f(X)$, passing through x with direction ξ;

(b) $\mathcal{P}$: there is no (ω, θ)-trajectory for $f(X)$ having different parallel segments;

(c) $\mathcal{S}$: $T_\gamma \neq T_\delta$ for any two different (ω, θ)-trajectories for $f(X)$.

Before going on let us make the following remark. Let T be a subset of $\mathbf{C}(X)$ and let $\{U_k\}$ be a sequence of open subsets of $\mathbf{R}^n$ such that $\cup_k U_k = \mathbf{R}^n$ and $X \subset U_k$ for any k. Then T is a residual subset of $\mathbf{C}(X)$ if and only if $T \cap C_{\text{emb}}^\infty(X, U_k)$ is residual in $C_{\text{emb}}^\infty(X, U_k)$ for each k. This follows easily from the fact that $C_{\text{emb}}^\infty(X, U_k)$ is open in $\mathbf{C}(X)$ for any k and

$$\mathbf{C}(X) = \cup_k C_{\text{emb}}^\infty(X, U_k).$$

Theorem 3.2.5: *Each of the sets $\mathcal{B}, \mathcal{P}, \mathcal{S}$, defined above, contains a residual subset of* $\mathbf{C}(X)$.

Proof: Fix an arbitrary open ball U_0 with radius a in $\mathbf{R}^n$ such that $X \subset U_0$.

For the sake of brevity set

$$Z_1 = Z_\omega, \quad Z_2 = Z_{-\theta}, \quad \pi_1 = \pi_\omega, \quad \pi_2 = \pi_{-\theta},$$

the notation Z_ξ, π_ξ being introduced in Section 2.4. According to the above remark, it is sufficient to establish that the intersection of each of the sets $\mathcal{B}, \mathcal{P}, \mathcal{S}$ with

$$\mathbf{C}(X, U_0) = C^\infty_{\text{emb}}(X, U_0)$$

contains a residual subset of $\mathbf{C}(X, U_0)$.

We need a modification of the notion of admissible map, introduced in the present section.

Let $k \geq s \geq 2$ be natural numbers and let α be a map of the form (3.21) such that (3.22) holds for $i = 1, \ldots, k-1$. We shall say that α is a *wns-map* if (3.23) holds for $1 \leq i < j \leq k-1$. If $k = 2m+1$, $\alpha(m-i+1) = \alpha(m+i+1)$ for $i = 0, 1, \ldots, m$ and (3.23) holds for $1 \leq i < j \leq m$, we shall say that α is a *ws-map*. By a *weakly admissible map* we mean a map (3.21) which is either a wns-map or a ws-map.

Next, we consider a fixed weakly admissible map (3.21) and set

$$\alpha(0) = 0, \quad \alpha(k+1) = s+1,$$

i.e. we regard α as a map $\alpha : \{0, 1, \ldots, k; \ k+1\} \to \{0, 1, \ldots, s, s+1\}$. Introduce the sets $I_i = I_i(\alpha)$ by (3.25) for $i = 1, \ldots, s$. Only in this proof we shall use the notation $y_0 = \pi_1(y_1)$ and $y_{s+1} = \pi_2(y_{\alpha(k)})$ for $y = (y_1, \ldots, y_s) \in (\mathbf{R}^n)^{(s)}$. Set $U^*_\alpha = U_\alpha \cap U_0^{(s)}$, U_α being defined at the beginning of this section. Then U^*_α is an open subset of $U_0^{(s)}$, and the function

$$F^* = F^*_\alpha : U^*_\alpha \to \mathbf{R}, \tag{3.33}$$

given by

$$F^*(y) = \sum_{i=0}^{k} \left\| y_{\alpha(i)} - y_{\alpha(i+1)} \right\|, \tag{3.34}$$

is smooth. If $y_1, \ldots, y_s$ are the different reflection points of an (ω, θ)-trajectory γ for X such that (3.28) are the successive reflection points of γ, we shall say that γ is of *type* α. In this case we have $y = (y_1, \ldots, y_s) \in U^*_\alpha$ and $F^*(y) = T_\gamma$. Moreover, y is a critical point of the map.

$$F^*_{|X_s} : X^s \to \mathbf{R}.$$

It is also clear that for every (ω, θ)-trajectory γ there exists a surjective weakly admissible map α such that γ is of type α.

The following lemma can be proved in the same way as Lemma 3.2.2. We leave the details to the reader.

Lemma 3.2.6: *For every* $i = 1, \ldots, s$ *and for every* $y \in U_\alpha^*$ *there exists* $j = 1, \ldots, n$ *such that* $\partial F_\alpha^* / \partial y_i^{(j)} \neq 0.$ ♠

Next, assume that (3.21) is a surjective wns-map such that $k > s$. Denote by $\mathcal{B}_\alpha$ the set of those $f \in C(X, U_0)$ such that there does not exist an (ω, θ)-trajectory of type α for $f(X)$. To show that $\mathcal{B}_\alpha$ contains a residual subset of $C(X, U_0)$ we use again Theorem 3.1.1 (B); this time with $U = U_\alpha^*, p = 1, H = F^*$.

Since $k > s$, there exists $i = 1, \ldots, s$ such that $\alpha^{-1}(i)$ contains more than one element. Take two arbitrary distinct elements j_1, j_2 of $\alpha^{-1}(i)$. Then $q = \alpha(j_1 - 1), j = \alpha(j_1 + 1), r = \alpha(j_2 - 1), t = \alpha(j_2 + 1)$ are distinct elements of I_i. Since $\{q, j\} \neq \{0, s+1\}$, either q or j is not in $\{0, s+1\}$. We may assume that $q \notin \{0, s+1\}$; otherwise we can exchange our notation by setting $q = \alpha(j_1 + 1), j = \alpha(j_1 - 1)$. Similarly, we may assume that $r \notin \{0, s+1\}$. For $u = 1, 2$ and $y \in U_\alpha^*$ set

$$
L_u(y) = \left\langle \frac{y_q - y_i}{\|y_q - y_i\|} + (-1)^u \frac{y_j - y_i}{\|y_j - y_i\|}, \frac{y_r - y_i}{\|y_r - y_i\|} - (-1)^u \frac{y_t - y_i}{\|y_t - y_i\|} \right\rangle
$$

and define $L : U_\alpha^* \to \mathbf{R}^2$ by $L(y) = (L_1(y), L_2(y))$. Now, repeating the corresponding part from the proof of Theorem 3.2.3, we deduce that $\mathcal{B}_\alpha$ contains a residual subset of $C(X, U_0)$.

Further, suppose that $\theta = -\omega$ and α is a surjective ws-map of the form (3.21) with $k > 2s - 1$. Let $k = 2m + 1$ and let $y_1, \ldots, y_s$ be the distinct reflection points of a (ω, θ)-trajectory γ for X of type α, i.e. (3.28) are the successive reflection points of γ. Then one gets easily that $y' = (y_1, \ldots, y_m, y_{m+1})$ is a critical point of the map $G_{|X^{(m+1)}}^*$, where

$$
G^* : (\mathbf{R}^n)^{(m+1)} \to \mathbf{R}
$$

is defined by

$$
G^*(y) = \sum_{i=0}^{m} \|y_{\alpha(i)} - y_{\alpha(i+1)}\|.
$$

Define $\mathcal{B}_\alpha$ as in the previous case. Now, using a similar argument as those above, replacing F^* by G^* and applying again Theorem 3.1.1 (B), we see that $\mathcal{B}_\alpha$ contains a residual subset of $C(X, U_0)$.

In this way we have established that the set $\mathcal{B}_1 = \cap_\alpha \mathcal{B}_\alpha$, where α runs over the surjective weakly admissible maps (3.21) with $k > s$ for wns-maps α and with $k > 2s - 1$ for ws-maps α, contains a residual subset of $C(X, U_0)$. To conclude the consideration of $\mathcal{B}$, we have to prove that the set $\mathcal{B}_2$ of those $f \in C(X, U_0)$ such that any two different (ω, θ)-trajectories for $f(X)$ have no common reflection point, contains a residual subset of $C(X, U_0)$. This can be established using a modification of the corresponding argument from the proof of Theorem 3.2.3. Here we omit the details.

Since $\mathcal{B}_1 \cap \mathcal{B}_2 \subset \mathcal{B}$, we deduce that $\mathcal{B}$ also contains a residual subset of $C(X, U_0)$.

Next, we proceed with the set $\mathcal{S}$. Let

$$f \in \mathcal{B} \cap \mathbf{C}(X, U_0),$$

and let γ and δ be different non-symmetric (ω, θ)-trajectories for $Y = f(X)$ such that $T_\gamma = T_\delta$. Let $y_i = f(x_i)$ be all distinct reflection points of γ and δ taken together. Since $f \in \mathcal{B}$, we may assume that $y_1, \ldots, y_k (k < s)$ are the successive reflection points of γ, while $y_{k+1}, y_{k+2}, \ldots, y_s$ are these of δ. Consider the functions $F^{**}, G^{**} : U \to \mathbf{R}$, defined by

$$F^{**}(z) = \|\pi_1(z_1) - z_1\| + \sum_{i=1}^{k-1} \|z_i - z_{i+1}\| + \|z_k - \pi_2(z_k)\|, \qquad (3.35)$$

$$G^{**}(z) = \|\pi_1(z_{k+1}) - z_{k+1}\| + \sum_{i=k+1}^{s-1} \|z_i - z_{i+1}\| + \|z_s - \pi_2(z_s)\|. \, (3.36)$$

Here U is the set of all $z \in (\mathbf{R}^n)^{(s)}$ such that $z_i \notin [z_{i-1}, z_{i+1}]$ for all $i = 2, 3, \ldots, s-1, i \neq k$, $z_1 \notin [\pi_1(z_1), z_2]$, $z_k \notin [z_{k-1}, \pi_2(z_k)]$, $z_{k+1} \notin [\pi_1(z_{k+1}), z_{k+2}]$, $z_s \notin [z_{s-1}, \pi_2(z_s)]$. Clearly, $F^{**}(y) = G^{**}(y)$. Applying Theorem 3.1.1 (B) with $H = (F^{**}, G^{**})$ and $L : U \to \mathbf{R}, L(z) = F^{**}(z) - G^{**}(z)$, we find that the set

$$S_1(k, s) = \{f \in \mathbf{C}(X, U_0) : \mathrm{grad}_x(H \circ f^s)(x) = 0 \Rightarrow L(f^s(x)) \neq 0\}$$

contains a residual subset of $\mathbf{C}(X, U_0)$. Hence

$$\mathcal{S}_1 = \cap_{k<s} \mathcal{S}_1(k, s) \cap \mathcal{B}$$

has the same property. Moreover, it follows from our considerations above that for $f \in \mathcal{S}_1$ we have $T_\gamma \neq T_\delta$ for any two different non-symmetric (ω, θ)-trajectories γ and δ for $f(X)$.

In a similar way one constructs $\mathcal{S}_2, \mathcal{S}_3$, containing residual subsets of $\mathbf{C}(X, U_0)$, such that for $f \in \mathcal{S}_2$ (resp. $f \in \mathcal{S}_3$) we have $T_\gamma \neq T_\delta$ for any two different symmetric (ω, θ)-trajectories γ and δ for $f(X)$ (resp. for any symmetric γ and non-symmetric δ). Since

$$\mathcal{S}_1 \cap \mathcal{S}_2 \cap \mathcal{S}_3 \subset \mathcal{S},$$

we obtain that $\mathcal{S}$ contains a residual subset of $\mathbf{C}(X, U_0)$.

Finally, we have to deal with $\mathcal{P}$. To this end we use an argument similar to these above. One can express analytically the fact that two segments $[y_i, y_{i+1}]$ and $[y_j, y_{j+1}]$ of an (ω, θ)-trajectory are parallel, by introducing a map $L : U \to \mathbf{R}^n$,

$$L(y) = \frac{y_i - y_{i+1}}{\|y_i - y_{i+1}\|} + \epsilon \frac{y_j - y_{j+1}}{\|y_j - y_{j+1}\|}.$$

Here $\epsilon = \pm 1$ and the set $U \subset (\mathbf{R}^n)^{(s)}$ are defined appropriately. A standard application of Theorem 3.1.1 (B) shows that $\mathcal{P}$ contains a residual subset of $\mathbf{C}(X, U_0)$. The details are left to the reader. ♠

3.3. Absence of tangent segments

In this section X will be again a fixed smooth $(n-1)$-dimensional submanifold of $\mathbf{R}^n, n \geq 2$. Recall that a reflecting ray for X is called *ordinary* if it has no segments tangent to X.

Theorem 3.3.1: *Let $\mathcal{T}$ be the set of those $f \in \mathbf{C}(X)$ such that every periodic reflecting ray for $f(X)$ is ordinary. Then $\mathcal{T}$ contains a residual subset of $\mathbf{C}(X)$.*
    To prove the theorem we need the following technical lemma.

Lemma 3.3.2: *Let $p = 1, \dots, n$ be fixed and let Q_p be the set of those $v = (v_0, v_1, v_2) \in (\mathbf{R}^n)^{(3)}$ such that $v_1^{(p)} \neq v_0^{(p)}$. Let*

$$d^{(m)} : Q_p \to \mathbf{R}, \quad 1 \leq m \leq n,$$

be defined by

$$d^{(m)}(v) = \frac{v_1^{(m)} - v_0^{(m)}}{\|v_1 - v_0\|} + \frac{v_2^{(m)} - v_0^{(m)}}{\|v_2 - v_0\|}.$$

Then the vectors grad $d^{(m)}(v), m = 1, \dots, n, m \neq p$, are linearly independent over $\mathbf{R}$ for every $v \in Q_p$.

Proof: Fix an arbitrary $v \in Q_p$, and assume that there exist real constants $D^{(m)}$ such that

$$\sum_{1 \leq m \leq n, m \neq p} D^{(m)} \operatorname{grad} d^{(m)}(v) = 0. \tag{3.37}$$

Set $D^{(p)} = 0$ and $D = (D^{(1)}, \dots, D^{(n)}) \in \mathbf{R}^n$. For convenience introduce the notation

$$\omega_i = \frac{v_i - v_0}{\|v_i - v_0\|}, \quad z_i = \frac{1}{\|v_i - v_0\|}, \quad i = 1, 2.$$

By direct calculations we obtain

$$\frac{\partial d^{(m)}}{\partial v_0^{(t)}}(v) = z_1 \omega_1^{(m)} \omega_1^{(t)} + z_2 \omega_2^{(m)} \omega_2^{(t)}, \quad t = 1, \dots, n, t \neq m,$$

and

$$\frac{\partial d^{(t)}}{\partial v_0^{(t)}}(v) = -(z_1 + z_2) + z_1 (\omega_1^{(t)})^2 + z_2 (\omega_2^{(t)})^2, \quad t = 1, \dots, n.$$

It then follows by (3.37) that for any t we have

$$0 = \sum_{m=1}^{n} D^{(m)} \frac{\partial d^{(t)}}{\partial v_0^{(t)}}(v)$$

$$= \sum_{m=1}^{n} D^{(m)}(z_1 \omega_1^{(m)} \omega_1^{(t)} + z_2 \omega_2^{(m)} \omega_2^{(t)}) - D_t(z_1 + z_2),$$

which is equivalent to

$$(z_1 + z_2)D^{(t)} = z_1 \langle D, \omega_1 \rangle \omega_1^{(t)} + z_2 \langle D, \omega_2 \rangle \omega_2^{(t)}.$$

Since the latter holds for any $t = 1, \ldots, n$, we get

$$(z_1 + z_2)D = z_1 \langle D, \omega_1 \rangle \omega_1 + z_2 \langle D, \omega_2 \rangle \omega_2. \tag{3.38}$$

Considering the inner product of (3.38) with ω_1, we find $\langle D, \omega_1 \rangle = \langle D, \omega_2 \rangle \langle \omega_1, \omega_2 \rangle$. In the same way we see that $\langle D, \omega_2 \rangle = \langle D, \omega_1 \rangle \langle \omega_1, \omega_2 \rangle$. Combining the latter two equalities one gets

$$\langle D, \omega_1 \rangle(1 - \langle \omega_1, \omega_2 \rangle^2) = 0, \quad \langle D, \omega_2 \rangle(1 - \langle \omega_1, \omega_2 \rangle^2) = 0. \tag{3.39}$$

First, assume that $\langle \omega_1, \omega_2 \rangle^2 \neq 1$. Then (3.39) implies $\langle D, \omega_1 \rangle = \langle D, \omega_2 \rangle = 0$, and by (3.38) we deduce $D = 0$. Let $\langle \omega_1, \omega_2 \rangle^2 = 1$. Then $\omega_2 = \epsilon \omega_1$ for $\epsilon = \pm 1$, and (3.38) becomes

$$(z_1 + z_2)D = (z_1 + z_2)\langle D, \omega_1 \rangle \omega_1.$$

Hence $D = \langle D, \omega_1 \rangle \omega_1$, and comparing the p-components of the vectors in the latter equality, we find $0 = D^{(p)} = \langle D, \omega_1 \rangle \omega_1^{(p)}$. Since $v \in Q_p$, we have $\omega_1^{(p)} \neq 0$, therefore $\langle D, \omega_1 \rangle = 0$. In the case under consideration this clearly implies $\langle D, \omega_2 \rangle = 0$, and by (3.38) we deduce $D = 0$ again. ♠

Proof of Theorem 3.3.1: Let $\mathcal{A}$ be the subset of $C(X)$, defined in Section 3.2 (cf. the text before Theorem 3.2.3). Let $f \in \mathcal{A}$ and let $s \geq 2$ be a natural number. Suppose that there exists a periodic reflecting ray γ for $f(X)$ with s reflection points and such that some segment of γ is tangent to $f(X)$. We may assume that $y_1 = f(x_1), \ldots, y_s = f(x_s)$ are the successive reflection points of γ, and the segment $[y_1, y_2]$ is tangent to $f(X)$ at some of its interior points $y_0 = f(x_0)$. The latter condition is equivalent to

$$\frac{y_1 - y_0}{\|y_1 - y_0\|} + \frac{y_2 - y_0}{\|y_2 - y_0\|} = 0$$

and $\langle y_1 - y_2, \nu(y_0) \rangle = 0$, where $\nu(y)$ denotes a unit normal vector to $f(X)$ at y.

Fix $s \geq 2$, and define U_s and $F = F_s$ as in the proof of Theorem 3.2.3. Denote by $\mathcal{T}_s$ the set of those $f \in C(X)$ such that there does not exist a critical point $x = (x_0, x_1, \ldots, x_s) \in X^{s+1}$ for which $x' = (x_1, \ldots, x_s) \in X^{(s)}$, $f^s(x') \in U_s$, x' is a critical point of the map $F \circ f^s$, and the segment $[f(x_1), f(x_2)]$ is tangent to $f(X)$ at $f(x_0)$. We shall prove that $\mathcal{T}_s$ contains a residual subset of $C(X)$. To this end we are going to use an argument similar to that in the proof of Theorem 3.1.1.

Set

$$V = \{j^1 f(x) \in J^1(X, \mathbf{R}^n) : \operatorname{rank} df(x) = n - 1\}.$$

Denote by M the set of those

$$\sigma = (j^1 f_0(x_0), j^1 f_1(x_1), \ldots, j^1 f_s(x_s)) \in V^{s+1} \qquad (3.40)$$

such that $x = (x_0, x_1, \ldots, x_s) \in X^{s+1}$, $x' = (x_1, \ldots, x_s) \in X^{(s)}$, $f^s(x') \in U_s$. Clearly, M is an open subset of $J^1_{s+1}(X, \mathbf{R}^n)$.

Introduce the singular set Σ as the set of all elements σ of M having the form (3.40) such that x' is a critical point of $F \circ (f_1 \times \ldots \times f_s)$ and the segment $[f_1(x_1), f_2(x_2)]$ is tangent to $f_0(X)$ at $f_0(x_0)$. Then we have

$$\mathcal{T}_s = \{f \in C(X) : j^1_{s+1}(X^{(s+1)}) \cap \Sigma = \emptyset\}. \qquad (3.41)$$

We shall establish that Σ can be covered by a countable family of smooth submanifolds of M of codimension greater than $\dim X^{(s+1)} = (s+1)(n-1)$. To this end consider a coordinate neighbourhood D of an element of Σ. We may take D of the same form as in the proof of Lemma 3.1.2. Namely, fix arbitrary charts $\varphi_i : V_i \to \mathbf{R}^{n-1}$, $i = 0, 1, \ldots, s$, where V_i are open subsets of X with $V_i \cap V_j = \emptyset$ for $1 \leq i, j \leq s, i \neq j$. Then we define D by (3.9), (3.11), (3.12), (3,13), (3.14), the only difference being that i runs from 0 to s (instead of $1 \leq i \leq s$). Note that the vector $N_i = (N_i^{(1)}, \ldots, N_i^{(n)})$, determined by

$$N_i^{(t)} = (-1)^t \det \begin{pmatrix} a_{i1}^{(1)} & \cdots & a_{i1}^{(t-1)} & a_{i1}^{(t+1)} & \cdots & a_{i1}^{(n)} \\ \cdots & \cdots & \cdots & \cdots & \cdots & \cdots \\ a_{in-1}^{(1)} & \cdots & a_{in-1}^{(t-1)} & a_{in-1}^{(t+1)} & \cdots & a_{in-1}^{(n)} \end{pmatrix}, \qquad (3.42)$$

is normal to $f_i(X)$ at $f_i(x_i)$.

We shall show that $\varphi(D \cap \Sigma)$ is contained in the union of a finite family of smooth submanifolds of $\varphi(D)$ of codimension $(s+1)(n-1)+1$. The elements ξ of $\varphi(D)$ will be written in the form

$$\xi = (u; v; a) \in (\mathbf{R}^{n-1})^{(s+1)} \times (\mathbf{R}^n)^{(s+1)} \times \mathbf{R}^{n(n-1)(s+1)},$$

where u, v, a are determined as above. For $p = 1, \ldots, n$ consider the open subset

$$G_p = \{\xi \in \varphi(D) : v_1^{(p)} \neq v_0^{(p)}\}$$

of $\varphi(D)$. Since $\varphi(D) = \cup_{p=1}^n G_p$, it is sufficient to show that for any p, $G_p \cap \varphi(D \cap \Sigma)$ is contained in a smooth submanifold of G_p of codimension $(s+1)(n-1)+1$.

Fix an arbitrary p. For $\xi \in \varphi(D)$ set

$$N(\xi) = (N_0^{(1)}(\xi), \ldots, N_0^{(n)}(\xi)),$$

where the components $N_0^{(t)}(\xi) = N_0^{(t)}$ are given by (3.42). Since $D \subset M \subset V^{s+1}$, we have $N(\xi) \neq 0$ for any $\xi \in \varphi(D)$. Set

$$c_{ij}(\xi) = \sum_{t=1}^n \frac{\partial F}{\partial y_i^{(t)}}(v) a_{ij}^{(t)}$$

for $\xi \in \varphi(D)$, $1 \le i \le s$, $1 \le j \le n-1$. Note that if $\xi = \varphi(\sigma)$ and σ has the form (3.40), then $x' = (x_1, \ldots, x_s)$ is a critical point of the map $F \circ (f_1 \times \ldots \times f_s)$ if and only if $c_{ij}(\xi) = 0$ for all $i = 1, \ldots, s$ and $j = 1, \ldots, n-1$. Consider the map

$$K : G_p \to \mathbf{R}^{s(n-1)} \times \mathbf{R}^{n-1} \times \mathbf{R},$$

defined by

$$K(\xi) = ((c_{ij}(\xi))_{1 \le i \le s, 1 \le j \le n-1}; (d^{(m)})_{1 \le m \le n, m \neq p}; L(\xi)),$$

where

$$d^{(m)}(\xi) = \frac{v_1^{(m)} - v_0^{(m)}}{\|v_1 - v_0\|} + \frac{v_2^{(m)} - v_0^{(m)}}{\|v_2 - v_0\|},$$

and $L(\xi) = \langle v_1 - v_2, N(\xi) \rangle$. It is straightforward to check that

$$G_p \cap \varphi(D \cap \Sigma) \subset K^{-1}(0).$$

Now it is sufficient to show that K is a submersion at any point of G_p. This will imply that $K^{-1}(0)$ is a submanifold of G_p of codimension $(s+1)(n-1)+1$.

Fix an arbitrary $\xi \in G_p$ and assume that there exist real constants C_{ij}, D_m, E such that

$$\sum_{i=1}^s \sum_{j=1}^{n-1} C_{ij} \operatorname{grad} c_{ij}(\xi) + \sum_{1 \le m \le n, m \neq p} D_m \operatorname{grad} d^{(m)}(\xi) + E \operatorname{grad} L(\xi) = 0.$$

$$(3.43)$$

It follows by Lemma 3.2.2 that for any $i = 1, \ldots, s$ there exists $t = 1, \ldots, n$ such that (3.29) holds with $y = v$. Considering in (3.43) the derivatives with respect to $a_{ij}^{(t)}$, we find $C_{ij} = 0$ for all i and j. Consequently, the first double sum in (3.43) is zero. Since $\frac{\partial L}{\partial v_0^{(t)}}(\xi) = 0$ for any $t = 1, \ldots, n$, we can use

(3.43) to apply Lemma 3.3.2. It then follows that $D_m = 0$ for every $m \neq p$. Finally, consider arbitrary $t = 1, \ldots, n$ with $N^{(t)}(\xi) \neq 0$. Then by (3.43) and

$$\frac{\partial L}{\partial v_1^{(t)}}(\xi) = N^{(t)}(\xi) \neq 0$$

it follows that $E = 0$. Hence K is a submersion at ξ.

In this way we have shown that $K^{-1}(0)$ is a smooth submanifold of G_p of codimension $(s+1)(n-1)+1$. As we have already mentioned above, this implies that $D \cap \Sigma$ is contained in the union of finitely many smooth submanifolds of M of codimension $(s+1)(n-1)+1$.

Since Σ can be covered with a countable family of coordinate neighbourhoods D, we conclude that Σ is contained in the union of a countable family $\{W_m\}$ of smooth submanifolds W_m of M with codim $W_m = (s+1)(n-1)+1$. Then (3.41) yields

$$\cap_m T_s^{(m)} \subset T_s, \tag{3.44}$$

where

$$T_s^{(m)} = \{f \in \mathbf{C}(X) : j_{s+1}^1 f(X^{s+1}) \cap W_m = \emptyset\}.$$

Since M is open in $J_{s+1}^1(X, \mathbf{R}^n)$, W_m is a smooth submanifold of $J_{s+1}^1(X, \mathbf{R}^n)$ with

$$\text{codim } W_m = (s+1)(n-1)+1 > \dim X_{s+1}.$$

On the other hand, $j_{s+1}^1 f : X^{s+1} \to J_{s+1}^1(X, \mathbf{R}^n)$, therefore

$$j_{s+1}^1 f(X^{s+1}) \cap W_m = \emptyset \Leftrightarrow j_{s+1}^1 f \pitchfork W_m.$$

It then follows by the multijet transversality theorem (cf. Section 1.1) that $T_s^{(m)}$ is a residual subset of $\mathbf{C}(X)$ for any m. Consequently, according to (3.44), we get that T_s contains a residual subset of $\mathbf{C}(X)$.

Since $\cap_{s=2}^\infty T_s \cap \mathcal{A} \subset T$, according to Theorem 3.2.3, $\mathcal{A}$ contains a residual subset of $\mathbf{C}(X)$, we deduce that T has the same property. ♠

The next theorem can be proved applying the previous argument with slight modifications. We leave the details to the reader.

Theorem 3.3.3: Let ω and θ be two fixed unit vectors in $\mathbf{R}^n$ and let X be a compact smooth $(n-1)$-dimensional submanifold of $\mathbf{R}^n$. Let $T(\omega, \theta)$ be the set of those $f \in \mathbf{C}(X)$ such that every reflecting (ω, θ)-ray for $f(X)$ is ordinary. Then $T(\omega, \theta)$ contains a residual subset of $\mathbf{C}(X)$. ♠

3.4. Non-degeneracy of reflecting rays

In this section it is proved that for generic domains Ω in $\mathbf{R}^n$ all periodic reflecting rays are non-degenerate. It is established also that, given two fixed unit vectors ω and θ, for generic Ω with bounded complements, every reflecting (ω, θ)-ray γ in Ω is non-degenerate, i.e. $\det dJ_\gamma \neq 0$.

Hereafter X will be a fixed smooth compact $(n-1)$-dimensional submanifold of $\mathbf{R}^n$, $n \geq 2$.

Theorem 3.4.1: *Let Λ be an arbitrary countable set of complex numbers and let T_Λ be the set of those $f \in \mathbf{C}(X)$ such that every periodic reflecting ray γ for $f(X)$ is ordinary and spec P_γ does not contain elements of Λ. Then T_Λ contains a residual subset of $\mathbf{C}(X)$.*

Since the definition of the Poincaré map involves second derivatives, there is no chance to prove the above theorem applying somehow Theorem 3.1.1. However, as in Section 3.3, an appropriate modification of the scheme of the proof of the latter theorem is useful again.

We begin with some preliminary considerations.

Let γ be an ordinary primitive non-symmetric periodic reflecting ray for X with successive reflection points $q_1, \ldots, q_m, q_{m+1} = q_1$. We shall suppose that the points $q_1, \ldots, q_m$ are all different. Introduce the notation Π_i, σ_i, λ_i, ν_i, G_i, $\tilde{\psi}_i$, ψ_i, s_i as in Section 2.3. Then P_γ has the representation (2.26). For the sake of brevity we set

$$A_j = \begin{pmatrix} I & \lambda_j I \\ \psi_j & I + \lambda_j \psi_j \end{pmatrix}, \quad j = 1, \ldots, m. \tag{3.45}$$

Since P_γ is a symplectic matrix, we have $0 \notin \operatorname{spec} P_\gamma$. Let $1/\mu \in \mathbf{C}$ be an eigenvalue of P_γ. Then by (2.26) we find

$$0 = \det \left(\frac{1}{\mu} I - \begin{pmatrix} s_m & 0 \\ 0 & s_m \end{pmatrix} A_m \ldots A_1 \right)$$

$$= \det \left(I - A_m \ldots A_1 \begin{pmatrix} \mu s_m & 0 \\ 0 & \mu s_m \end{pmatrix} \right).$$

Set

$$E = \det \left(I - A_m \ldots A_1 \begin{pmatrix} \mu s_m & 0 \\ 0 & \mu s_m \end{pmatrix} \right), \tag{3.46}$$

and denote by $d_{ij}^{(t)}$ the elements of the matrix ψ_t, $t = 1, \ldots, m$, $i, j = 1, \ldots, n-1$. Then clearly E can be expressed as a polynomial of the elements $d_{ij}^{(t)}$. The terms in E, involving only products of elements $d_{ij}^{(1)}$, $i, j = 1, \ldots, n-1$, are contained in the determinant

$$D = \det \left(I - \begin{pmatrix} I & \lambda_m I \\ 0 & I \end{pmatrix} \ldots \begin{pmatrix} I & \lambda_2 I \\ 0 & I \end{pmatrix} A_1 \begin{pmatrix} \mu s_m & 0 \\ 0 & \mu s_m \end{pmatrix} \right)$$

$$= \det \left(A_1^{-1} - \begin{pmatrix} \mu s_m & \mu \left(\sum_{i=2}^m \lambda_i \right) s_m \\ 0 & \mu s_m \end{pmatrix} \right).$$

Since

$$A_1^{-1} = \begin{pmatrix} I + \lambda_1 \psi_1 & -\lambda_1 I \\ -\psi_1 & I \end{pmatrix},$$

we find

$$D = \det \begin{pmatrix} I + \lambda_1 \psi_1 - \mu s_m & -\mu \left(\sum_{i=2}^m \lambda_i\right) s_m - \lambda_1 \\ -\psi_1 & I - \mu s_m \end{pmatrix}$$

$$= \det \begin{pmatrix} I - \mu s_m & -\mu \left(\sum_{i=1}^m \lambda_i\right) s_m \\ -\psi_1 & I - \mu s_m \end{pmatrix}.$$

It is clear now that the product $d_{11}^{(1)} d_{22}^{(1)} \ldots d_{n-1n-1}^{(1)}$ has in E a non-zero coefficient $\epsilon \mu (\sum_{i=1}^m \lambda_i) \det s_m$, where $\epsilon = \pm 1$. Consequently, E is a non-trivial polynomial of $d_{ij}^{(t)} (1 \le i, j \le n - 1, 1 \le t \le m)$ with coefficients depending on $\mu, \lambda_i, \sigma_i (i = 1, \ldots, m)$.

Introduce the multiindices

$$\tau = ((i_1, j_1, t_1), \ldots, (i_l, j_l, t_l)), \tag{3.47}$$

consisting of triples (i_s, j_s, t_s) of integers such that

$$1 \le i_s, j_s \le n - 1, \ 1 \le t_s \le m, \ t_1 \ge t_2 \ge \ldots \ge t_l \ge 1. \tag{3.48}$$

For $i \le l$ denote by $p_i(\tau)$ the number of those triples (i_s, j_s, t_s) in τ such that $t_s = t_i$. Further, set

$$|\tau| = \sum_{i=1}^l t_i, \ d^\tau = d_{i_1 j_1}^{(t_1)} d_{i_2 j_2}^{(t_2)} \ldots d_{i_l j_l}^{(t_l)},$$

and define the function $\partial^\tau E$ by

$$\partial^\tau E = \frac{\partial^{|\tau|} E}{\partial d_{i_1 j_1}^{(t_1)} \partial d_{i_2 j_2}^{(t_2)} \ldots \partial d_{i_l j_l}^{(t_l)}}.$$

It follows from our arguments above that

$$E = \sum_\tau c_\tau d^\tau,$$

where τ runs over the set of the multiindices (3.47), satisfying (3.48), such that $|\tau| \le m(n-1)$, $p_i(\tau) \le n - 1$. Here

$$c_\tau = c_\tau(\mu, \lambda_1, \ldots, \lambda_m, \sigma_1 \ldots, \sigma_m)$$

are real coefficients.

Next, we define an open subset M of the m-fold bundle of 2-jets $J_m^2(X, \mathbf{R}^n)$. First, consider the open subset

$$V = \{j^2 f(x) \in J^2(X, \mathbf{R}^n) : \text{ rank } df(x) = n - 1\}$$

of $J^2(X, \mathbf{R}^n)$. Denote by U_m the set of those $y = (y_1, \ldots, y_m) \in (\mathbf{R}^n)^{(m)}$ such that for every $i = 1, \ldots, m$ the point y_i does not belong to the segment $[y_{i-1}, y_{i+1}]$. As before, we set for convenience $y_0 = y_m$ and $y_{m+1} = y_1$. We need also the function

$$F = F_m : U_m \to \mathbf{R},$$

given by

$$F(y) = \sum_{i=1}^{m} \| y_i - y_{i+1} \|.$$

Finally, set

$$M = (\alpha^m)^{-1}(X^{(m)}) \cap (\beta^m)^{-1}(U_m) \cap V^m,$$

where α and β are the source and the target maps, respectively, defined in Section 1.1.

An atlas for M can be described in a similar way as in Sections 3.1 and 3.3. Namely, consider arbitrary coordinate neighbourhoods $V_1, \ldots, V_m$ of different elements of X such that $V_i \cap V_j = \emptyset$ whenever $i \neq j$, and let $\varphi_i : V_i \to \mathbf{R}^{n-1}$ be arbitrary smooth charts. Set

$$D = M \cap \prod_{i=1}^{m} J^2(V_i, \mathbf{R}^n) \tag{3.49}$$

and define the chart

$$\varphi : D \to (\mathbf{R}^{n-1})^{(s)} \times (\mathbf{R}^n)^{(s)} \times \mathbf{R}^{s(n-1)n} \times \mathbf{R}^{s(n-1)(n-2)n/2}$$

by

$$\varphi(\sigma) = (u; v; a; b)$$

for every element

$$\sigma = (j^2 f_1(x_1), \ldots, j^2 f_m(x_m)) \tag{3.50}$$

of D, where u, v, a are defined by (3.12) (replacing s by m), (3.13) and (3.14), while

$$b = (b_{ijl}^{(t)})_{1 \leq i \leq m, 1 \leq j, l \leq n-1, 1 \leq t \leq n} \tag{3.51}$$

is given by

$$b_{ijl}^{(t)} = \frac{\partial^2 (f_i^{(t)} \circ \varphi_i^{-1})}{\partial u_i^{(j)} \partial u_i^{(l)}}(u_i). \tag{3.52}$$

As we mentioned in Section 3.3, the vector

$$N_i = (N_i^{(1)}, \ldots, N_i^{(n)}), \tag{3.53}$$

determined by (3.42), is orthogonal to $f_i(X)$ at the point $f_i(x_i)$.

Let σ be an element of D of the form (3.50), and let $q_i = f_i(x_i)$, $i = 1, \ldots, m$. Now we can define the numbers λ_i and the maps σ_i as in the case of a periodic reflecting ray for a given submanifold $f(X)$. Namely, we first define λ_i by (2.14) and N_i by (3.53) and (3.42). Next, denote by Π_i the hyperplane, passing through $f_i(x_i)$ and orthogonal to $\overrightarrow{f_i(x_i)f_{i+1}(x_{i+1})}$, and by α_i the hyperplane, passing through $f_i(x_i)$ and orthogonal to N_i. Now the symmetry σ_i is defined as in Section 2.4. Till now we have used only the data $j^1 f(x_1), \ldots, j^1 f(x_m)$. Further, we define the differential of the Gauss map

$$G_i : \alpha_i \rightarrow \alpha_i$$

by means of $j^2 f_i(x_i)$ and determine ψ_i and $\tilde{\psi}_i$ by (2.16), (2.24) and (2.25). Finally, define the matrices $A_i = A_i(\sigma)$ by (3.45) and the function $E = E(\sigma)$ by (3.46), where μ is a fixed complex number. Clearly, E is completely determined by σ, therefore E can be viewed as a function

$$E : D \rightarrow \mathbf{R}.$$

Denote by Σ the set of all elements σ of M having the form (3.50) and such that $x = (x_1, \ldots, x_m)$ is a critical point of $F_m \circ f^m$, $f^m(x) \in U_m$, and $E(\sigma) = 0$.

Lemma 3.4.2: Σ is contained in the union of a countable family of smooth submanifolds of M of codimension $s(n-1) + 1$.

Proof: Consider a coordinate neighbourhood D of the form (3.49) of an element of Σ, and let the chart φ on D be defined as above. To prove the assertion it is sufficient to show that $\varphi(D \cap \Sigma)$ is contained in the union of finitely many smooth submanifolds of $\varphi(D)$ of codimension $s(n-1) + 1$.

Now the elements ξ of $\varphi(D)$ have the form

$$\xi = (u; v; a; b),$$

where u, v, a, b are given by (3.12) (with s replaced by m), (3.13), (3.14), (3.51), (3.52). For the sake of brevity we shall use the notation $E(\xi) = E(\sigma)$ for $\varphi(\sigma) = \xi$, i.e. $E(\xi) = E(\varphi^{-1}(\xi))$. Set

$$c_{ij}(\xi) = \sum_{t=1}^{n} \frac{\partial F_m}{\partial y_i^{(t)}}(v) a_{ij}^{(t)}$$

for $1 \leq i \leq m$, $1 \leq j \leq n-1$. Consider the multiindices

$$\delta_p = ((1,1,1), (2,2,1), \ldots, (p,p,1)),$$

and denote by M_p the set of those $\xi \in \varphi(D)$ such that $c_{ij}(\xi) = 0$ for all $i = 1, \ldots m$, $j = 1, \ldots, n-1$, and

$$\partial^{\delta_p} E(\xi) = 0, \quad \partial^{\delta_p + 1} E(\xi) \neq 0$$

for $p = 0, 1, \ldots, n - 2$. Here we set for convenience $\partial^{\delta_0} E = E$.

Given two multiindices τ and τ' of the form (3.47) with (3.48), we shall write $\tau < \tau'$ if $|\tau| < |\tau'|$ and τ' contains all triples in τ. Denote by $\mathcal{M}$ the set of all multiindices τ of the form (3.47) with (3.48) such that $|\tau| \leq m(n-1)$, $\tau > \delta_{n-1}$, and τ contains exactly $n-1$ triples (i, j, t) with $t = 1$. Let $\mathcal{M}_2$ be the set of all pairs $(\tau, \tau') \in \mathcal{M}^2$ such that $\tau < \tau'$ and $|\tau'| = |\tau| + 1$. Given $(\tau, \tau') \in \mathcal{M}_2$, set

$$M(\tau, \tau') = \{\xi \in \varphi(D) : c_{ij}(\xi) = 0 \ \forall i = 1, \ldots, m,$$
$$j = 1, \ldots, n-1, \partial^{\tau} E(\xi) = 0, \partial^{\tau'} E(\xi) \neq 0\}.$$

It then follows from above that

$$\varphi(D \cap \Sigma) \subset \cup_{p=0}^{n-2} M_p \cup \cup_{(\tau, \tau') \in \mathcal{M}_2} M(\tau, \tau').$$

Therefore the lemma will be proved if we establish that each of the sets M_p and $M(\tau, \tau')$ is a smooth submanifold of $\varphi(D)$ with codimension $s(n-1) + 1$.

Before going on let us make the following remark. Let $A = (A_{ij})$ and $B = (B_{ij})$ be $n \times n$ symmetric matrices such that

$$UAV = B \tag{3.54}$$

for some invertible matrices U and V. Let $\psi(B_{11}, B_{12}, \ldots, B_{nn})$ be a smooth function of the elements B_{ij} of the matrix B. If the elements of the matrices U and V do not depend on (A_{ij}) and (B_{ij}), then $\partial \psi / \partial B_{ij} = 0$ for all $i, j = 1, \ldots, n$ is equivalent to $\partial \psi / \partial A_{ij} = 0$ for all $i, j = 1, \ldots, n$. Here ψ is a considered as a function of (A_{ij}), replacing each B_{ij} by the corresponding function of (A_{ij}) according to (3.54).

For every $k = 1, \ldots, m$ we have

$$\psi_k = \sigma_1 \ldots \sigma_k \pi_k^* G_k \pi_k \sigma_k \ldots \sigma_1,$$

where the projection π_k is defined as in Section 2.4. Let $G_k = (g_{ij}^{(k)})_{i,j=1}^{n-1}$. It then follows from the above remark that if

$$\frac{\partial}{\partial d_{ij}^{(k)}} (\partial^{\tau} E)(\xi) \neq 0 \tag{3.55}$$

for some (i, j, k), then there exist i' and j' such that

$$\frac{\partial}{\partial g_{i'j'}^{(k)}} (\partial^{\tau} E)(\xi) \neq 0.$$

Note that

$$g_{jl}^{(k)} = -\sum_{t=1}^{n} b_{kjl}^{(t)} N_k^{(t)},$$

where $b_{kjl}^{(t)}$ are given by (3.52) and $N_k^{(t)}$ by (3.42). Consequently, $E(\xi)$ is a polynomial of $b_{kjl}^{(t)}$, $1 \le k \le m$, $1 \le j \le n-1$, $1 \le t \le n$. Therefore if

$$\frac{\partial}{\partial g_{jl}^{(k)}}(\partial^\tau E)(\xi) \ne 0$$

for some (k, j, l), then there exists $t = 1, \ldots, n$ such that

$$\frac{\partial}{\partial b_{kjl}^{(t)}}(\partial^\tau E)(\xi) \ne 0. \tag{3.56}$$

Fix an arbitrary $(\tau, \tau') \in \mathcal{M}_2$, and denote by $\mathcal{O}_{\tau'}$ the set of those $\xi \in \varphi(D)$ such that $\partial^{\tau'} E(\xi) \ne 0$. Define the map

$$L : \mathcal{O}_{\tau'} \to \mathbf{R}^{s(n-1)+1}$$

by

$$L(\xi) = ((c_{ij}(\xi))_{1 \le i \le m, 1 \le j \le n-1}; \quad \partial^\tau E(\xi)).$$

Clearly,

$$M(\tau, \tau') = L^{-1}(0) \subset \mathcal{O}_{\tau'}.$$

We are going to show that L is a submersion in $\mathcal{O}_{\tau'}$; this will imply that $M(\tau, \tau')$ is a smooth submanifold of $\varphi(D)$ of codimension $s(n-1)+1$.

Let $\xi \in \mathcal{O}_{\tau'}$ and assume that

$$\sum_{i=1}^{m} \sum_{j=1}^{n-1} C_{ij} \operatorname{grad} c_{ij}(\xi) + A \operatorname{grad}(\partial_\tau E)(\xi) = 0 \tag{3.57}$$

for some constants C_{ij}, A. Since $\xi \in M(\tau, \tau')$, there exists (i_0, j_0, k_0) such that (3.55) holds with $i = i_0$, $j = j_0$, $k = k_0$. It then follows from our reasonings above that there exist i, j and t such that (3.56) holds with $k = k_0$. Since the functions $c_{ij}(\xi)$ do not depend on the variables $b_{kjl}^{(t)}$, we get $A = 0$. Next, fix arbitrary i and j and consider in (3.57) the derivatives with respect to $a_{ij}^{(t)}$. Since $v \in U_m$, according to Lemma 3.2.2 and using the same argument as that in the proof of Lemma 3.1.2, we find $C_{ij} = 0$. Therefore L is a submersion at ξ.

This shows that $M(\tau, \tau')$ is a smooth submanifold of $\varphi(D)$ of codimension $s(n-1)+1$. Applying the same argument with a slight modification, we see also that M_p has the same property for any $p = 0, 1, \ldots, n-2$. This concludes the proof of the lemma. ♠

Proof of Theorem 3.4.1:   For a complex number μ and an integer $m \geq 2$ set

$$T'(\mu, m) = \{f \in \mathbf{C}(X) : j_m^2 f(X^{(m)}) \cap \Sigma = \emptyset\}.$$

Then $T'(\mu, m)$ contains a residual subset of $\mathbf{C}(X)$. This follows easily from Lemma 3.4.2, applying the multijet transversality theorem and an argument, similar to these in the proofs of Theorems 3.1.1 and 3.3.1.

Let $f \in T'(\mu, m)$. Then if $x = (x_1, \ldots, x_m) \in X^{(m)}$ is a critical point of the map $F_m \circ f^m$ with $f^m(x) \in U_m$, we have $E(\sigma) \neq 0$, where σ is defined by (3.50) with $f_1 = \ldots = f_m = f$. According to our considerations at the beginning of the section, this implies that for any ordinary primitive non-symmetric periodic reflecting ray γ for $f(X)$ we have $\mu \notin \operatorname{spec} P_\gamma$.
Set

$$\tilde{\Lambda} = \{z \in \mathbf{C} : \exists k \in \mathbf{N} \text{ with } z^k \in \Lambda\}.$$

Since Λ is countable, $\tilde{\Lambda}$ is also countable. Let $\mathcal{A}, \mathcal{T}$ be the subsets of $\mathbf{C}(X)$ from Theorems 3.2.3 and 3.3.1, respectively. It follows from the considerations above and the theorems just cited that

$$T' = \cap_{\mu \in \tilde{\Lambda}, m \geq 2} T'(\mu, m)$$

contains a residual subset of $\mathbf{C}(X)$. Moreover, the properties of the sets $T'(\mu, m)$, mentioned above, yield that for every $f \in T'$ and every ordinary non-symmetric periodic reflecting ray γ for $f(X)$, spec P_γ does not contain elements of Λ.

In a similar way one constructs $T'' \subset \mathbf{C}(X)$, containing a residual subset of $\mathbf{C}(X)$, and such that for every $f \in T'$ and every ordinary symmetric periodic reflecting ray γ for $f(X)$, spec P_γ does not contain elements of Λ. To this end one can repeat most of the considerations in this section with minor changes. Let us mention that if $q_i = f(x_i)$, $i = 1, \ldots, m$ are the successive reflection points of a symmetric primitive periodic reflecting ray γ for $f(X)$, $f \in \mathcal{A} \cap \mathcal{T}$, we may always assume that the segment $[q_1, q_2]$ is orthogonal to $f(X)$ at q_1. Moreover, $f \in \mathcal{A} \cap \mathcal{T}$ implies that γ is ordinary and the points $q_1, \ldots, q_k$, $k = (m-2)/2$, are all different. Finally, $x = (x_1, \ldots, x_k)$ is a critical point of the function $G_k \circ f^k$, where

$$G_k : U_k' \to \mathbf{R}$$

is given by

$$G_k(y) = \sum_{i=1}^{k-1} \|y_i - y_{i+1}\|,$$

and $U_k' \subset (\mathbf{R}^n)^{(k)}$ is defined appropriately. After these preparatory remarks, one can define the function E in the same way and repeat the argument from the non-symmetric case, considered above. The necessary modifications are very easy and we leave them to the reader.

Since $T' \cap T'' \subset T_\Lambda$, we obtain that T_Λ contains a residual subset of $\mathbf{C}(X)$. ♠

Combining Theorems 3.2.3, 3.3.1 and 3.4.1, we deduce that the set

$$\mathcal{F} = \mathcal{A} \cap \mathcal{T} \cap T_{Q/Z}$$

contains a residual subset of $\mathbf{C}(X)$. Here

$$\mathbf{Q/Z} = \{z \in \mathbf{C} : \exists k \in \mathbf{N} \text{ with } z^k = 1\},$$

and, as before, X is a compact smooth $(n-1)$-dimensional submanifold of $\mathbf{R}^n$, $n \geq 2$.

The next theorem shows that for generic domains Ω in $\mathbf{R}^n$ there exist at most countably many periodic reflecting rays in Ω.

Theorem 3.4.3: *For every $f \in \mathcal{F}$ and every integer $s \geq 2$, there exist only finitely many periodic reflecting rays for $f(X)$ with exactly s reflection points.*

Proof: Fix arbitrary $f \in \mathcal{F}$ and $s \geq 2$. Without loss of generality we may assume that $f = \mathrm{id}$; otherwise we can replace X by $f(X)$. Denote by K_s the set of all $x = (x_1, \dots, x_s) \in X^s$ such that $x_1, \dots, x_s$ are the successive reflection points of a periodic reflecting ray for X. We are going to prove that K_s is finite.

Assume that K_s is infinite. It then follows by the compactness of X that there exists a sequence

$$\{(x_{1,m}, \dots, x_{s,m})\}_{m=1}^{\infty}$$

of different elements of K_s such that $x_i = \lim_{m \to \infty} x_{i,m}$ exists for every $i = 1, \dots, s$. As before, we set for convenience $x_{s+1,m} = x_{1,m}, x_{s+1} = x_1$.

Lemma 3.4.4: *There exists $i \neq j$ with $x_i \neq x_j$.*

Proof of Lemma 3.4.4: Set

$$e_{i,m} = \frac{x_{i+1,m} - x_{i,m}}{\|x_{i+1,m} - x_{i,m}\|}, \qquad a_{i,m} = \frac{\|x_{i+1,m} - x_{i,m}\|}{\sum_{j=1}^{s} \|x_{j+1,m} - x_{j,m}\|}.$$

Then $\|e_{i,m}\| = 1$ and $\sum_{j=1}^{s} a_{i,m} = 1$. Without loss of generality we may assume that there exist

$$\lim_{m \to \infty} e_{i,m} = e_i, \qquad \lim_{m \to \infty} a_{i,m} = a_i.$$

Then $\|e_i\| = 1$ for all $i = 1, \dots, s$ and $\sum_{i=1}^{s} a_i = 1$.

First, assume that $x_1 = \dots = x_s$. Then using

$$\lim_{m \to \infty} x_{1,m} = \lim_{m \to \infty} x_{2,m} = \lim_{m \to \infty} x_{3,m},$$

we find that $e_2 = e_1$. In the same way we get $e_{i+1} = e_i$ for any i, therefore

$e_1 = e_2 = \ldots = e_s$. Then

$$\sum_{i=1}^{s}(x_{i+1,m} - x_{i,m}) = 0$$

implies $\Sigma_{i=1}^{s}a_{i,m}e_{i,m} = 0$. Letting $m \to \infty$, we get $e_1 = (\Sigma_{i=1}^{s}a_i)e_i = 0$, which is a contradiction with $\|e_1\| = 1$. This proves the assertion. ♠

We continue the proof of Theorem 3.4.3. Without loss of generality, according to the above lemma, we may assume that $x_1 \neq x_2$. There exists a uniquely determined sequence

$$i_1 = 1 < i_2 < \ldots < i_k \leq s, i_{k+1} = s+1$$

of integers such that every $j = 2, \ldots, k, i_j$ is the maximal index $i > i_{j-1}$ such that the points

$$x_{i_{j-1}}, x_{i_{j-1}+1}, \ldots x_i$$

are collinear. It then follows that the points

$$x_{i_1}, x_{i_2}, \ldots, x_{i_k}$$

are the successive reflection points of a periodic reflecting ray γ for X.

Lemma 3.4.5: *We have $k = s$ and $i_j = j$ for every $j = 1, \ldots, s$.*

Proof of Lemma 3.4.5: Suppose that $i_2 > 2$; then $i_2 \geq 3$. There are two cases.
Case 1. There exists i with $1 < i < i_2$ and $x_i \neq x_{i_2}$. In this case the segment $[x_1, x_{i_2}]$ of γ is tangent to X at the point x_i, which is a contradiction with id $\in \mathcal{F} \subset \mathcal{T}$.
Case 2. $x_2 = x_3 = \ldots = x_{i_2}$. Denote by θ_m the measure of the angle between the vector $x_{3,m} - x_{2,m}$ and the tangential hyperplane to X at $x_{2,m}$. Since

$$\lim_{m \to \infty} x_{3,m} = \lim_{m \to \infty} x_{2,m} = x_2,$$

we obtain $\lim_{m \to \infty} \theta_m = 0$. On the other hand, θ_m coincides with the measure of the angle between the vector $x_{1,m} - x_{2,m}$ and the tangential hyperplane to X at $x_{2,m}$. This is why the vector

$$x_1 - x_2 = \lim_{m \to \infty}(x_{1,m} - x_{2,m})$$

is tangent to X at $x_2 = x_{i_2}$. The latter implies that x_{i_2} is contained in the segment $[x_1, x_{i_3}]$, which is a contradiction with the choice of i_2.
In this way we have shown that $i_2 = 2$. Proceeding in the same way, we find $i_3 = 3, \ldots, i_s = s$. Consequently, $k = s$. ♠

It follows from the two lemmas above that $x_i \neq x_{i+1}$ for any $i = 1, \ldots, s$. Moreover, it is clear that $x_1, \ldots, x_s$ are the successive reflection points of a periodic reflecting ray γ for X. Let us remark that in general some of the reflection points of γ could coincide even if $x_{1,m}, \ldots, x_{s,m}$ are different for any m. For example, γ could be a symmetric periodic reflecting ray with $1 + s/2$ different reflection points.

Denote by γ_m the periodic reflecting ray for X with successive reflection points $x_{1,m}, \ldots, x_{s,m}$. Let Ω be one of the domains in $\mathbf{R}^n$ with boundary $\partial \Omega = X$, chosen in such a way that $\gamma_m \subset \bar{\Omega}$ for infinitely many m. We assume that this is true for any m. Set

$$\eta = \frac{x_2 - x_1}{\|x_2 - x_1\|}, \quad \eta_m = \frac{x_{2,m} - x_{1,m}}{\|x_{2,m} - x_{1,m}\|}.$$

Then for every m, $(x_{1,m}, \eta_m)$ is a periodic point of period s of the billiard ball map B, related to Ω (see Section 2.1). The same is true for (x_1, η). It follows easily from the definition of B and our considerations in Section 2.3 that the map $L = dB^s(x_1, \eta)$ is conjugated to the Poincaré map P_γ. Consequently, by $\mathrm{id} \in \mathcal{F} \subset T_{Q/Z}$, we find that $\mathrm{spec}\, L$ does not contain roots of 1. On the other hand, there exists a sequence of different fixed points $(x_{1,m}, \eta_m)$ of B^s. Therefore 1 is an eigenvalue of $L = dB^s(x_1, \eta)$, which is a contradiction.

In this way we have shown that K^s is finite and this proves the theorem. ♠

Next, we consider reflecting (ω, θ)-rays for X, where ω and θ are two fixed unit vectors in $\mathbf{R}^n$.

Theorem 3.4.6: *Let $T(\omega, \theta)$ be the set of those $f \in \mathbf{C}(X)$ such that every reflecting (ω, θ)-ray for $f(X)$ is ordinary and $\det dJ_\gamma \neq 0$. Then $T(\omega, \theta)$ contains a residual subset of $\mathbf{C}(X)$.*

Proof: Fix an open ball U_0, containing X, and set

$$Z_1 = Z_\omega, \quad Z_2 = Z_{-\theta}, \quad \pi_1 = \pi_\omega, \quad \pi_2 = \pi_{-\theta},$$

the notation Z_ξ, π_ξ being introduced in Section 2.4. According to the remark before Theorem 3.2.5, it is sufficient to establish that

$$T(\omega, \theta) \cap \mathbf{C}(X, U_0)$$

contains a residual subset of $\mathbf{C}(X, U_0)$.

Let $f \in \mathcal{B} \cap T(\omega, \theta) \cap \mathbf{C}(X, U_0)$, where $\mathcal{B}$ and $T(\omega, \theta)$ are the sets from Theorems 3.2.6 and 3.3.3, respectively. Let γ be a reflecting (ω, θ)-ray for $f(X)$ with successive reflection points $q_1, \ldots, q_k$. Set $u_\gamma = q_0 = \pi_1(q_1)$, $q_{k+1} = \pi_2(q_k)$, and define $\Pi_i, \lambda_i, \sigma_i, \tilde{\psi}_i$, etc. as in Section 2.4. Then for $dJ_\gamma(u_\gamma)$ we have the representation (2.32).

Next, introduce the function $E = \det(dJ_\gamma)(u_\gamma)$. Clearly, E is a polynomial of the elements $(\tilde{d}_{ij}^{(t)})_{i,j=1}^{n-1}$ of the matrices $\tilde{\psi}_t$, $t = 1, \ldots, k$.

First, assume that γ is non-symmetric. Since $f \in \mathcal{B} \cap T(\omega, \theta)$, we have that γ is ordinary and the reflection points $q_1, \ldots, q_k$ of γ are all different. We shall verify that the coefficient in front of

$$\tilde{d}^{(1)}_{11}\tilde{d}^{(1)}_{22}\cdots\tilde{d}^{(1)}_{n-1n-1} \tag{3.58}$$

in E is not zero. Let s_i be given by (2.24). Then we have

$$\begin{pmatrix} \sigma_k & \lambda_k\sigma_k \\ 0 & \sigma_k \end{pmatrix}\cdots\begin{pmatrix} \sigma_2 & \lambda_2\sigma_2 \\ 0 & \sigma_2 \end{pmatrix}\begin{pmatrix} \sigma_1 & \lambda_1\sigma_1 \\ \tilde{\psi}_1\sigma_1 & \sigma_1+\lambda_1\tilde{\psi}_1\sigma_1 \end{pmatrix}$$

$$= \begin{pmatrix} s_k & 0 \\ 0 & s_k \end{pmatrix}\begin{pmatrix} I & \sum_{i=2}^{k}\lambda_i I \\ 0 & I \end{pmatrix}\begin{pmatrix} I & \lambda_1 I \\ \psi_1 & I+\lambda_1\psi_1 \end{pmatrix}.$$

Therefore

$$pr_2\begin{pmatrix} s_k & 0 \\ 0 & s_k \end{pmatrix}\begin{pmatrix} I & \sum_{i=2}^{k}\lambda_i I \\ 0 & I \end{pmatrix}\begin{pmatrix} I & \lambda_1 I \\ \psi_1 & I+\lambda_1\psi_1 \end{pmatrix}\begin{pmatrix} u \\ 0 \end{pmatrix} = s_k\psi_1(u),$$

where

$$pr_2\begin{pmatrix} u \\ v \end{pmatrix} = v.$$

This clearly implies that the coefficient in front of the product (3.58) in E is 1. Consequently, E is a non-trivial polynomial of the variables $\tilde{d}^{(t)}_{ij}$ with coefficients depending on λ_i, σ_i, $i = 1,\ldots,k$.

Further, repeating most of the considerations from the proof of Theorem 3.4.1, replacing F by the function F^*, determined by (3.33) and (3.34), we prove that there exists a residual subset R' of $\mathbf{C}(X)$, $R' \subset \mathcal{B}\cap\mathcal{T}(\omega,\theta)$, such that $f \in R'$ yields $\det dJ_\gamma(u_\gamma) \neq 0$ for any non-symmetric reflecting (ω,θ)-ray γ for X.

To deal with the symmetric case, assume $\theta = -\omega$, and let again $f \in \mathcal{B} \cap\mathcal{T}(\omega,\theta)\cap\mathbf{C}(X,U_0)$. Let $q_1,\ldots,q_k$ be the successive reflection points of a symmetric reflecting (ω,θ)-ray γ. Then γ is ordinary, $k = 2m+1$ and the points $q_1,\ldots,q_m$ are all different. Clearly, $q_{m+i+1} = q_{m-i+1}$ for $i = 0,1,\ldots,m$.

We have

$$M = \begin{pmatrix} \sigma_1 & \lambda_1\sigma_1 \\ 0 & \sigma_1 \end{pmatrix}\cdots\begin{pmatrix} \sigma_{m-1} & \lambda_{m-1}\sigma_{m-1} \\ 0 & \sigma_{m-1} \end{pmatrix}\begin{pmatrix} \sigma_m & \lambda_m\sigma_m \\ \tilde{\psi}_m\sigma_m & \sigma_m+\lambda_m\tilde{\psi}_m\sigma_m \end{pmatrix}$$

$$\times\begin{pmatrix} \sigma_{m-1} & \lambda_{m-1}\sigma_{m-1} \\ 0 & \sigma_{m-1} \end{pmatrix}\cdots\begin{pmatrix} \sigma_1 & \lambda_1\sigma_1 \\ 0 & \sigma_1 \end{pmatrix}$$

$$= \begin{pmatrix} s^{-1}_{m-1} & 0 \\ 0 & s^{-1}_{m-1} \end{pmatrix}\begin{pmatrix} I & \sum_{i=1}^{m-1}\lambda_i I \\ 0 & I \end{pmatrix}\begin{pmatrix} I & \lambda_m I \\ \psi_m & I+\lambda_m\psi_m \end{pmatrix}$$

$$\times\begin{pmatrix} s_m & 0 \\ 0 & s_m \end{pmatrix}\begin{pmatrix} I & \sum_{i=1}^{m-1}\lambda_i I \\ 0 & I \end{pmatrix}.$$

Therefore

$$pr_2\left(M\begin{pmatrix} u \\ 0 \end{pmatrix}\right) = s^{-1}_m\psi_m s_m(u),$$

which shows that the coefficient in front of the product $\tilde{d}_{11}^{(m)}\ \tilde{d}_{22}^{(m)}\ \ldots\ \tilde{d}_{n-1\,n-1}^{(m)}$ in E is not zero. Consequently, E is a non-trivial polynomial, and we can apply the argument from the proof of Theorem 3.4.1, replacing F by the map

$$G^* : U_m^* \to \mathbf{R},$$

defined by

$$G^*(y) = \|y_1 - \pi_1(y_1)\| + \sum_{i=1}^{m-1} \|y_i - y_{i+1}\|,$$

where $U_m^* \subset (\mathbf{R}^n)^{(m)}$ is to be defined appropriately. In this way we prove that there exists a residual subset R'' of $\mathbf{C}(X)$, $R'' \subset \mathcal{B} \cap T(\omega,\theta)$, such that $f \in R''$ implies $\det dJ_\gamma(u_\gamma) \neq 0$ for any symmetric reflecting (ω,θ)-ray γ for X.

Finally, mention that $R' \cap R'' \subset T(\omega,\theta)$. Since each of the sets R' and R'' contains a residual subset of $\mathbf{C}(X)$, the same is true for $T(\omega,\theta)$. This concludes the proof of the theorem. ♠.

Consider the subset

$$\mathcal{F}(\omega,\theta) = \mathcal{B} \cap T(\omega,\theta) \cap T(\omega,\theta)$$

of $\mathbf{C}(X)$, where $\mathcal{B}$ and $T(\omega,\theta)$ are the sets from Theorems 3.2.6 and 3.3.3, respectively. Then, according to Theorems 3.2.6, 3.3.3 and 3.4.6, we see that $\mathcal{F}(\omega,\theta)$ contains a residual subset of $\mathbf{C}(X)$. Applying the argument from the proof of Theorem 3.4.3, we get the following.

Corollary 3.4.7: *For every $f \in \mathcal{F}(\omega,\theta)$ and every integer $m \geq 1$ there exist only finitely many reflecting (ω,θ)-rays for $f(X)$ with m reflection points.* ♠

3.5. Notes

Theorem 3.1.1 was proved in a slightly different form in [PS2] (see also [PS1]). The material in Section 3.2 is a modification of parts of [PS2, S1] and [CPS], while that in Section 3.3 is taken from [PS4]. The case $n = 2$ of Theorem 3.4.1 was proved in [PS2] (see also [PS1]). In its present form this theorem, as well as Theorem 3.4.6, were established in [PS3]. Finally, Theorem 3.4.3 and Corollary 3.4.7 are taken from [PS4].

Generic properties of reflecting rays were first considered by Lazutkin [L1, L2], who proved an analogue of the Kupka–Smale theorem for billiards in strictly convex planar domains. Note that Theorem 3.4.1 may be considered as a first part of a Kupka–Smale-type theorem for billiards.

4 BUMPY METRICS

In this chapter we study again generic properties of compact smooth submanifolds M of $\mathbf{R}^n$ of positive codimension. This time they concern the behaviour of the geodesic flow on M determined by the standard metric on M inherited from the Euclidean structure of $\mathbf{R}^n$. These properties, together with the generic properties of the periodic reflecting rays established in the previous chapter, will be important when we study the Poisson relation for generic domains in $\mathbf{R}^n$.

Our aim in the present chapter is to establish the existence of a residual set of smooth embeddings F of M into $\mathbf{R}^n$ such that the standard metric on $M' = F(M)$ is a *bumpy metric*, i.e. all closed geodesics on M' are non-degenerate. As a consequence, the classical bumpy metric theorem of Abraham–Klingenberg–Takens– Anosov is obtained:for every compact smooth manifold M there exists a residual set in the space of all smooth Riemannian metrics on M consisting of bumpy metrics.

4.1. Poincaré map for closed geodesics

We begin with some standard facts from the theory of ordinary differential equations which will be useful later.

Let $\Delta = [0, a] \subset \mathbf{R}$, $a > 0$, and let

$$X = (X^{(1)}, \dots, X^{(k)}) : \Delta \times U \to \mathbf{R}^k$$

be a C^1 map, where U is an open neighbourhood of 0 in $\mathbf{R}^k$. For $u \in U$ close to 0, let $x(t; u)$ be a solution of the differential equation

$$\dot{x}(t; u) = X(t; x(t; u)). \tag{4.1}$$

Here $\dot{x}$ denotes the derivative with respect to t. We shall assume that $x(t; u)$ exists for all $t \in \Delta$, provided u belongs to a small neighbourhood V of 0 in $\mathbf{R}^k$, $V \subset U$. Define the map

$$\mathcal{P}_t : V \to \mathbf{R}^k$$

by $\mathcal{P}_t(u) = x(t; u)$.

Proposition 4.1.1: *Under the assumptions above, the map $\mathcal{P}_t$ is differentiable* at 0, and for any $t \in \Delta$ the matrix $P_t = dP_t(0)$ is a solution of the problem

$$\begin{cases} \dot{P}_t = d_x X(t; x(t; 0)).P_t, & t \in \Delta, \\ P_0 = I, \end{cases} \tag{4.2}$$

where I is the identity matrix and

$$d_x X = \begin{pmatrix} \dfrac{\partial X^{(1)}}{\partial x_1} & \cdots & \dfrac{\partial X^{(1)}}{\partial x_k} \\ \cdots & \cdots & \cdots \\ \dfrac{\partial X^{(k)}}{\partial x_1} & \cdots & \dfrac{\partial X^{(k)}}{\partial x_k} \end{pmatrix}.$$

Proof: It is convenient to write X as a row-vector and

$$x = \begin{pmatrix} x_1 \\ \vdots \\ x_k \end{pmatrix}, \quad u = \begin{pmatrix} u_1 \\ \vdots \\ u_k \end{pmatrix}$$

as column vectors. For $i = 1, \dots,$ set

$$y_i(t) = \frac{\partial x}{\partial u_i}(t; u)_{|u=0}.$$

Then, by the variational equations for (4.1) (cf. [Pon]), we find that for every $j = 1, \dots, k$ the jth component $y_i^{(j)}$ of y_i is a solution of the problem

$$\begin{cases} \dot{y}_i^{(j)}(t) = \displaystyle\sum_{j=1}^{k} \frac{\partial X^{(j)}}{\partial x_k}(t; x(t; 0)) y_i^{(j)}(t), & t \in \Delta, \\ y_i^{(j)}(0) = \delta_{ij}. \end{cases}$$

Hence for the vector function $y_i(t)$ we obtain

$$\begin{cases} \dot{y}_i(t) = d_x X(t; x(t; 0)) - y_i(t), & t \in \Delta, \\ y_i(0) = e_i, \end{cases}$$

where e_i is the ith basis vector in $\mathbf{R}^k$. Since

$$P_t = \begin{pmatrix} y_1^{(1)}(t) & \cdots & y_k^{(1)}(t) \\ \cdots & \cdots & \cdots \\ y_1^{(k)}(t) & \cdots & y_k^{(k)}(t) \end{pmatrix},$$

we deduce that P_t is a solution of the problem (4.2). ♠

Corollary 4.1.2:   *Let*

$$Y : \Delta \times U \to \mathbf{R}^k$$

be another C^1 map. Consider the differential equation

$$\dot{y}(t; u) = Y(t; y(t; u)), \tag{4.3}$$

and suppose that $x(t; 0)$, $t \in \Delta$, is a solution of both (4.1) and (4.3), and

$$\frac{\partial X}{\partial x_i}(t; x(t; 0)) = \frac{\partial Y}{\partial x_i}(t; x(t; 0))$$

for all $i = 1, \ldots, k$ and all $t \in \Delta$. Define the map

$$\mathcal{Q}_t : V \to \mathbf{R}^k$$

(eventually with a smaller V) by $\mathcal{Q}_t(u) = y(t; u)$. Then $\mathrm{d}\mathcal{P}_t(0) = \mathrm{d}\mathcal{Q}_t(0)$ for every $t \in \Delta$.

Proof: Clearly, $\mathrm{d}_x X(t; x(t; 0)) = \mathrm{d}_x Y(t; x(t; 0))$ for all t, and therefore both $\mathrm{d}\mathcal{P}_t(0)$ and $\mathrm{d}\mathcal{Q}_1(0)$ are solutions of (4.2). ♠

Let M be a smooth manifold and let $\pi : T^*M \to \mathbf{R}$ be the *cotangent bundle* of M. In this chapter we shall denote the dimension of M by $m + 1$, i.e. we set

$$m = \dim M - 1,$$

and we assume that $m \geq 1$.

Let ω be the *canonical symplectic form* on T^*M (cf. [AbM], for example), and let g be a fixed smooth Riemannian metric on M. Define the *energy function*

$$H = H_g : T^*M \to \mathbf{R}$$

by $H(q) = \frac{1}{2}\langle q, q \rangle_g$, where $< ., . >_g$ is the inner product in T^*M, related to g. The *Hamiltonian vector field*, determined by H (and therefore by g), is the unique smooth vector field $X = X_g$ on T^*M such that

$$\omega(X, Y) = \mathrm{d}H.Y$$

for any smooth vector field Y on T^*M. The flow on T^*M, determined by X, is called the *geodesic flow*. A curve c in T^*M is an integral curve of X if and only if the curve $\pi \circ c$ in M is a geodesic with respect to the metric g. We refer the reader to [AbM, Ch. 3] for the basic facts concerning the Hamiltonian dynamics.

Consider a closed integral curve

$$c : [0, \theta] \to T^*M, \quad \theta > 0, \tag{4.4}$$

of X, i.e. such that $c(0) = c(\theta)$, and $\dot{c}(t) = X(c(t))$ for all $t \in [0, \theta]$. Then θ is called *period* of c. If $c(t) \neq c(0)$ for every $t \in (0, \theta)$, we shall say that

θ is the *minimal period* of c. The same terminology will be used for the *closed geodesic*

$$\gamma : [0,\theta] \to M, \quad \gamma = \pi \circ c \qquad (4.5)$$

on (M, g).

To define the Poincaré map of c, set $q = c(0)$, $p = \pi(q)$, and consider a smooth m-submanifold Σ^* of M, containing p and such that $\dot{c}(0) = X(q)$ is transversal to the $(2m+1)$-submanifold $\Sigma = \pi^{-1}(\Sigma^*)$ of T^*M at q. For $q' \in \Sigma$ close to q, the integral curve of X, passing through q' at $t = 0$, after time t close to θ, intersects Σ at some point $q'' \in \Sigma$. Thus we obtain a map

$$\mathcal{P} = \mathcal{P}^{(g)} : \Sigma \ni q' \to q'' \in \Sigma,$$

defined in a small neighbourhood of q in Σ. This map leaves the $2m$-submanifold

$$\tilde{\Sigma} = \{q' \in \Sigma : H(q') = H(q)\}$$

invariant and preserves the natural symplectic form on $\tilde{\Sigma}$, induced by ω. In this way $\mathcal{P}$ induces a local symplectic diffeomorphism

$$\mathcal{P} : (\tilde{\Sigma}, q) \to (\tilde{\Sigma}, q).$$

The linear map

$$P = P^{(g)} = d\mathcal{P}^{(g)}(q) : T_q\tilde{\Sigma} \to T_q\tilde{\Sigma}$$

is called *(linear) Poincaré map* of the integral curve c (or of the closed geodesic γ). The reader may consult Chapters 7 and 8 in [AbM] for more details on the definition of the Poincaré map, as well as for the proof of the fact that, up to conjugacy, it does not depend on the choice of the initial point q and the submanifold Σ^*. The latter shows that the *spectrum* spec P does not depend on the choice of q and Σ^*. We shall say that c (resp. γ) is *non-degenerate* as an integral curve (resp. closed geodesic) of period θ, if $1 \notin$ spec P. If all closed geodesics on (M, g) are non-degenerate, then g is called a *bumpy metric* on M.

Hereafter, we assume that (4.1) is a fixed closed integral curve of X with **minimal period $\theta > 0$**, M being endowed with the fixed smooth Riemannian metric g. Define γ by (4.2). There exist *Fermi coordinates* in a neighbourhood of Im γ in M (cf. for example Section 1.12 in [K1]). This means that there exist an *open neighbourhood* U of Im γ in M and a *local diffeomorphism*

$$r : V = (-\alpha, \theta+\alpha) \times B_\alpha(0) \to U, \qquad (4.6)$$

where $\alpha > 0$ and

$$B_\alpha(0) = \{x \in \mathbf{R}^m : \|x\| < \alpha\},$$

such that the following conditions are satisfied:

(i) $\gamma(t) = r(t, 0, \ldots, 0)$ for every $t \in [0, \theta]$;

(ii) the 1-jet of g_{00} coincides with the 1-jet of the constant 1 at all points of

$$\gamma_0 = \{(t,0,\ldots,0) \in \mathbf{R}^{m+1} : 0 \leq t \leq \theta\};$$

(iii) $g_{0i} = 0$ on γ_0 for all $i = 1,\ldots,m$.

Here g_{ij}, i, $j = 0,1,\ldots,m$, are the components of the metric g with respect to the coordinates $x_0, x_1, \ldots, x_m$ provided by r. To be more precise we should say that $x_i : U \to \mathbf{R}$ are smooth functions such that

$$r(x_0(\xi),\ldots,x_m(\xi)) = \xi$$

for any $\xi \in U$. In fact, r is only a local diffeomorphism, therefore x_i are coordinates only locally, that is, every point in U has a neighbourhood $W \subset U$ such that the restriction of r onto $r^{-1}(W)$ is a diffeomorphism between $r^{-1}(W)$ and W. For our next considerations this is already sufficient in order to treat x_i in the same way as if they were coordinates in the whole U.

Let $y_0, y_1, \ldots, y_m$ be the coordinates dual to $x_0, x_1, \ldots, x_m$; then

$$x_0, x_1, \ldots x_m, y_0, y_1, \ldots, y_m$$

are coordinates in T^*M in a neighbourhood of Im c (in the same sense as above). With respect to these coordinates we have $\omega = \sum_{i=0}^m \mathrm{d}x_i \wedge \mathrm{d}y_i$ and

$$H(x_0,\ldots,x_m,y_0,\ldots,y_m) = \frac{1}{2} \sum_{i,j=0}^m g_{ij}(x_0,\ldots,x_m) y_i y_j$$

(cf. Chapter 3 in [AbM]).

Introduce the following abbreviations:

$$x = (x_0; x'), \quad x' = (x_1,\ldots,x_m), \quad y = (y_0; y'), \quad y' = (y_1,\ldots,y_m).$$

For $0 \leq t \leq 0$ define

$$\Sigma(t) = \{(x;y) : x_0 = t\}, \quad \tilde{\Sigma}(t) = \{(x;y) : x_0 = t, y_0 = 1\}.$$

Given $t \geq 0$, let

$$\mathcal{P}_t : \Sigma(0) \to \Sigma(t)$$

be the map, defined in a small neighbourhood of $q = c(0)$, which assigns to each $q' \in \Sigma(0)$ the first point of intersection of the positive integral curve of X through q' with $\Sigma(t)$.

Further, we consider a small perturbation g' of g so that

$$\tilde{g} = g + g'$$

is another smooth Riemannian metric on M. We assume that g' satisfies the following conditions:

(ii′) the 1-jet of g'_{00} is zero on γ_0;

(iii′) $g'_{0i} = 0$ on γ_0 for each $i = 1, \ldots, m$.

Then $\tilde{X} = X_{\tilde{g}}$ can be written in the form $\tilde{X} = X + X'$, where X' is the Hamiltonian vector field on T^*M (defined only locally around $\operatorname{Im} c$), determined by the Hamiltonian function

$$H'(x, y) = \frac{1}{2} \sum_{i,j=0}^{m} g'_{ij}(x) y_i y_j.$$

Note that $c(t)$ is an integral curve not only for X but for $\tilde{X}$ as well, that is $\gamma(t)$ is a geodesic on $(M, \tilde{g})$. This follows immediately from the conditions (ii′) and (iii′), writing down the corresponding Hamiltonian system of differential equations for an integral curve of X.

Define the maps

$$\mathcal{P}'_t : \Sigma(0) \rightarrow \Sigma(t)$$

in the same way as $\mathcal{P}_t$, using the vector field $\tilde{X}$ instead of X.

Written explicitly, in a neighbourhood of $\operatorname{Im} c$ the vector fields X and X' have the form

$$X = \sum_{i=0}^{m} \left(\frac{\partial H}{\partial y_i} \frac{\partial}{\partial x_i} - \frac{\partial H}{\partial x_i} \frac{\partial}{\partial y_i} \right),$$

$$X' = \sum_{i=0}^{m} \left(\frac{\partial H'}{\partial y_i} \frac{\partial}{\partial x_i} - \frac{\partial H'}{\partial x_i} \frac{\partial}{\partial y_i} \right).$$

Now we define the time dependent vector fields X_t and X'_t on $\Sigma(0)$ by

$$X_t(\zeta) = \sum_{i=1}^{m} \left(\frac{\partial H}{\partial y_i}(t; \zeta) \frac{\partial}{\partial x_i} - \frac{\partial H}{\partial x_i}(t; \zeta) \frac{\partial}{\partial y_i} \right),$$

$$X'_t(\zeta) = \sum_{i=1}^{m} \left(\frac{\partial H'}{\partial y_i}(t; \zeta) \frac{\partial}{\partial x_i} - \frac{\partial H'}{\partial x_i}(t; \zeta) \frac{\partial}{\partial y_i} \right)$$

for $t \in [0, \theta]$ and

$$\zeta = (x_1, \ldots, x_m; \ y_0, y_1, \ldots, y_m) \in \mathbf{R}^{2m+1}$$

close to 0. To these vector fields there correspond the local diffeomorphisms

$$\tilde{\mathcal{P}}_t, \tilde{\mathcal{P}}'_t : (\Sigma(0), c(0)) \rightarrow (\Sigma(t), c(t)),$$

determined in the same way as $\mathcal{P}_t$ and $\mathcal{P}'_t$, replacing X by X_t and X' by X'_t, respectively. More precisely, given $\zeta \in \Sigma(0)$ close to $c(0)$, $\tilde{\mathcal{P}}_t(\zeta)$ is the first intersection point of the positive integral curve of the time dependent vector field X_t starting at ζ with $\Sigma(t)$. The definition of $\tilde{\mathcal{P}}'_t(\zeta)$ is similar.

Lemma 4.1.3:   *For every t we have*

$$d\tilde{\mathcal{P}}_t(0) = d\mathcal{P}_t(0), \quad d\tilde{\mathcal{P}}'_t(0) = d\mathcal{P}'_t(0).$$

Proof: We shall verify the first of the above equalities, the proof of the second is the same.

Notice that the zeroth component of X, corresponding to the coordinate x_0, has the form

$$X^{(0)}(x;y) = \frac{\partial H}{\partial y_0}(x;y) = \sum_{i=0}^{m} g_{i0}(x)y_i.$$

Therefore

$$\frac{\partial X^{(0)}}{\partial x_k}(c(t)) = \frac{g_{00}}{\partial x_k}(t;0) = 0,$$

and

$$\frac{\partial X^{(0)}}{\partial y_k}(c(t)) = g_{k0}(t;0) = 0$$

for every $k = 0, 1, \ldots, m$. Here we have used the fact that the metric g satisfies the conditions (i), (ii) and (iii).

Similarly, for the $(m+1)$th component $X^{(m+1)}$ of X, corresponding to the coordinate y_0, we find

$$X^{(m+1)}(x;y) = -\frac{\partial H}{\partial x_0}(x;y) = -\frac{1}{2}\sum_{i,j=0}^{m} \frac{\partial g_{ij}(x)}{\partial x_0}y_iy_j.$$

Therefore

$$\frac{\partial X^{(m+1)}}{\partial x_k}(c(t)) = -\frac{1}{2}\frac{\partial^2 g_{00}}{\partial x_k \partial x_0}(t;0) = 0,$$

and

$$\frac{\partial X^{(m+1)}}{\partial y_k}(c(t)) = -\frac{\partial g_{k0}}{\partial x_0}(t;0) = 0$$

for every $k = 0, 1, \ldots, m$.

Now the assertion follows from Corollary 4.1.2. ♠.

The definition of X_t implies that for any integral curve $\xi(t)$ of X_t we have $y_0(t) = $ const. The same is true for the integral curves of X'_t, therefore

$$\tilde{\mathcal{P}}_t(\tilde{\Sigma}(0)) \subset \tilde{\Sigma}(t), \quad \tilde{\mathcal{P}}'_t(\tilde{\Sigma}(0)) \subset \tilde{\Sigma}(t) \tag{4.7}$$

for every t. Note that similar inclusions are not satisfied in general by $\mathcal{P}_t$, and $\mathcal{P}'_t$. However, according to Lemma 4.1.3. and (4.7), we get

$$\mathrm{d}\mathcal{P}_t(c(0))(T_{c(0)}\tilde{\Sigma}(0)) \subset T_{c(t)}\tilde{\Sigma}(t), \quad \mathrm{d}\mathcal{P}'_t(c(0))(T_{c(0)}\tilde{\Sigma}(0)) \subset T_{c(t)}\tilde{\Sigma}(t)$$

for every $t \in [0, \theta]$.

Let P_t and P'_t be the *matrices of the restrictions* of the linear maps $\mathrm{d}\mathcal{P}_t(c(0))$ and $\mathrm{d}\mathcal{P}'_t(c(0))$, respectively, on $T_{c(0)}\tilde{\Sigma}(0)$. Here, using the coordinates $x_1, \ldots, x_m, y_1, \ldots, y_m$, we identify $\tilde{\Sigma}(t)$ for all t with an open neighbourhood of 0 in $\mathbf{R}^{2m}$. Correspondingly, $T_{c(t)}\tilde{\Sigma}(t)$ are identified with $\mathbf{R}^{2m} \times \mathbf{R}^{2m}$, so P_t and P'_t are $2m \times 2m$ symplectic matrices smoothly depending on t. In particular,

$$R_t = P_t^{-1} P'_t \tag{4.8}$$

is also a symplectic matrix smoothly depending on t.

We are going to show that the matrix function R_t is the solution of a certain matrix differential equation. To this end we need the following simple fact.

Lemma 4.1.4:   *Let Δ be a subinterval of $\mathbf{R}$ and let Q_t, Q'_t, Y_t, Y'_t ($t \in \Delta$) be $k \times k$ real matrices, differentiable with respect to t in Δ and such that Q_t is invertible for every $t \in \Delta$. If*

$$\dot{Q}_t = Y_t Q_t, \quad \dot{Q}'_t = (Y_t + Y'_t)Q'_t, \quad t \in \Delta,$$

then for $S_t = Q_t^{-1} Q'_t$ we have

$$\dot{S}_t = (Q_t^{-1} Y'_t Q_t)S_t, \quad t \in \Delta.$$

Proof: It follows from $Q_t^{-1}Q_t = I$ that

$$(Q_t^{-1})^{\cdot}Q_t + Q_t^{-1}\dot{Q}_t = 0,$$

and therefore

$$(Q_t^{-1})^{\cdot} = -Q_t^{-1}\dot{Q}_t Q_t^{-1}$$

for any $t \in \Delta$. Then

$$\begin{aligned}
\dot{S}_t &= (Q_t^{-1})^{\cdot}Q'_t + Q_t^{-1}\dot{Q}'_t = -Q_t^{-1}\dot{Q}_t Q_t^{-1}Q'_t + Q_t^{-1}(Y_t + Y'_t)Q'_t \\
&= -Q_t^{-1}Y_t Q'_t + Q_t^{-1}Y_t Q'_t + (Q_t^{-1}Y'_t Q_t)Q_t^{-1}Q'_t = (Q_t^{-1}Y'_t Q_t)S_t
\end{aligned}$$

for every $t \in \Delta$. ♠.

Consider $\tilde{X}_t(\xi)$ as a time dependent vector field, defined for

$$\xi = (x'; y') = (x_1, \ldots, x_m; y_1, \ldots, y_m) \in \tilde{\Sigma}(0).$$

Then

$$d_\xi \tilde{X}_t(\xi) = J \begin{pmatrix} d^2_{x'x'} H(t;\xi) & d^2_{x'y'} H(t;\xi) \\ d^2_{x'y'} H(t;\xi) & d^2_{y'y'} H(t;\xi) \end{pmatrix},$$

where

$$J = \begin{pmatrix} 0 & I \\ -I & 0 \end{pmatrix}$$

is the *canonical* $2m \times 2m$ *symplectic matrix*, and

$$d^2_{x'x'} H(t;\xi) = \left(\frac{\partial^2 H}{\partial x_i \partial x_j}(t,\xi) \right)^m_{i,j=1}, d^2_{x'y'} H(t;\xi) = \left(\frac{\partial^2 H}{\partial x_i \partial y_j}(t,\xi) \right)^m_{i,j=1},$$

etc. In the same way one obtains

$$d_\xi \tilde{X}'_t(\xi) = J \begin{pmatrix} d^2_{x'x'} H'(t;\xi) & d^2_{x'y'} H'(t;\xi) \\ d^2_{x'y'} H'(t;\xi) & d^2_{y'y'} H'(t;\xi) \end{pmatrix}. \tag{4.9}$$

Now, applying Proposition 4.1.1 and Lemma 4.1.3, we deduce that P_t and P'_t are solutions of the problems:

$$\begin{cases} \dot{P}_t = d_\xi \tilde{X}_t(0) P_t, & t \in [0,\theta], \\ P_0 = I, \end{cases}$$
$$\begin{cases} \dot{P}'_t = (d_\xi \tilde{X}_t(0) + d_\xi \tilde{X}'_t(0)) P_t, & t \in [0,\theta], \\ P'_0 = I. \end{cases}$$

It then follows by Lemma 4.1.4 that the matrix function R_t, determined by (4.8), is a solution of the problem:

$$\begin{cases} \dot{R}_t = (P_t^{-1} d_\xi \tilde{X}'_t(0) P_t) R_t, & t \in [0,\theta], \\ R_0 = I. \end{cases} \tag{4.10}$$

Further, observe that

$$\frac{\partial H'}{\partial x_i} = \frac{1}{2} \sum_{j,k=0}^m \frac{\partial g'_{kj}}{\partial x_i}(x) y_k y_j, \qquad \frac{\partial H'}{\partial y_i} = \sum_{j=0}^m g'_{ij}(x) y_j.$$

Since $x_0 = t$, $y_0 = 1$, $x' = y' = 0$ on $c(t)$, we find

$$\frac{\partial^2 H'}{\partial x_i \partial x_j}(c(t)) = \frac{1}{2} \frac{\partial^2 g'_{00}}{\partial x_i \partial x_j}(t;0), \qquad \frac{\partial^2 H'}{\partial x_i \partial y_j}(c(t)) = \frac{\partial^2 g'_{0j}}{\partial x_i}(t;0),$$

$$\frac{\partial^2 H'}{\partial y_i \partial y_j}(c(t)) = g'_{ij}(t;0).$$

Introduce the *homogeneous polynomial*

$$\tilde{H}_t(x';y') = \sum_{i,j=1}^{m} \left(\frac{1}{2}a_{ij}(t)x_ix_j + b_{ij}(t)x_iy_j + \frac{1}{2}c_{ij}(t)y_iy_j \right), \qquad (4.11)$$

where

$$a_{ij}(t) = \frac{\partial^2 g'_{00}}{\partial x_i \partial x_j}(t;0), \quad b_{ij}(t) = \frac{\partial g'_{0j}}{\partial x_i}(t;0), \quad c_{ij}(t) = \frac{1}{2}g'_{ij}(t;0). \quad (4.12)$$

Clearly, $\tilde{H}_t(x'_i;y')$ is the sum of those terms in the Taylor series of the function $H'_{|\tilde{\Sigma}(t)}$ which involve second derivatives.

It follows from (4.9) and the expressions for the second derivatives of H' along $c(t)$ that $d_\xi \tilde{X}'_t(0) = J \cdot D(t)$, where

$$D(t) = \begin{pmatrix} A(t) & B(t) \\ B(t)^{\mathrm{T}} & C(t) \end{pmatrix}, \qquad (4.13)$$

with

$$A(t) = (a_{ij}(t)), \quad B(t) = (b_{ij}(t)), \quad C(t) = (c_{ij}(t)). \qquad (4.14)$$

Clearly, for every t, $J \cdot D(t)$ belongs to the *Lie algebra* sp$(2m)$ of the *symplectic Lie group* Sp$(2m)$ (cf. [AbM]). Finally, by (4.10) and the last expression for $d_\xi \tilde{X}'_t(0)$ we obtain the following.

Proposition 4.1.5: *The matrix function* $R_t = P_t^{-1}P'_t$ *is a solution of the problem*

$$\begin{cases} \dot{R}_t = P_t^{-1}JD(t)P_tR_t, & t \in [0,\theta], \\ R_0 = I, \end{cases} \qquad (4.15)$$

where $D(t)$ is given by (4.13), (4.14), (4.12). ♠.

4.2. Local perturbations of smooth surfaces

Let M be a smooth submanifold of $\mathbf{R}^n$, $n \geq 3$. As in the previous section, we set

$$\dim M = m+1,$$

and assume that $1 \leq m \leq n-2$, i.e. $\dim M \leq n-1$. Consider the *standard Riemannian metric* g on M, inherited from the Euclidean structure of $\mathbf{R}^n$, and let $H = H_g$, $X = X_g$ (cf. the notation in Section 4.1). Given $F \in \mathbf{C}(M)$ we denote by g_F the Riemannian metric on M with respect to which the map

$$F : M \to F(M) \subset \mathbf{R}^n$$

is isometric, $F(M)$ being considered with the standard metric.

Throughout this section we consider a fixed closed integral curve (4.4) of X with minimal period $\theta > 0$ and the corresponding closed geodesic (4.5) on M.

Theorem 4.2.1: *Let Λ be an arbitrary countable set of complex numbers. Under the assumptions above, there exists $t_0 \in (0, \theta)$ such that for every neighbourhood $\mathcal{U}$ of 0 in $C^\infty(M, \mathbf{R}^n)$ and every neighbourhood $\mathcal{W}$ of $\gamma(t_0)$ in M there exists $f \in \mathcal{U}$ with supp $f \subset \mathcal{W}$ such that $F = \mathrm{id} + f \in C(M)$, γ is a geodesic on (M, g_F) and the spectrum of the Poincaré map, related to γ with respect of the metric g_F, does not contain elements of Λ.*

The proof of this theorem is rather lengthy, and is broken into several lemmas.

As in Section 4.1, consider a local diffeomorphism (4.6) satisfying the conditions (i), (ii), (iii), and introduce the coordinates

$$x_0, x_1, \ldots, x_m, \quad y_0, y_1, \ldots, y_m$$

in a neighbourhood of Im c in T^*M. Notice that the components g_{ij} of the metric g have the form

$$g_{ij}(x) = \left\langle \frac{\partial r}{\partial x_i}(x), \frac{\partial r}{\partial x_j}(x) \right\rangle, \quad x \in V, \quad i, j = 0, 1, \ldots, m,$$

where r is considered as a map $r: V \to \mathbf{R}^n$, and $\langle ., . \rangle$ is the *standard inner product* in $\mathbf{R}^n$.

Given an embedding $F \in C(M)$, we can write it in the form $F = \mathrm{id} + f$, with $f \in C^\infty(M, \mathbf{R}^n)$. Next, consider the corresponding perturbed metric

$$\tilde{g} = g_F = g + g'.$$

Clearly, in the special case under consideration, we have

$$g'_{ij}(x) = \left\langle \frac{\partial r}{\partial x_i}(x), \frac{\partial (f \circ r)}{\partial x_i}(x) \right\rangle + \left\langle \frac{\partial r}{\partial x_j}(x), \frac{\partial (f \circ r)}{\partial x_i}(x) \right\rangle$$
$$+ \left\langle \frac{\partial (f \circ r)}{\partial x_i}(x), \frac{\partial (f \circ r)}{\partial x_j}(x) \right\rangle \qquad (4.16)$$

for all $x \in V$ and $i, j = 0, 1, \ldots, m$.

Further, introduce the symplectic matrices P_t, P'_t and R_t and the homogeneous polynomial $\tilde{H}_t$ as in Section 4.1. We shall consider only perturbations g' of g with supports in a small neighbourhood of $\gamma(t)$ for some $t \in (0, \theta)$, so that the matrix function $D(t)$, defined by (4.13), (4.14) and (4.12), will have a compact support in $(0, \theta)$.

Now the problem is to find a perturbation $F = \mathrm{id} + f$ with small f such that, if R_t is the solution of the problem (4.15) for the corresponding matrix function $D(t)$, then the spectrum of the matrix $P'_\theta = P_\theta R_\theta$ does not contain elements of the set Λ.

Since γ is a closed curve, the vector $\partial r/\partial x_0$ is not constant along γ_0, therefore there exists $t_0 \in (0, \theta)$ such that $(\partial^2 r/\partial x_0^2)(t; 0) \neq 0$. Fix such a t_0 and an arbitrary neighbourhood $\mathcal{W}$ of $\gamma(t_0)$ in M. There exist real numbers a, b with

$$\begin{cases} 0 < a < t_0 < b < \theta_0, \\ \dfrac{\partial^2 r}{\partial x_0^2}(t; 0) \neq 0 \quad \text{for every } t \in [a, b], \\ r(t; 0) \in \mathcal{W} \quad \text{for every } t \in [a, b]. \end{cases} \tag{4.17}$$

Choose an arbitrary β with

$$0 < \beta < \min\{a, \theta - b, \alpha\},$$

and consider an arbitrary smooth function

$$\rho : \mathbf{R}^{m+1} \to [0, 1]$$

such that $\operatorname{supp} \rho \subset (0, \theta) \times (-\beta, \beta)^m$ and $\rho(x) = 1$ for all $x \in [a, b] \times [-\beta/2, \beta/2]^m$.

In what follows the numbers t_0, a, b, β and the function ρ with the above properties will be fixed.

Define the map $h : V \to \mathbf{R}^n$ by

$$h(x) = \frac{1}{2} \sum_{i,j=1}^{m} v_{ij}(x_0) x_i x_j, \tag{4.18}$$

where $v_{ij} : [0, \theta] \to \mathbf{R}^n$ are smooth maps with

$$v_{ji} = v_{ij}, \quad \operatorname{supp} v_{ij} \subset [a, b] \quad (i, j = 1, \dots, m) \tag{4.19}$$

which will be constructed later. We set

$$f(\xi) = \begin{cases} \rho(r^{-1}(\xi)) h(r^{-1}(\xi)), & \xi \in U, \\ 0, & \xi \in M \backslash U. \end{cases} \tag{4.20}$$

Then $f : M \to \mathbf{R}^n$ is a smooth map with

$$\operatorname{supp} f \subset r([a, b] \times [-\beta, \beta]^m).$$

Clearly, $\operatorname{supp} f \subset W$, provided β is chosen sufficiently small. Moreover, if the maps $v_{ij} \in C^\infty([0, \theta], \mathbf{R}^n)$ are sufficiently close to 0 in the C^∞ topology, then $f \in \mathcal{U}$ and $F = \operatorname{id} + f \in \mathbf{C}(M)$.

As above we set $g' = g - g_F$. Note that for x close to γ_0 we have $f(r(x)) =$

$h(x)$, therefore (4.16) implies

$$
\begin{aligned}
g'_{ij}(x) &= \left\langle \frac{\partial r}{\partial x_i}(x), \frac{\partial h}{\partial x_j}(x) \right\rangle + \left\langle \frac{\partial r}{\partial x_j}(x), \frac{\partial h}{\partial x_i}(x) \right\rangle \\
&\quad + \left\langle \frac{\partial h}{\partial x_i}(x), \frac{\partial h}{\partial x_j}(x) \right\rangle.
\end{aligned}
\tag{4.21}
$$

Using the form (4.18) of h, by direct calculations one checks that g' satisfies the conditions (ii′) and (iii′) from Section 4.1. Therefore γ is a geodesic on (M, g_F).

We are going to show now that choosing the maps v_{ij} with (4.19) in a special way, one can obtain at least a special kind of perturbations g' of the metric g.

Lemma 4.2.2: Let $a_{ij}, b_{ij} : [0, \theta] \to \mathbf{R}^n$ be *smooth maps such that for all* i, $j = 1, \ldots, m$

$$
\begin{cases}
a_{ij} = a_{ji}, \quad b_{ij} = b_{ji}, \\
\operatorname{supp} a_{ij} \subset [a, b], \quad \operatorname{supp} b_{ij} \subset [a, b].
\end{cases}
\tag{4.22}
$$

Then there exists a neighbourhood $\mathcal{V}$ of 0 in $C^\infty([0, \theta], \mathbf{R}^n)$ such that if all a_{ij}, b_{ij} are in $\mathcal{V}$, then there exists a smooth map h of the form (4.18) with (4.19) for which the map f, defined by (4.20), belongs to $\mathcal{U}$, and for $g' = g - g_F$, $F = \mathrm{id} + f$ the corresponding polynomial $\tilde{H}_t$ has the form

$$
\tilde{H}_t(x'; y') = \sum_{i,j=1}^{m} \left(\frac{1}{2} a_{ij}(t) x_i x_j + b_{ij}(t) x_i y_j \right).
\tag{4.23}
$$

Proof: We have to choose the maps v_{ij} in such a way that

$$
\frac{\partial^2 g'_{00}}{\partial x_i \partial x_j}(t; 0) = a_{ij}(t), \quad \frac{\partial g'_{0i}}{\partial x_j}(t; 0) = b_{ij}(t), \quad g'_{ij}(t; 0) = 0,
\tag{4.24}
$$

for all $i, j = 1, \ldots, n$ and all $t \in [0, \theta]$.

Let $v_{ij} : [0, \theta] \to \mathbf{R}^n$ be arbitrary smooth maps satisfying (4.19). Then for $i = 1 \ldots, n$ and $t \in [0, \theta]$ we have

$$
\frac{\partial h}{\partial x_i}(t; 0) = 0,
$$

therefore $g'_{ij} = 0$ by (4.21). Then for $x \in V$ close to γ_0 and $i, j \geq 1$ we find

$$
\frac{\partial^2 h}{\partial x_0 \partial x_i}(x) = \sum_{j=1}^{n} \frac{\partial v_{ij}}{\partial x_0}(x_0) x_j, \quad \frac{\partial^3 h}{\partial x_0 \partial x_i \partial x_j}(x) = \frac{\partial v_{ij}}{\partial x_0}(x_0).
$$

Further, differentiating (4.21) one gets

$$\frac{\partial^2 g'_{00}}{\partial x_i \partial x_j}(t;0) = 2\left\langle \frac{\partial r}{\partial x_0}(t;0), v'_{ij}(t)\right\rangle, \quad \frac{\partial g'_{0i}}{\partial x_j}(t;0) = \left\langle \frac{\partial r}{\partial x_0}(t;0), v_{ij}(t)\right\rangle$$

for all $i, j \geq 1, t \in [0, \theta]$.

Set $w(t) = \dfrac{\partial r}{\partial x_0}(t;0)$. Then by (4.17)

$$\|w(t)\| = \left\|\frac{\partial^2 r}{\partial^2 x_0}(t;0)\right\| > 0$$

whenever $t \in [a, b]$.

Fix arbitrary $i, j = 1, \ldots, n$. According to (4.24), we have to choose the maps v_{ij} in such a way that

$$\begin{cases} \langle w(t), v_{ij}(t)\rangle = b_{ij}(t) \\ 2\langle w(t), \dot{v}_{ij}(t)\rangle = a_{ij}(t) \end{cases} \quad (t \in [0, \theta]). \tag{4.25}$$

Define $v_{ij} : [0, \theta] \to \mathbf{R}^n$ by $v_{ij}(t) = 0$ for $t \notin [a, b]$ and

$$v_{ij}(t) = b_{ij}(t)w(t) + \frac{\dot{b}_{ij}(t) - \frac{1}{2}a_{ij}(t)}{\|w(t)\|^2}w(t)$$

for $t \in [a, b]$. It follows above that v_{ij} are well-defined smooth maps with supp $v_{ij} \subset [a, b]$. A straightforward verification shows that (4.25) and (4.19) hold. Moreover, if all a_{ij}, and b_{ij}, are taken in a small neighbourhood $\mathcal{V}$ of 0 in $C^\infty([0, \theta].\mathbf{R}^n)$, then h is C^∞ close to 0 and the map f, defined by (4.20), belongs to $\mathcal{U}$. This proves the assertion. ♠.

Our interest to Hamiltonians of the form (4.23) with $c_{ij} = 0$ for all i, j and symmetric matrices $A = (a_{ij})$, $B = (b_{ij})$, leads naturally to the consideration of a special subset of sp$(2m)$. Denote by **a** the *linear subspace of* sp$(2m)$ *consisting of all matrices of the form*

$$\begin{pmatrix} B & 0 \\ A & -B \end{pmatrix},$$

where A and B are symmetric $m \times m$ real matrices. The following lemma shows that **a** is sufficiently big for our aims.

Lemma 4.2.3:   *Let*

$$P = \begin{pmatrix} X & Y \\ Z & T \end{pmatrix}$$

*be an arbitrary $2m \times 2m$ real matrix. For every $\epsilon > 0$ there exists $N \in$ **a** such that $\|N\| < \epsilon$ and $\det(P - \exp N) \neq 0$.*

Proof: Fix an arbitrary $\epsilon > 0$. There exists $\delta > 0$ such that if $N_i \in$ **a**, $\|N_i\| < \delta$, $i = 1, 2$, and $\exp N = (\exp N_1)(\exp N_2)$, then $\|N\| < \epsilon$.

Denote by **a**′ the set of all matrices of the form

$$\begin{pmatrix} 0 & 0 \\ A & 0 \end{pmatrix},$$

where A is a symmetric $m \times m$ real matrix. A straightforward verification shows that **a**′ is a Lie subalgebra of $sp(2m)$, **a**′ $\subset$ **a**, and the commutator $[\mathbf{a}, \mathbf{a}']$ is contained in **a**′. Therefore for every $N_1 \in$ **a** the set

$$\mathbf{a}(N_1) = \{tN_1 + X : t \in \mathbf{R}, X \in \mathbf{a}'\}$$

is a Lie subalgebra of $sp(2m)$. Consequently, if $N_1 \in$ **a**, $N_2 \in$ **a**′ are sufficiently close to 0 and $\exp N = (\exp N_1)(\exp N_2)$, then $N \in \mathbf{a}(N_1)$, and in particular $N \in$ **a**.

Next, we consider matrices N_1, N_2 of the form

$$N_1 = \begin{pmatrix} -B & 0 \\ 0 & B \end{pmatrix}, \quad N_2 = \begin{pmatrix} 0 & 0 \\ A & 0 \end{pmatrix},$$

where A, B are symmetric $m \times m$ real matrices. Then $N_i \in$ **a** and there exists $\delta' > 0$ such that if $\|A\| < \delta'$, $\|B\| < \delta'$, then $\|N_i\| < \delta$ for $i = 1, 2$. Define N by $\exp N = (\exp N_1)(\exp N_2)$, then $N \in$ **a** and setting $D = \exp B$ we obtain

$$\exp N = \begin{pmatrix} D^{-1} & 0 \\ 0 & D \end{pmatrix} \begin{pmatrix} I & 0 \\ A & I \end{pmatrix} = \begin{pmatrix} D^{-1} & 0 \\ DA & D \end{pmatrix}.$$

Choose $\tau > 0$ such that if D is a symmetric positive definite $m \times m$ matrix with $\|D - I\| < \tau$, then $D = \exp B$ for some symmetric $m \times m$ matrix B with $\|B\| < \delta'$.

Assume that

$$\det(P - \exp N) = 0 \qquad (4.26)$$

for every choice of the symmetric matrices A and B with $\|A\| < \delta'$, $\|B\| < \delta'$. Consider an arbitrary symmetric positive definite matrix D with $\|D - I\| < \tau$. We may write D in the form $D = E^{-1} D_1 E$, where E is an orthogonal matrix and

$$D_1 = \begin{pmatrix} y_1 & 0 & \cdots & 0 \\ 0 & y_2 & \cdots & 0 \\ \cdots & \cdots & \cdots & \cdots \\ 0 & 0 & \cdots & y_m \end{pmatrix}, \quad |y_i - 1| < \tau, \quad i = 1, \ldots, m.$$

Fix E, A with $\|A\| < \delta'$ and $y_2, \ldots, y_m$ with $|y_i - 1| < \tau$ for $i = 2, \ldots, m$.

Then by (4.26) we get that

$$\det \begin{pmatrix} X - E^{-1}D_1^{-1}E & Y \\ Z - E^{-1}D_1EA & T - E^{-1}D_1E \end{pmatrix} = 0 \qquad (4.27)$$

holds for every $y_1 \in \mathbf{R}$ with $|y_1 - 1| < \tau$. The left-hand side of (4.27) is a rational function of y_1, determined for all $y_1 \neq 0$. Since it vanishes for infinitely many values of y_1, it is zero for all $y_1 \neq 0$. Therefore, for fixed E, A, $y_2 \ldots, y_m$, (4.27) holds for all $y_1 \neq 0$. Using the same argument, we get by induction that, for fixed E and A, (4.27) holds for all $y_1 \neq 0, \ldots, y_m \neq 0$. This is true for any choice of the orthogonal matrix E, hence

$$\det \begin{pmatrix} X - D^{-1} & Y \\ Z - DA & T - D \end{pmatrix} = 0 \qquad (4.28)$$

holds for every non-singular symmetric matrix D. Next, for a fixed non-singular symmetric matrix D, we can apply an argument similar to the previous one to show that (4.28) holds for any symmetric matrix A.

In this way we have established that (4.28) holds for all symmetric matrices A and D with $\det D \neq 0$. On the other hand, it is well-known from the linear algebra that any square matrix Z can be written in the form $Z = D_0 A_0$, where A_0 and D_0 are symmetric matrices and $\det D_0 \neq 0$. Let A_0 and D_0 be such matrices. Set $D = yD_0$, $A = \frac{1}{y}A_0$ with an arbitrary real $y \neq 0$. Then $Z = DA$, and (4.28) implies

$$\det(X - yD_0^{-1})\det(T - \frac{1}{y}D_0) = \det \begin{pmatrix} X - D^{-1} & Y \\ 0 & T - D \end{pmatrix} = 0.$$

Since each of the equalities $\det(X - yD_0^{-1}) = 0$ and $\det(T - \frac{1}{y}D_0) = 0$ is satisfied only for a finite number of values of y, we get a contradiction. This proves the assertion. ♠

The next lemma concerns analytical dependence of solutions of a special kind of systems of ordinary differential equations on parameters.

Lemma 4.2.4: *Let L be a smooth map of $[0, \theta]$ into the space of $k \times k$ real matrices such that*

$$\|L(t)\| \leq c\psi(t)$$

for each $t \in [0, \theta]$, where $c \in (0, 1)$ is a constant, and

$$\psi : [0, \theta] \rightarrow [0, N],$$

$N > 0$, *is an integrable function with*

$$\int_0^\theta \psi(t)\,dt \leq 1.$$

For $\epsilon \in \mathbf{R}$ let $Y_\epsilon(t)$ be the $k \times k$ matrix function which is the solution of the problem:

$$\begin{cases} \dot{Y}_\epsilon(t) = \epsilon L(t)Y_\epsilon(t), & t \in [0,\theta], \\ Y_\epsilon(0) = I. \end{cases} \tag{4.29}$$

Then there exist smooth matrix functions $B_p(t)$, $p = 0, 1, \ldots$, such that

$$\|B_p(t)\| \le c^p, \quad \|\dot{B}_p(t)\| \le Nc^p \quad (p = 0, 1, \ldots; \quad t \in [0,\theta]) \tag{4.30}$$

and

$$Y_\epsilon(t) = \sum_{p=0}^{\infty} \epsilon^p B_p(t) \quad \left(t \in [0,\theta], |\epsilon| < \frac{1}{c} \right). \tag{4.31}$$

Proof: Set

$$Z^{(p)}(t,\epsilon) = \frac{d^p}{d\epsilon^p} Y_\epsilon(t).$$

Writing the variational equations of (4.29), we get by induction

$$\begin{cases} \dot{Z}^{(p)}(t,\epsilon) = \epsilon L(t)Z^{(p)}(t,\epsilon) + pL(t)Z^{(p-1)}(t,\epsilon), \\ Z^{(p)}(0,\epsilon) = 0, \end{cases}$$

for all $t \in [0,\theta]$, $\epsilon \in \mathbf{R}$, $p = 1, 2, \ldots$. Then for $\epsilon = 0$ one finds

$$\dot{Z}^{(p)}(t,0) = pL(t)Z^{(p-1)}(t,0), \quad Z^{(p)}(0,0) = 0$$

for $t \in [0,\theta]$, $p = 1, 2, \ldots$. Clearly, $\|Z^{(0)}(t,0)\| = \|Y_0(t)\| = 1$. Suppose that for some integer $p > 0$ we have

$$\|Z^{(p-1)}(t,0)\| \le (p-1)!c^{p-1}$$

for all $t \in [0,\theta]$. Then it follows from the above that

$$\|Z^{(p)}(0,t)\| = \left\| p \int_0^t L(s)Z^{(p-1)}(0,s)ds \right\| \le p!c^{p-1} \int_0^t \|L(s)\| \, ds \le p!c^p.$$

Moreover, $\|Z^{(o)}(0,t)\| = 0$ and for $p \ge 1$ we get

$$\|\dot{Z}^{(p)}(t,0)\| \le p\|L(t)\| \|Z^{(p-1)}(t,0)\|$$
$$\le pcn(p-1)!c^{p-1} = Np!c^p.$$

Therefore the maps

$$B_p(t) = \frac{1}{p!} Z^{(p)}(t,0) = \frac{1}{p!} \frac{d^p}{d\epsilon^p} Y_\epsilon(t)_{|\epsilon=0}$$

satisfy the conditions (4.30). Consequently, the series

$$G(t,\epsilon) = \sum_{p=0}^{\infty} \epsilon^p B_p(t)$$

is uniformly and absolutely convergent for $t \in [0, \theta]$, provided $|\epsilon| \leq \text{const} < \frac{1}{c}$. Then, for ϵ satisfying the latter condition, we have

$$
\dot{G}(t, \epsilon) = \sum_{p=0}^{\infty} \epsilon^p \dot{B}_p(t) = \sum_{p=0}^{\infty} \frac{\epsilon^p}{p!} \dot{Z}^{(p)}(t, 0)
$$

$$
= \epsilon L(t) \sum_{p=1}^{\infty} \frac{\epsilon^{p-1}}{(p-1)!} Z^{(p-1)}(t, 0)
$$

$$
= \epsilon L(t) \sum_{p=1}^{\infty} \epsilon^{p-1} B_{p-1}(t) = \epsilon L(t) G(t, \epsilon).
$$

Moreover, $G(0, \epsilon) = B_0(0) = Y_\epsilon(0) = I$. Therefore $G(t, \epsilon)$ coincides with the solution $Y_\epsilon(t)$ of (4.29). This is true for every ϵ with $|\epsilon| < \frac{1}{c}$, and hence (4.31) is satisfied. ♠

We are now ready to prove the main result of this section.

Proof of Theorem 4.2.1: We shall use again the map r, the coordinates x_i, y_i and the numbers a, b, t_0, satisfying (4.17). Take an arbitrary $\mu > 0$ such that

$$
c = \mu \max\{\|P_t\|^2 : t \in [0, \theta]\} < 1. \tag{4.32}
$$

In what follows the numbers a, b, t_0, μ, c will be fixed.

Fix an arbitrary neighbourhood $\mathcal{W}$ of $\gamma(t_0)$ in M and an arbitrary neighbourhood $\mathcal{U}$ of 0 in $C^\infty(M, \mathbf{R}^n)$. Given a neighbourhood $\mathcal{O}$ of 0 in $C^\infty([0, \theta], \mathbf{a})$, denote by $\mathcal{A}$ the set of all $N \in \mathcal{O}$ with supp $N \subset [a, b]$. We shall consider $\mathcal{A}$ with the topology induced by $\mathcal{O} \subset C^\infty([0, \theta], \mathbf{a})$. Then, as an open subset of a closed linear subspace of $C^\infty([0, \theta], \mathbf{a})$, $\mathcal{A}$ is a Baire topological space. For $N \in \mathcal{A}$ we have

$$
N(t) = \begin{pmatrix} (b_{ij}(t)) & 0 \\ (a_{ij}(t)) & -(b_{ij}(t)) \end{pmatrix}, \tag{4.33}
$$

where the functions $a_{ij}(t)$ and $b_{ij}(t)$ satisfy (4.22). Let $\mathcal{V}$ be a neighbourhood of 0 in $C^\infty([0, \theta], \mathbf{R}^n)$ as in Lemma 4.2.2. We fix $\mathcal{O}$ in such a way that $a_{ij}, b_{ij} \in \mathcal{V}$ for all $i, j = 1, \ldots, m$. Then for every $N \in \mathcal{A}$ we construct $h = h_N$ and $f = f_N$ as in Lemma 4.2.2 and set $F_n = \text{id} + f_N$. Denote by $R_N(t)$ the smooth $2m \times 2m$ matrix function which is the solution of the problem

$$
\begin{cases} \dot{R}_N(t) = P_t^{-1} N(t) P_t R_N(t), & t \in [0, \theta], \\ R_N(0) = I. \end{cases} \tag{4.34}
$$

For a given $\lambda \in \mathbf{C}$ set

$$
\mathcal{A}_\lambda = \{N \in \mathcal{A} : \det(P_\theta R_N(\theta) - \lambda) \neq 0\}.
$$

The verification of the fact that $\mathcal{A}_\lambda$ is open in $\mathcal{A}$ is standard, and we leave it to the reader. We shall show that $\mathcal{A}_\lambda$ is dense in $\mathcal{A}$. In fact, it is sufficient to show that $\mathcal{A}_\lambda$ contains elements arbitrarily close to 0. Indeed, for $N \in \mathcal{A}$ we can apply this to the submanifold $M' = F_N(M)$ to show that there exist elements of $\mathcal{A}_\lambda$ arbitrarily close to N.

By Lemma 4.2.3 there exists $A \in \mathbf{a}$ such that $\|A\| < \mu$ and

$$\det(\exp A - \lambda P_{t_0} P_\theta^{-1} P_{t_0}^{-1}) \neq 0.$$

Fix an A with these properties, and take an arbitrary smooth function

$$\varphi : \mathbf{R} \rightarrow [0,1]$$

such that supp $\varphi \subset [-1,1]$ and $\int_\mathbf{R} \varphi(t)\, dt = 1$. For $\delta > 0$ set

$$\varphi_\delta(t) = \frac{1}{\delta} \varphi\left(\frac{t-t_0}{\delta}\right).$$

Then supp $\varphi_\delta \subset [t_0 - \delta, t_0 + \delta]$, $0 \leq \varphi_\delta \leq \frac{1}{\delta}$, $\int_\mathbf{R} \varphi_\delta(t)\, dt = 1$. Denote by $R_\delta(t)$ the matrix function which is the solution of (4.34) for $N(t) = \varphi_\delta(t)A$ (Note that $\varphi_\delta A \notin \mathcal{A}$ for small $\delta > 0$). Then for $t > t_0$ we have

$$R_\delta(t) \rightarrow \exp(P_{t_0}^{-1} A P_{t_0}) = P_{t_0}^{-1}(\exp A)P_{t_0} \quad \text{as } \delta \rightarrow 0.$$

The choice of A now implies the existence of $\delta > 0$ with

$$\det(R_\delta(\theta) - \lambda P_\theta^{-1}) \neq 0. \tag{4.35}$$

Fix an arbitrary $\delta > 0$ with (4.35) and set

$$L(t) = \varphi_\delta(t) P_t^{-1} A P_t, \quad t \in [0,\theta].$$

Then

$$\|L(t)\| = \varphi_\delta(t) \|J P_t^\mathsf{T} J A P_t\| \leq \mu \varphi_\delta \max_s \|P_s\|^2 = c\varphi_\delta(t),$$

which yields that the assumptions of Lemma 4.2.4 are fulfilled for $L(t)$, c, $N = \frac{1}{\delta}$, $\psi = \varphi_\delta$. Therefore the solution $Y_\epsilon(t)$ of (4.29) satisfies (4.31) with (4.30). In particular

$$\chi_\lambda(\epsilon) = \det(Y_\epsilon(\theta) - \lambda P_\theta^{-1}) \tag{4.36}$$

is an analytic function of $\epsilon \in \mathbf{R}$ for $|\epsilon| < \frac{1}{c}$. Note that for $\epsilon = 1$ the solution of (4.29) coincides with the solution of (4.34) for $N = \varphi_\delta A$. Consequently, $Y_1(t) = R_\delta(t)$ for every t, and (4.36) and (4.35) imply

$$\chi_\lambda(1) = \det(R_\delta(\theta) - \lambda P_\theta^{-1}) \neq 0.$$

On the other hand, $c < 1$ by (4.33), therefore $1 < \frac{1}{c}$. This shows that $\chi_\lambda(\epsilon)$ is an analytic function for $|\epsilon| < \frac{1}{c}$ which is not trivially zero. In particular, 0 is not a cluster point of the set

$$E_\lambda = \left\{ \epsilon \in \left(0, \frac{1}{c} \right) : \chi_\lambda(\epsilon) = 0 \right\}.$$

Hence for all sufficiently small $\epsilon > 0$ the map N_ϵ, defined by $N_\epsilon(t) = \epsilon\varphi(t)A$, belongs to $\mathcal{A}$ and $\chi_\lambda(\epsilon) \neq 0$, i.e. $N_\epsilon \in \mathcal{A}_\lambda$. Since $N_\epsilon \to 0$ as $\epsilon \to 0$, we deduce that $\mathcal{A}_\lambda$ contains elements arbitrarily close to 0.

In this way we have established that $\mathcal{A}_\lambda$ is open and dense in $\mathcal{A}$ for each complex number λ. Since Λ is countable,

$$\mathcal{A}_\Lambda = \cap_{\lambda \in \Lambda} \mathcal{A}_\lambda$$

is a residual subset of $\mathcal{A}$. In particular, there exists $N \in \mathcal{A}_\Lambda$ such that $f = f_N \in \mathcal{U}$ and $F = F_N$ satisfy the requirements of the theorem. ♠

4.3. Non-degeneracy and transversality

Our aim in this section is to describe the notion of non-degeneracy of a closed geodesic by means of some transversality condition. This will allow us in the next section to apply the transversality theorem of Abraham and to establish the existence of a residual set of embeddings inducing bumpy metrics on M.

Throughout M will be a compact smooth $(m+1)$-dimensional submanifold of $\mathbf{R}^n$, $1 \leq m \leq n-2$. Let

$$\sigma : S^*M \to M$$

be the *cospherical bundle* over M. That is,

$$S^*M = \cup_{x \in M} S^*_x M,$$

where for every $x \in M$, $S^*_x M$ is the space of all straight lines through 0 in $T^*_x M$ endowed with the quotient topology, and $\sigma(v) = x$ for each $v \in S^*_x M$. Let

$$p : T^*M \to S^*M$$

be the *canonical map*. Consider the map

$$\mathcal{H} : T^*M \times \mathbf{R}^+ \times C(M) \to T^*M,$$

which assigns to $(X, t, F) \in T^*M \times \mathbf{R}^+ \times C(M)$ the shift of X after time t under the action of the geodesic flow on T^*M generated by the metric g_F. There exists a unique map

$$\mathcal{F} : S^*M \times \mathbf{R}^+ \times C(M) \to S^*M$$

such that $\mathcal{F} \circ (p \times \mathrm{id} \times \mathrm{id}) = p \circ \mathcal{H}$.

In what follows we shall use not only the topological structure of $\mathbf{C}(M)$ but its structure of a Banach manifold as well. The latter is inherited naturally from the Banach space structure of $C^\infty(M, \mathbf{R}^n)$. We are going to describe it briefly. As a general reference on infinite dimensional manifolds, we refer the reader to [Lang].

Let $2 \leq s \leq \infty$, and fix a finite number of smooth charts $\varphi_i : B_r \to U_i$, $i = 1, \ldots, k$, where U_i are open subsets of M, B_r is the open ball with centre 0 and radius $r > 0$ in $\mathbf{R}^{m+1}$ and $\cup_{i=1}^k K_i = M$, where $K_i = \varphi_i(\overline{B_{r/2}})$. For $f \in C^s(M, \mathbf{R}^n)$ define

$$\|f\|_s = \sum_{j=0}^s \frac{1}{j!} \max_i \sup_{x \in K_i} \left\| D^j (f \circ \varphi_i^{-1})(x) \right\|.$$

Then $\| \cdot \|_s$ is a norm in $C^s(M, \mathbf{R}^n)$ with respect to which the latter is a Banach space. This norm generates the Whitney C^s topology (cf. Chapter 2, Section 1 of [Hir]). Denote by $\mathbf{C}^s M$ the subset of $C^s(M, \mathbf{R}^n)$ consisting of all C^s embeddings of M into $\mathbf{R}^n$. Then $\mathbf{C}^s M$ is an open subset of $C^s(M, \mathbf{R}^n)$ (cf. Chapter 2, Section 1 of [Hir]) and therefore has a natural structure of a Banach manifold with model space $C^s(M, \mathbf{R}^n)$. Next, we consider $\mathbf{C}^s M$ equipped with this structure. As before the space $\mathbf{C}^\infty M$ will be denoted briefly by $\mathbf{C}(M)$.

The following lemma is a consequence of Chapter 5 in [AbR], and in fact can be derived easily from some standard facts concerning smooth dependence of solutions of systems of differential equations on parameters. We omit the details.

Lemma 4.3.1: *The maps $\mathcal{F}$ and $\mathcal{H}$ are smooth.* ♠

The notion of non-degeneracy of a closed geodesic, introduced in the previous section, has a local character, and at a first glance seems to be inconvenient when we try to perturb the whole manifold M in order to make all closed geodesics on it non-degenerate. However, this notion can be described in terms of some global characteristics of the corresponding geodesic flow.

Given a fixed smooth Riemannian metric g on M, for $t \in \mathbf{R}$, $Y \in T^*M$ denote by $\varphi_t(Y)$ the shift of Y after time t under the action of the geodesic flow on T^*M corresponding to g. Let ψ_t be the projection of φ_t on S^*M, that is

$$\psi_t : S^*M \to S^*M$$

is such that $\psi_t \circ p = p \circ \varphi_t$ for any $t \in \mathbf{R}$. Let $Y \in T^*M \setminus \{0\}$ generate a periodic trajectory with period $\theta > 0$, i.e. $\varphi_\theta(Y) = Y$. Consider the linear map

$$T\varphi_\theta : T_Y(T^*M) \to T_Y(T^*M).$$

The following proposition is a particular case of a general fact in the theory of dynamical systems (see for example Sections 7.1 and 8.1 in [AbM]). We omit the proof.

Proposition 4.3.2: *Under the assumptions above, let P be a linear Poincaré* *map, related to the closed integral curve*

$$c = \{\varphi_t(Y) : t \in [0, \theta]\},$$

generated by Y. Then

$$\operatorname{spec} T\varphi_\theta = \{1\} \cup \operatorname{spec} P,$$

and the multiplicity of 1 in spec $T\varphi_\theta$ *is* $k + 2$, k *being the multiplicity of 1 in spec P.* ♠

According to the definition in Section 4.1, c (and so the corresponding geodesic on M) is non-degenerate as a curve of period θ, if $k = 0$ in the notation of the above proposition. This is equivalent to the fact that 1 occurs with multiplicity exactly 2 in spec $T\varphi_\theta$. Setting $X = p(Y)$, we see that c is non-degenerate if and only if 1 occurs with multiplicity exactly 1 in the spectrum of the linear map $T\psi_\theta$.

Denote by **D** the *diagonal* of $S^*M \times S^*M$, i.e.

$$\mathbf{D} = \{(X, X) : X \in S^*M\}.$$

The following lemma is the main point in this section.

Lemma 4.3.3: *Let*

$$(Y_0, \theta, F_0) \in (T^*M \backslash \{0\}) \times \mathbf{R}^+ \times \mathbf{C}(M)$$

be such that

$$c : [0, \theta] \to T^*M, \quad c(t) = \mathcal{H}(Y_0, t, F_0),$$

is a closed geodesic with respect to the metric g_{F_0} of period θ, and let $X_0 = p(Y_0)$.

(a) If c is non-degenerate as a curve of period θ, then the map

$$\mathcal{F}_1 : S^*M \times \mathbf{R}^+ \to S^*M \times S^*M,$$

*defined by $\mathcal{F}_1(X, t) = (X, \mathcal{F}(X, t, F_0))$, is transversal to **D** at (X_0, θ);*

(b) If θ is the minimal period of c, then the map

$$\mathcal{F}_2 : \mathbf{R}^+ \times \mathbf{C}(M) \to S^*M \times S^*M,$$

*defined by $\mathcal{F}_2(t, F) = (X_0, \mathcal{F}(X_0, t, F))$, is transversal to **D** at (θ, F_0).*

Proof: It is sufficient to consider only the case $F_0 = $ id; in the general one we can replace M by $F_0(M)$ and g by g_{F_0}.

Introduce coordinates $x_0 \ldots, x_m, y_0, \ldots, y_m$ in a neighbourhood of Im c in T^*M by means of a map (4.6) as in Section 4.1 such that

$$Y_0 = (0, \ldots, 0; \; 1, 0, \ldots, 0).$$

Using these coordinates, we shall identify $T_{Y_0}(T^*M)$ with

$$\mathbf{R}^{2m+2} = (\mathbf{R} \times \mathbf{R}^m) \times (\mathbf{R} \times \mathbf{R}^m).$$

Note that we can use $x_0, \ldots, x_m, y_1, \ldots, y_m$ as coordinates in a neighbourhood of $p \circ c$ in S^*M.

(a) Let c be non-degenerate as a curve of period θ. Define

$$\mathcal{H}_1 : T^*M \times \mathbf{R}^+ \to T^*M \times T^*M$$

by $\mathcal{H}_1(Y, t) = (Y, \mathcal{H}(Y, t, \mathrm{id}))$. For $s, t \in \mathbf{R}^+$, $Y = (s, 0; 1, 0)$ we have

$$\mathcal{H}_1(Y, T) = (Y, (s+t, 0; 1, 0)).$$

Therefore

$$\operatorname{Im} T\mathcal{H}_1(Y_0, \theta) \supset (\mathbf{R} \times \{0\}) \times (\mathbf{R} \times \{0\}). \tag{4.37}$$

Fix $t = \theta$ and consider $Y \in T^*M$ of the form $Y = (0, x; 0, y)$, x, $y \in \mathbf{R}^{2m}$. Then $Y \in \Sigma(0)$ (see the notation in Section 4.1) and $\mathcal{H}_1(Y, \theta) = (Y, Y')$, where $Y' = (0, x'; 0, y') \in \Sigma(\theta)$, $(x', y') = \mathcal{P}_\theta(x, y)$. Since $1 \notin \operatorname{spec} d\mathcal{P}_\theta(0, 0)$, the map

$$(x, y) \to ((x, y), \mathcal{P}_\theta(x, y))$$

of $\mathbf{R}^{2m}$ into $\mathbf{R}^{2m}$ is transversal to the diagonal at $(0, 0)$. Therefore

$$\mathbf{D}' + \operatorname{Im} T\mathcal{H}_1(Y_0, \theta) \supset (\{0\} \times \mathbf{R}^m)^4,$$

$\mathbf{D}'$ being the diagonal of $\mathbf{R}^{2m+2} \times \mathbf{R}^{2m+2} = (T_{Y_0}(T^*M))^2$. Combining this with (4.37), we find

$$\mathbf{D}' + \operatorname{Im} T\mathcal{H}_1(Y_0, \theta) \supset (\mathbf{R}^{m+1} \times (\{0\} \times \mathbf{R}^m))^2,$$

which proves (a).

(b) Let θ be the minimal period of c. Fix the numbers a, b, t_0 with (4.17) and a smooth map $\rho : \mathbf{R}^{m+1} \to [0, 1]$ as in Section 4.2. Let $\varphi : \mathbf{R} \to [0, 1]$ be a smooth function with

$$\operatorname{supp} \varphi \subset [-1, 1], \quad \int_{\mathbf{R}} \varphi(t) \, dt = 1.$$

For $\epsilon, \delta > 0$ set

$$\varphi_\delta(t) = \frac{1}{\delta} \varphi\left(\frac{t - t_0}{\delta}\right), \quad \chi(t) = \epsilon \varphi_\delta(t).$$

Later on we shall see how small the numbers ϵ and δ should be. Clearly, if δ is sufficiently small, then $\operatorname{supp} \chi \subset [a, b]$.

It follows by (4.17) that for $\lambda(t) = \frac{\partial r}{\partial x_0}(t; 0)$ we have $\dot{\lambda}(t) \neq 0$ whenever $t \in [a, b]$. As in the proof of Lemma 4.2.2 we construct smooth maps e, d : $[0, \theta] \to \mathbf{R}^n$ with supports in $[a, b]$ such that

$$\begin{cases} \langle \lambda(t), e(t) \rangle = \chi(t), & \langle \lambda(t), \dot{e}(t) \rangle = 0, \\ \langle \lambda(t), d(t) \rangle = 0, & 2\langle \lambda(t), \dot{d}(t) \rangle = -\chi(t) \end{cases} \tag{4.38}$$

for every $t \in [0, \theta]$. Namely, we set

$$e(t) = \chi(t)\lambda(t) + \frac{\dot{\chi}(t)}{\|\dot{\lambda}(t)\|^2}\dot{\lambda}(t),$$

and

$$d(t) = \frac{\chi(t)}{2\|\dot{\lambda}(t)\|^2}\dot{\lambda}(t)$$

for every $t \in [0, \theta]$. Next, define

$$W = \{\omega = (u_1, \ldots, u_m; v_1, \ldots, v_m) \in \mathbf{R}^{2m} : \|\omega\| < 1\},$$

and consider the map

$$\Omega : W \to C(M), \quad \Omega(\omega) = F = \mathrm{id} + f,$$

where f is determined by (4.20) by means of the map h, given by

$$h(x) = \left(\sum_{i=1}^m u_i x_i\right) e(x_0) + \left(\sum_{i=1}^m v_i x_i\right) d(x_0) \tag{4.39}$$

for $x = (x_0, x_1, \ldots, x_m) \in V$. If e and d are sufficiently close to 0 in the C^∞ topology (this means that χ is C^∞ close to 0), then Ω is well-defined. In fact, Ω coincides with the restriction of a linear map $\mathbf{R}^{2m} \to C^\infty(M, \mathbf{R}^n)$.

To prove (b) it is sufficient to show that the map

$$(t, F) \mapsto \mathcal{F}(X_0, t, F)$$

of $\mathbf{R}^+ \times C(M)$ into S^*M is a submersion at (θ, id). Consider the map

$$\mathcal{H}_2 : \mathbf{R}^+ \times W \to T^*M,$$

defined by $\mathcal{H}_2(t, w) = \mathcal{H}(Y_0, t, \Omega(w))$. The assertion will be proved if we show that

$$\mathrm{Im}\, T\mathcal{H}_2(\theta, 0) = (\mathbf{R} \times \mathbf{R}^m) \times (\{0\} \times \mathbf{R}^m). \tag{4.40}$$

Let $\omega = 0$, then $\mathcal{H}_2(t, 0) = (t, 0; 1, 0)$ and therefore

$$\operatorname{Im} T\mathcal{H}_2(\theta, 0) \supset (\mathbf{R} \times \{0\}) \times (\{0\} \times \{0\}). \tag{4.41}$$

Let $\omega = (u; v) \in W$. Define h by (4.39) and set $\tilde{g} = g_F$, $F = \Omega(\omega)$, $g' = \tilde{g} - g$. Then $\tilde{g}_{ij} = g_{ij} + g'_{ij}$, where g'_{ij} are determined by (4.21). In a small neighbourhood of $\operatorname{Im} c$ in T^*M the Hamiltonian function, corresponding to $\tilde{g}$, has the form

$$\tilde{H}(x; y) = \frac{1}{2} \sum_{i,j=0}^{m} \tilde{g}_{ij}(x) y_i y_j,$$

where $x = (x_0, \ldots, x_m)$, $y = (y_0, \ldots, y_m)$. Therefore the coordinate functions $\tilde{x}(t; \omega)$, $\tilde{y}(t; w)$ of $\mathcal{H}(Y_0, t, \Omega(\omega))$, $t \in [0, \theta]$, satisfy the Hamiltonian equations

$$\dot{\tilde{x}}_k = \frac{\partial \tilde{H}}{\partial y_k}(\tilde{x}; \tilde{y}; \omega), \quad \dot{\tilde{y}}_k = -\frac{\partial \tilde{H}}{\partial x_k}(\tilde{x}; \tilde{y}; \omega). \tag{4.42}$$

Set $x(t) = \tilde{x}(t; 0)$, $y(t) = \tilde{y}(t; 0)$, then $x(t) = (t, 0, \ldots, 0)$ and $y(t) = (1, 0, \ldots, 0)$ are the coordinates of $c(t)$.

Next, we consider $\tilde{x}(t; \omega)$ and $\tilde{y}(t; \omega)$ as column vectors. For $q = 1, \ldots, m$ and $t \in [0, \theta]$ set

$$\xi_q(t) = \frac{d}{du_q}\bigg|_{\omega=0} \begin{pmatrix} \tilde{x}(t; \omega) \\ \tilde{y}(t; \omega) \end{pmatrix}, \quad \eta_q(t) = \frac{d}{dv_q}\bigg|_{\omega=0} \begin{pmatrix} \tilde{x}(t; \omega) \\ \tilde{y}(t; \omega) \end{pmatrix}.$$

Writing the variational equations of (4.42), we find $\xi_q(0) = 0$ and

$$\dot{\xi}_q(t) = S(t)\xi_q(t) + R_q(t), \quad t \in [0, \theta], \tag{4.43}$$

where

$$S(t) = \begin{pmatrix} S_1(t) & S_2(t) \\ S_3(t) & S_4(t) \end{pmatrix},$$

$$S_1(t) = \left(\frac{\partial^2 \tilde{H}}{\partial y_k \partial x_i}(x(t); y(t); 0) \right), \quad S_2(t) = \left(\frac{\partial^2 \tilde{H}}{\partial y_k \partial y_i}(x(t); y(t); 0) \right),$$

$$S_3(t) = \left(-\frac{\partial^2 \tilde{H}}{\partial x_k \partial x_i}(x(t); y(t); 0) \right), \quad S_4(t) = \left(-\frac{\partial^2 \tilde{H}}{\partial x_k \partial y_i}(x(t); y(t); 0) \right),$$

and

$$R_q(t) = \left(\left(-\frac{\partial^2 \tilde{H}}{\partial y_k \partial u_q}(x(t); y(t); 0) \right)_k ; \left(-\frac{\partial^2 \tilde{H}}{\partial x_k \partial u_q}(x(t); y(t); 0) \right)_k \right)^T.$$

By $y(t) = (1, 0, \ldots, 0)$, we get

$$\frac{\partial^2 \tilde{H}}{\partial y_k \partial x_i}(x(t); y(t); 0) = \sum_{j=0}^{m} \frac{\partial \tilde{g}_{ki}}{\partial x_i}(x(t); 0) y_j(t) = \frac{\partial \tilde{g}_{k0}}{\partial x_i}(x(t); 0).$$

On the other hand, (4.21) and (4.39) imply

$$\tilde{g}_{k0} = g_{k0} + \left\langle \frac{\partial r}{\partial x_0}, \frac{\partial h}{\partial x_k} \right\rangle + \left\langle \frac{\partial r}{\partial x_k}, \frac{\partial h}{\partial x_0} \right\rangle + O(\|\omega\|^2).$$

Differentiating the latter equality with respect to x_i and evaluating at $x_0 = t$, $x_1 = \ldots = x_m = 0$, we find

$$\frac{\partial \tilde{g}_{k0}}{\partial x_i}(x(t); 0) = \frac{\partial g_{k0}}{\partial x_i}(t; 0).$$

Therefore

$$S_1(t) = \left(\frac{\partial g_{k0}}{\partial x_i}(t; 0) \right)_{k,i}.$$

By similar calculations one obtains $S_2(t) = (g_{ki}(t; 0))_{k,i}$,

$$S_3(t) = \left(-\frac{1}{2} \frac{\partial^2 g_{00}}{\partial x_k \partial x_i}(t; 0) \right)_{k,i}, \quad S_4(t) = \left(-\frac{\partial g_{i0}}{\partial x_k}(t; 0) \right)_{k,i}.$$

Moreover, (4.38) yields

$$\frac{\partial^2 \tilde{H}}{\partial y_k \partial u_q}(x(t); y(t); 0) = \frac{\partial \tilde{g}_{k0}}{\partial u_q}(x(t); 0) = \begin{cases} 0, & k \neq q, \\ \langle \lambda(t), e(t) \rangle = \chi(t), & k = q, \end{cases}$$

and

$$\frac{\partial^2 \tilde{H}}{\partial x_k \partial u_q}(x(t); y(t); 0) = \frac{\partial^2 \tilde{g}_{00}}{\partial x_k \partial u_q}(x(t); 0) = 0$$

for $k = 0, 1, \ldots, m$ and $q = 1, \ldots, m$. Thus, the column-vector $R_q(t)$ has the form

$$R_q(t) = (0, \ldots, 0, \chi(t), 0, \ldots, 0; 0, \ldots, 0)^{\mathrm{T}},$$

where $\chi(t)$ is the qth component. Since $g_{00}(t; 0) = 1$,

$$\frac{\partial g_{00}}{\partial x_i}(t; 0) = \frac{\partial^2 g_{00}}{\partial x_0 \partial x_i}(t; 0) = \frac{\partial g_{i0}}{\partial x_0}(t; 0) = 0$$

for $i = 0, 1, \ldots, m$, and $g_{i0}(t; 0) = 0$ for $i = 1, \ldots, m$, we see that the 0th row of $S(t)$ has the form $(0, \ldots, 0; 1, 0, \ldots, 0)$, while the $(m+1)$th one consists only of zeros. Combining this with (4.43), we deduce

$$\dot{\xi}_q^{(0)}(t) = \xi_q^{(m+1)}(t), \quad \dot{\xi}_q^{(m+1)}(t) = 0, \quad t \in [0, \theta].$$

Here $\xi_q^{(i)}$ is the ith component of the vector ξ_q. Consequently, we get $\xi_q^{(0)}(t) = \xi_q^{(m+1)}(t) = 0$ for all $t \in [0, \theta]$ and every $q = 1, \ldots, m$. In particular,

$$\xi_q(\theta) \in (\{0\} \times \mathbf{R}^m) \times (\{0\} \times \mathbf{R}^m), \quad q = 1, \ldots, m.$$

In the same way one gets similar inclusions for the vectors $\eta_q(\theta), q = 1, \ldots, m$.

Next, notice that the subspace $T\mathcal{H}_2(\theta, 0)(\{0\} \times T_0 W)$ is generated by the vectors

$$\xi_1(\theta), \ldots, \xi_m(\theta), \eta_1(\theta), \ldots, \eta_m(\theta) \tag{4.44}$$

(see the beginning of the proof of (b)). Therefore

$$T\mathcal{H}_2(\theta, 0)(\{0\} \times T_0 W) \subset (\{0\} \times \mathbf{R}^m) \times (\{0\} \times \mathbf{R}^m). \tag{4.45}$$

Now we shall show that if δ and ϵ are sufficiently small, then (4.45) becomes an equality. In fact, it is sufficient to choose δ so small that the vectors (4.44) are linearly independent. To see that this is possible, introduce the fundamental solution $Z(t)$ of (4.43). That is $Z(t)$ is the $(2m + 2) \times (2m + 2)$ smooth matrix function with $Z(0) = I$ and $\dot{Z}(t) = S(t)Z(t)$ for $t \in \mathbf{R}$. Since $S(t) \in \mathrm{sp}(2m)$, we have $Z(t) \in \mathrm{Sp}(2m)$ for every t. Now by (4.43) we have

$$\xi_q(t) = Z(t) \int_0^t Z(s)^{-1} R_q(s) \, ds, \quad t \in \mathbf{R}. \tag{4.46}$$

Let $\lambda_0(t), \ldots, \lambda_m(t), \mu_0(t), \ldots, \mu_m(t)$ be the successive column vectors of $Z(t)^{-1}$. Then

$$\int_0^\theta \varphi_\delta(s) \lambda_q(s) \, ds \to \lambda_q(t_0)$$

and

$$\int_0^\theta \varphi_\delta(s) \mu_q(s) \, ds \to \mu_q(t_0)$$

as $\delta \to 0$. Hence, if δ is sufficiently small, the vectors

$$\lambda_1, \ldots, \lambda_m, \quad \mu_1, \ldots, \mu_m,$$

given by

$$\lambda_q = \int_0^\theta \varphi_\delta(s) \lambda_q(s) \, ds, \quad \mu_q = \int_0^\theta \varphi_\delta(s) \mu_q(s) \, ds,$$

are linearly independent. Fix such a $\delta > 0$ and choose $\epsilon > 0$ so small that $\Omega(\omega) \in \mathbf{C}(M)$ for every $\omega \in W$ (cf. (4.38) and (4.39)). According to (4.46), we find

$$\xi_q(\theta) = Z(\theta) \int_0^\theta Z(s)^{-1} R_q(s)\,\mathrm{d}s = \epsilon Z(\theta) \int_0^\theta \varphi_\delta(s)\lambda_q(s)\,\mathrm{d}s = \epsilon Z(\theta)\lambda_q.$$

In the same way one gets $\eta_q(\theta) = \epsilon Z(\theta)\mu_q$. The vectors (4.44) are therefore linearly independent, and consequently (4.45) becomes an equality by this choice of δ. This and (4.41) imply (4.40), which concludes the proof of the lemma. ♠

4.4. Global perturbations of smooth surfaces

Here we use the notation of the previous section.

The considerations in the present chapter culminate in the following theorem.

Theorem 4.4.1:   *Let M be a smooth compact submanifold of $\mathbf{R}^n$, $n \geq 3$, having a positive codimension. Then there exists a residual subset $\mathcal{B}$ of $\mathbf{C}(M)$ such that for every $F \in \mathcal{B}$ the standard metric on $F(M)$ is a bumpy metric.*

Proof: Given $F \in \mathbf{C}(M)$, $X \in S^*M$, $t \in \mathbf{R}$, set

$$\psi_t^F(X) = \mathcal{F}(X, t, F).$$

For $0 < a \leq b$ denote by $\mathbf{C}(a, b)$ the set of those $F \in \mathbf{C}(M)$ such that if $\tilde{c} = \{\psi_i^F(v) : 0 \leq t \leq \omega\}$ is a periodic trajectory of the flow $\{\psi_t^F\}$ of period $\omega \leq b$ and having minimal period $\theta \leq a$, then $\tilde{c}$ is non-degenerate as a curve of period ω. Notice that $\mathbf{C}(a', b') \subset \mathbf{C}(a, b)$, provided $a \leq a'$, $b \leq b'$, $a' \leq b'$. Now we set

$$\mathcal{B} = \cap_{k=1}^\infty \mathbf{C}(k, k).$$

Then evidently for every $F \in \mathcal{B}$ the standard metric on $F(M)$ is a bumpy metric. Thus, the theorem will be proved if we establish that $\mathbf{C}(k, k)$ is open and dense in $\mathbf{C}(M)$ for every $k = 1, 2, \dots$.

First, notice that for every $F \in \mathbf{C}(M)$ there exist a neighbourhood $\mathcal{U}$ of F in $\mathbf{C}(M)$ and $\alpha > 0$ such that for every $G \in \mathcal{U}$ the flow $\{\psi_t^G\}$ has no periodic trajectories with period $\leq \alpha$. Indeed, suppose this is not true. Then there exist sequences $\{F_k\} \subset \mathbf{C}(M)$, $\{t_k\} \subset \mathbf{R}^+$, $\{v_k\} \subset S^*M$ such that $F_k \to F, t_k \to 0$, $\psi_{t_k}^{F_k} v_k = v_k$ for every k. Due to the compactness of S^*M we may assume that $v_k \to v \in S^*M$. Fix an arbitrary $t > 0$. For each k write $t = m_k t_k + s_k$ with $m_k \in N$ and $0 \leq s_k < t_k$. Then $s_k \to 0$, so we have

$$\psi_t^{F_k} v_k = \psi_{s_k}^{F_k} v_k \to_{k \to \infty} v.$$

On the other hand, clearly $\psi_t^{F_k} v_k \to \psi_t^F V$. Therefore $\psi_t^F v = v$. This is true for every $t > 0$ which is a contradiction with the well known fact that the geodesic flow ψ_t^F has no fixed points.

Now let $0 < a \leq b$. We show that $\mathbf{C}(a, b)$ is open in $\mathbf{C}(M)$. Assume $F_k \to F$, $F_k \notin \mathbf{C}(a, b)$ for every k. Then there exist $v_k \in S^*M$, $t_k \in (0, a]$ and $l_k \in \mathbf{N}$ such that $\psi_{t_k}^{F_k} v_k = v_k$, $l_k t_k \leq b$ and spec $T\psi_{l_k t_k}^{F_k}(v_k)$ contains 1 with multiplicity at least 2. According to the argument above, there exist a neighbourhood $\mathcal{U}$ of F in $C(M)$ and $\alpha > 0$ such that for every $G \in \mathcal{U}$ the flow $\{\psi_t^G\}$ has no periodic trajectories with period $\leq \alpha$. Without loss of generality we may assume that $F_k \in \mathcal{U}$ for every k and $t_k \to t$, $v_k \to v \in S^*M$. Since $t_k > \alpha$ for each k, we find $t \geq \alpha$. Now we see that the sequence $\{l_k\}$ is bounded; therefore we may assume $l_k = l$ for every k. It then follows that $\psi_t^F v = v$, $lt \leq b$ and spec $T\psi_{lt}^F(v)$ contains 1 with multiplicity at least 2. Consequently $F \notin \mathbf{C}(a, b)$, which proves the openness of $\mathbf{C}(a, b)$.

Given $a > 0$ consider the map

$$\mathcal{F}^a : S^*M \times (0, 2a) \times \mathbf{C}(a, 2a) \to S^*M \times S^*M,$$

defined by $\mathcal{F}^a(X, t, F) = (X, \mathcal{F}(X, t, F))$. It follows now by Lemma 4.3.3 that $\mathcal{F}^a \pitchfork \mathbf{D}$. Indeed, let $\mathcal{F}^a(X_0, \theta, F_0) \in \mathbf{D}$. Then $\theta \leq 2a$ and $\tilde{c} = \{\psi_t^{F_0}(X) : 0 \leq t \leq \theta\}$ is a periodic trajectory of period θ. If θ is the minimal period, then $\mathcal{F}^a(X_0, t, F) = \mathcal{F}_2(t, F)$ and Lemma 4.3.3 (b) imply that $\mathcal{F}^a \pitchfork \mathbf{D}$ at (X_0, θ, F_0). Assume that for the minimal period ω of $\tilde{c}$ we have $\omega < \theta$; then clearly $\omega \leq \theta/2 \leq a$. Since $F_0 \in \mathbf{C}(a, 2a)$, $\tilde{c}$ is non-degenerate as a trajectory with period θ. By Lemma 4.3.3 (a) and $\mathcal{F}^a(X, t, F_0) = \mathcal{F}_1(X, t)$ we see that $\mathcal{F}^a \pitchfork \mathbf{D}$ at (X_0, θ, F_0) again. Hence $\mathcal{F}^a \pitchfork \mathbf{D}$.

Next, for $F \in \mathbf{C}(M)$ define

$$\mathcal{F}_F^a : S^*M \times (0, 2a) \to S^*M \times S^*M$$

by $\mathcal{F}_F^a(X, t) = \mathcal{F}^a(X, t, F)$. Set

$$U = \{F \in \mathbf{C}(a, 2a) : \mathcal{F}_F^a \pitchfork \mathbf{D}\}.$$

Applying the Abraham transversality theorem (cf. Section 1.1) we see that U is open and dense in $\mathbf{C}(a, 2a)$. On the other hand, $U \subset \mathbf{C}(3a/2, 3a/2)$. Indeed, if $\mathcal{F}_F^a \pitchfork \mathbf{D}$, then every periodic trajectory of $\{\psi_t^F\}$ of period $\theta < 2a$ is non-degenerate as a trajectory with period θ, and therefore $F \in \mathbf{C}\left(\dfrac{3a}{2}, \dfrac{3a}{2}\right)$. Consequently,

$$\mathbf{C}\left(\frac{3a}{2}, \frac{3a}{2}\right) \cap \mathbf{C}(a, 2a)$$

is an open and dense subset of $\mathbf{C}(a, 2a)$.

Further, fix an arbitrary $a > 0$. We are going to show that $\mathbf{C}(a, 2a)$ is dense in $\mathbf{C}(a, a)$. Consider an arbitrary $F_0 \in \mathbf{C}(a, a)$ and an arbitrary neighbourhood

$\mathcal{U}$ of F_0 in $\mathbf{C}(a,a)$. Note that if $X \in S^*M$ generates a periodic trajectory $\tilde{c}$ of $\{\psi_t^{F_0}\}$ with period not greater than a, then $\tilde{c}$ is non-degenerate as a trajectory with period a. Therefore there exists a neighbourhood V of X in S^*M such that there is no $Y \in V \backslash \{X\}$ generating a periodic trajectory of period $\leq a$. Now the compactness of S^*M shows that there exist only finitely many periodic trajectories

$$\tilde{c}_i = \{\psi_t^{F_0} X_i : 0 \leq t \leq \theta_i\}, \quad i = 1,\dots,k$$

of the flow $\{\psi_t^{F_0}\}$ the minimal periods θ_i of which are $\leq a$. A standard result from the theory of differential equations implies that there exist a neighbourhood $\mathcal{V} \subset \mathcal{U}$ of F_0 and continuous maps

$$Z_i : \mathcal{V} \to S^*M, \quad \omega_i : \mathcal{V} \to \mathbf{R}^+, \quad i = 1,\dots,k,$$

with $Z_i(F_0) = X_i, \omega_i(F_0) = \theta_i$ for each i and such that for every $F \in \mathcal{V}$,

$$\tilde{c}_i(F) = \{\psi_t^F Z_i(F) : 0 \leq t \leq \omega_i(F)\}, \quad i = 1,\dots,k \qquad (4.47)$$

are periodic trajectories of the flow $\{\psi_t^F\}$ with minimal periods $\omega_i(F)$, respectively. Moreover, exploiting the non-degencracy of the trajectories $\tilde{c}_i$ and using a compactness argument, very similar to one of those already applied in this proof, we see that if $\mathcal{V}$ is sufficiently small, then for every $F \in \mathcal{V}$ if $\tilde{c} = \{\psi_t^F(X)\}$ is a periodic trajectory with period $\leq a$, then it coincides with some of the trajectories (4.47). It then follows by Theorem 4.2.1 that there exists $F \in \mathcal{V}$ such that the trajectories (4.47) are non-degenerate as trajectories with periods, respectively, $k\omega_i(F)$ for any integer $k > 0$. But, as we have already mentioned, the trajectories (4.47) are the only periodic trajectories of $\{\psi_i^F\}$ having minimal periods $\leq a$. Therefore $F \in \mathbf{C}(a,2a)$.

In this way we have proved that $\mathbf{C}(a,2a)$ is dense in $\mathbf{C}(a,a)$. Consequently, $C(3a/2,3a/2)$ is dense in $\mathbf{C}(a,a)$. The latter clearly implies that $\mathbf{C}((3a/2)^k,(3a/2)^k)$ is dense in $\mathbf{C}(a,a)$ for every integer $k > 0$. Thus for every $b \geq a$, $\mathbf{C}(b,b)$ is dense in $\mathbf{C}(a,a)$.

Consider again an arbitrary fixed $a > 0$. To show that $\mathbf{C}(a,a)$ is dense in $\mathbf{C}(M)$, fix an arbitrary $F \in \mathbf{C}(M)$ and an arbitrary neighbourhood $\mathcal{U}$ of F in $\mathbf{C}(M)$. We may assume $\mathcal{U}$ is so small that there exists $\alpha \in (0,a)$ such that for every $G \in \mathcal{U}$ the flow $\{\psi_t^G\}$ has no periodic trajectories with period $\leq \alpha$. Then obviously $\mathcal{U} \subset \mathbf{C}(\alpha,\alpha)$. Now the density of $\mathbf{C}(a,a)$ in $\mathbf{C}(\alpha,\alpha)$ yields that $\mathcal{U} \cap \mathbf{C}(a,a)$ is non-empty. Hence $\mathbf{C}(a,a)$ is dense in $\mathbf{C}(M)$.

We have established that each of the sets $\mathbf{C}(k,k)$ is open and dense in $\mathbf{C}(M)$. Thus $\mathcal{B}$ is residual in $\mathbf{C}(M)$, which proves the theorem. ♠

The classical bumpy metric theorem concerns the space $\mathcal{G}M$ of all smooth Riemannian metrics on a given smooth compact manifold M. More precisely, $\mathcal{G}M$ is the set of all smooth symmetric and positive definite tensors $g \in \mathcal{T}_2^0(M)$ (cf., for example Sections 1.7 and 2.5 in [AbM]), endowed with the Whitney C^∞ topology. Using the well-known fact that for every $g \in \mathcal{G}M$, the Riemannian manifold (M, g) can be isometrically embedded into some $\mathbf{R}^n$, the image of M being endowed with its standard metric, and applying Theorems 4.2.1 and 4.4.1 one gets the following.

Corollary 4.4.2 (bumpy metric theorem):   *There exists a residual subset of* $\mathcal{G}M$ *consisting of bumpy metrics on* M. ♠

4.5. Notes

The classical bumpy metric theorem (Corollary 4.4.2) was announced by Abraham [Ab] giving an idea of a proof. More general results were published by Klingenberg and Takens [KT]. Given a smooth manifold M with dim $M = m+1$, an integer $k > 0$ and an open dense and invariant subset Q of the Lie group of k-jets of smooth local symplectic maps

$$(\mathbf{R}^{2m}, 0) \rightarrow (\mathbf{R}^{2m}, 0),$$

it was established in [KT] that there exists a residual subset R of $\mathcal{G}M$ such that for $g \in R$ the k-jet of the Poincaré map of every closed geodesic on (M, g) belongs to Q. This global result is derived as a consequence of the following local result in [KT]: if $g \in \mathcal{G}M$ and γ is a closed geodesic on (M, g), then there exists $g' \in \mathcal{G}M$ arbitrarily close to g such that γ is a geodesic on (M, g') and the k-jet of the Poincaré map of γ with respect to g' belongs to Q. Different proofs of these results for $k = 1$ and 3 are given in [K2]. In fact, both [KT] and [K2] do not contain a detailed proof of the global theorem in the case $k = 1$ which includes the bumpy metric theorem. The latter theorem was proved in details by Anosov [An] using the local theorem of [KT].

The main results in this chapter, which are analogous to special cases of the results in [KT], are taken from [S3]. The material in Section 4.1 is a mixture of parts of [KT] and [S3], presented in a different form here. Sections 4.2 and 4.3 follow closely [S3], while the globalization argument in Section 4.4 is a modification of Section 4 in [An]. Lemma 4.3.3 is an analogue of the main Lemma 1 in [An].

Complete analogues of the results in [KT] for hypersurfaces in $\mathbf{R}^n$ are established in [ST].

Generic properties of more general Hamiltonian systems were studied by Robinson [R], Meyer and Palmore [MeyP], Takens [T] and others. The reader may consult [AbM] for more information in this direction. Let us note that the analogue of the bumpy metric theorem for general Hamiltonian systems is not true, as an example in [MeyP] shows.

5 POISSON RELATION FOR MANIFOLDS WITH BOUNDARY

This chapter is devoted to the analysis of the singularities of the distribution

$$\sigma(t) = \sum_j \cos \lambda_j t,$$

where $\{\lambda_j^2\}_{j=1}^\infty$ are the eigenvalues of the Laplacian in a bounded domain Ω with Dirichlet boundary condition on $\partial\Omega$. The main aim is to establish the so called *Poisson relation for manifolds with boundary*, namely that

$$\text{sing\,supp}\,\sigma(t) \subset \{0\} \cup \{\pm T_\gamma : \gamma \in \mathcal{L}_\Omega\},$$

where $\mathcal{L}_\Omega$ is the set of all generalized periodic bicharacteristics of $\Box$ in Ω and T_γ denotes the period of $\gamma \in \mathcal{L}_\Omega$. In Section 5.1 the fundamental solutions $e_0(t, x - y)$ and $h_0(t, x - y)$ of $\Box$ and $\Box^2$, respectively, are studied. We describe the singularities of e_0 and h_0 on the diagonal of $\partial\Omega \times \partial\Omega$. In Section 5.2 we introduce the distribution $\sigma(t)$ and show that it coincides with the trace of the fundamental solution $E(t, x, y)$ of the Dirichlet problem for $\Box$.

In Section 5.3 we reduce the proof of the Poisson relation to the analysis of the trace of a distribution $B(t, x, y)$, defined by (5.13), on the manifold without boundary $\partial\Omega$. For convex domains we examine the singularities of $B(t, x, y)$ and those of the trace $B(t, x, x)$, $x \in \partial\Omega$.

The analysis of the singularities of $B(t, x, y)$ for non-convex Ω leads to some difficulties. For this reason for general domains we study in Section 5.4 separately the singularities of $E(t, x, y)$ for $x, y \in \Omega^\circ$ and those for x, y close to $\partial\Omega$ provided $t \notin \{\pm T_\gamma : \gamma \in \mathcal{L}_\Omega\}$. For this purpose we apply a localization which will be exploited also in the next chapter. The advantage of the proof in Section 5.4 is that it works without any change for Neumann and Robin boundary conditions, according to the results on propagation of singularities in [MS1, MS2].

5.1. Traces of the fundamental solutions of □ and □²

Let $e_0(t, x - y)$ be the solution of the problem

$$\begin{cases} (\partial_t^2 - \Delta_x)e_0(t, x - y) = \delta(t)\delta(x - y), \\ \operatorname{supp} e_0(t, x - y) \subset \{(t, x, y) \in \mathbf{R}_t \times \mathbf{R}^{2n} : t \geq 0\}. \end{cases}$$

The second condition implies that the Fourier transform $\hat{e}_0(\tau, x - y)$ of $e_0(t, x - y)$ with respect to t admits an analytic continuation in

$$\{\tau \in \mathbf{C} : \operatorname{Im} \tau < 0\}.$$

This implies

$$\hat{e}_0(\tau, x - y) = (2\pi)^{-n} \int e^{i\langle x - y, \xi \rangle} [\xi^2 - (\tau - i0)^2]^{-1} \, d\xi,$$

and

$$e_0(t, x - y) = (2\pi)^{-n-1} \lim_{e \to +0} \iint e^{it\tau|\xi|} [1 - (\tau - i\epsilon)^2]^{-1} \cdot e^{i\langle x - y, \xi \rangle} |\xi|^{-1} \, d\tau d\xi.$$

For $\epsilon > 0$ the theorem of residues yields

$$\int_{-\infty}^{\infty} e^{it\tau|\xi|} [1 - (\tau - i\epsilon)^2]^{-1} \, d\tau = -i\pi \left(e^{it|\xi|(1+i\epsilon)} - e^{it|\xi|(i\epsilon - 1)} \right).$$

Letting $\epsilon \to +0$, for $t \geq 0$ we obtain the equality

$$e_0(t, x - y) = \frac{(2\pi)^{-n}}{2i} \int \left(e^{i(t|\xi| + \langle x - y, \xi \rangle)} - e^{i(-t|\xi| + \langle x - y, \xi \rangle)} \right) \frac{d\xi}{|\xi|}.$$

The latter integral can be considered for $|\xi| \geq 1$ as a sum of oscillatory integrals with phase functions

$$\Phi_{\pm}(t, x, y, \xi) = \pm t |\xi| + \langle x - y, \xi \rangle$$

(see Proposition 1.3.2). Moreover, by using Theorem 1.3.3, it is easy to find the wave front set

$$WF(e_0(t, x - y)) \subset \Big\{ (t, x, y, \tau, \xi, \eta) \in T^*(\mathbf{R}_t \times \mathbf{R}_x^n \times \mathbf{R}_y^n) \setminus \{0\} :$$

$$t > 0, \quad x = y \mp t\frac{\xi}{|\xi|}, \quad \tau = \pm |\xi|, \quad \xi + \eta = 0 \Big\}$$

$$\cup \{(t,x,y,\tau,\xi,\eta) \in T^*(\mathbf{R}_t \times \mathbf{R}_x^n \times \mathbf{R}_y^n)\backslash\{0\} :$$
$$t = 0, \quad x = y, \quad \xi + \eta = 0\}. \tag{5.1}$$

Now let $\Omega \subset \mathbf{R}$ be a bounded **closed** domain with smooth boundary $\partial\Omega$. We wish to define the *trace*

$$f_0(t,x-y) = e_0(t,x-y)_{|\mathbf{R}_t \times \partial\Omega \times \partial\Omega},$$

and to find $WF(f_0(t,x-y))$. To this end consider the inclusion map

$$j : \mathbf{R}_t \times \partial\Omega \times \partial\Omega \to \mathbf{R}_t \times \mathbf{R}_x^n \times \mathbf{R}_y^n,$$

$j(t,x,y) = (t,x,y)$. The *normal set* N_j of j has the form

$$N_j = \{(t,x,y,\tau,\xi,\eta) \in T^*(\mathbf{R}_t \times \mathbf{R}_x^n \times \mathbf{R}_y^n) : x \in \partial\Omega, \ y \in \partial\Omega,$$
$$\tau = 0, \ \xi_{|T_x(\partial\Omega)} = \eta_{|T_y(\partial\Omega)} = 0\}.$$

Obviously,

$$WF\left(e_0(t,x-y)_{|\mathbf{R}_t^+ \times \mathbf{R}_x^n \times \mathbf{R}_y^n}\right) \cap N_j = \emptyset,$$

where $\mathbf{R}_t^+ = \{t \in \mathbf{R} : t > 0\}$. According to Theorem 1.3.6, we can define $f_0(t,x-y)$ for $t > 0$ by setting

$$f_0(t,x-y) = j^* e_0(t,x-y),$$

where j^* is the *pull-back* of j. The same procedure works for $t = 0$, $x \neq y$.

The definition of $f_0(t,x-y)$ for $t = 0$, $x = y$, $x \in \partial\Omega$, leads to some difficulties, since the set N_j contains some points lying over the set

$$\{(0,x,x) \in \mathbf{R}_t \times \partial\Omega \times \partial\Omega\}.$$

To cover this case we shall make a more precise analysis of $\hat{f}_0$ for (t,x,y) close to $(0,x^0,x^0)$, $x^0 \in \partial\Omega$. For $x \in \Omega^\circ$, $y \in \partial\Omega$, we have

$$e_0(\tau,x-y) = (2\pi)^{-n} \int e^{i\langle x-y,\xi\rangle}(\xi^2 - \tau^2)^{-1}\,\mathrm{d}\xi, \quad \mathrm{Im}\,\tau < 0.$$

To find the limit of the left-hand side as $x \to x^0$, $\mathrm{Im}\,\tau \to -0$, we introduce near $x^0 \in \partial\Omega$ local coordinates (x',x_n), $x' = (x_1,\ldots,x_{n-1})$ so that in a neighbourhood of x^0, Ω is given by $x_n \geq g(x')$, while $\partial\Omega$ has the form $x_n = g(x')$, g being a smooth function. Let $H(x',y') \in C^\infty(\mathbf{R}^{n-1},\mathbf{R}^{n-1})$ be a smooth vector-valued function such that

$$g(x') - g(y') = \langle x' - y', H(x',y')\rangle$$

with $H(x',x') = dg(x')$. Here and below $\langle\cdot,\cdot\rangle$ denotes the *standard inner product* in $\mathbf{R}^{n-1}$. The standard metric on $T(\partial\Omega)$, inherited from the standard

Euclidean metric in $\mathbf{R}^n$, has the form

$$m^2(x', y', \nu') = \langle \nu', \nu' \rangle - \frac{\langle \nu', H(x', y') \rangle^2}{1 + |H|^2}.$$

In the local coordinates (x', x_n), for $x \in \Omega^\circ$, $y \in \partial\Omega$, we get

$$e_0(\tau, x, y') = (2\pi)^{-n} \int e^{i \langle x' - y', \xi' \rangle} e^{i\xi_n (x_n - g(y'))}$$

$$\times \left(1 + |dg(y')|^2\right)^{1/2} (\xi^2 - \tau^2)^{-1} \, d\xi. \tag{5.2}$$

Here the factor $(1 + |dg(y')|^2)^{1/2}$ appears, since in the coordinates (y', z_n) with $z_n = y_n - g(y')$ we have

$$\delta_{\partial\Omega} = (1 + |dg(y')|^2)^{1/2} \otimes \delta(z_n).$$

Next, introduce the coordinates

$$\begin{cases} \zeta' = \xi_n H(x', y'), \\ \zeta_n = \xi_n. \end{cases}$$

Then (5.2) becomes

$$e_0(\tau, x, y') = (2\pi)^{-n} \int e^{i \langle x' - y', \xi' \rangle + i\rho\xi_n} (1 + |dg(x')|^2)^{1/2} (\xi^2 - \tau^2)^{-1} \, d\xi d\xi_n,$$

with $\rho = x_n - g(x')$,

$$\xi^2 - \tau^2 = (\zeta')^2 - 2\zeta_n \langle \zeta', H \rangle + (1 + |H|^2)\zeta_n^2 - \tau^2.$$

The roots $z_\pm$ of the equation $\xi^2 - \tau^2 = 0$ with respect to ζ_n have the form

$$z_\pm = \frac{\langle \zeta', H \rangle}{1 + |H|^2} \pm i \left(\frac{m^2 - \tau^2}{1 + |H|^2} \right)^{1/2}.$$

Here we choose the square root so that $\mathrm{Re}(m^2 - \tau^2)^{1/2} > 0$. Hence $\mathrm{Im}\, z_+ > 0$, and by the theorem of residues we have

$$\int_{-\infty}^{\infty} e^{i\rho\zeta_n} \left((1 + |H|^2)(\zeta_n - z_+)(\zeta_n - z_-) \right)^{-1} \, d\zeta_n$$

$$= \pi e^{i\rho\zeta_n} \left((1 + |H|^2)(m^2 - \tau^2) \right)^{-1/2}.$$

Letting $\rho \to +0$, for $x' \in \partial\Omega$, $y' \in \partial\Omega$, we obtain

$$e_0(\tau, x', y') = \frac{(2\pi)^{-n+1}}{2} \int \frac{e^{i\langle x'-y', \zeta' \rangle}(1+|dg(y')|^2)^{1/2}}{[1+|H(x', y')|^2](m^2(x', y'; \zeta') - \tau^2]^{1/2}} \, d\zeta'.$$

The form $m^2(x', y', \zeta')$ is positively definite, so there exists a positively definite symmetric matrix $Q(x', y')$ such that

$$m^2(x', y', \zeta') = |Q(x', y')^{-1}\zeta'|^2, \quad Q(x', x') = \mathrm{Id}.$$

Changing the variables $\zeta' \mapsto Q(x', y')^{-1}\zeta'$, we deduce

$$e_0(\tau, x', y') = F(x', y') \int e^{i\langle Q(x', y')(x'-y'), \xi' \rangle}(\xi'^2 - (\tau - i0)^2)^{-1/2} \, d\xi'$$

with $F(x', y') \in C^\infty$.

Considering the inverse Fourier transform in τ, we have to examine the integrals

$$I_1 + I_2 = \iint_{|\tau| \le |\xi'|+1} e^{it\tau + i\langle Q(x', y')(x'-y'), \xi' \rangle}[\xi'^2 - (\tau - i0)^2]^{-1/2} \, d\tau d\xi'$$

$$+ \iint_{|\tau| > |\xi'|+1} e^{it\tau + i\langle Q(x', y')(x'-y'), \xi' \rangle}[\xi'^2 - (\tau - i0)^2]^{-1/2} \, d\tau d\xi'.$$

Setting $\tau = \sigma |\xi'|$, we can consider I_1, modulo smooth terms, as an oscillatory integral with a phase function

$$\Psi_1 = t\sigma |\xi'| + \langle Q(x', y')(x'-y'), \xi' \rangle,$$

since $\Psi_{1, x'} \ne 0$ for $x' = y'$, $|\xi'| \ge 1$. On the other hand, we can treat I_2 as an oscillatory integral with a phase function

$$\Psi_2 = t\tau + \langle Q(x', y')(x'-y'), \xi' \rangle,$$

since

$$|\Psi_{2,t}| + |\Psi_{2,x'}| \ne 0 \quad \text{for } x' = y'.$$

Consequently, we may define $e_0(t, x-y)_{|\mathbb{R} \times \partial\Omega \times \partial\Omega}$ for $t = 0$, and the analysis of the wave front sets of the integrals I_1 and I_2 yields

$$WF\left(e_0(t, x-y)_{|\mathbb{R} \times \partial\Omega \times \partial\Omega}\right) \cap \{t = 0\} \subset \{(t, x', y', \tau, \xi', \eta')$$

$$\in T^*(\mathbb{R} \times \partial\Omega \times \partial\Omega) \backslash \{0\} : t = 0, \quad x' = y', \quad \xi' + \eta' = 0\}. \quad (5.3)$$

In a similar way one studies the distribution $h_0(t, x - y)$ determined as the solution of the problem

$$\begin{cases} (\partial_t^2 - \Delta_x)^2 h_0(t, x - y) = \delta(t)\delta(x - y), \\ \operatorname{supp} h_0(t, x - y) \subset \{(t, x, y) \in \mathbf{R}_t \times \mathbf{R}_x^n \times \mathbf{R}_y^n : t \geq 0\}. \end{cases}$$

Write the Fourier transform $\hat{h}_0(\tau, x - y)$ in the form

$$\hat{h}_0(\tau, x - y) = (2\pi)^{-n} \int e^{i\langle x - y, \xi \rangle} [\xi^2 - (\tau - i0)^2]^{-2} \, d\xi,$$

then for $t \geq 0$ we obtain the oscillatory integral

$$h_0(t, x - y) = \frac{i(2\pi)^{-n}}{2} \int \left[e^{i\Phi^+}(it|\xi| - 1) + e^{i\Phi^-}(it|\xi| + 1) \right] |\xi|^{-3} \, d\xi.$$

This implies a relation, completely analogous to (5.1). Thus for $t > 0$ and for $t = 0$, $x \neq y$ we can define

$$f_1(t, x - y) = j^* h_0(t, x - y).$$

For the trace on $t = 0$, $x = y$, we repeat the above procedure. Since

$$\lim_{\rho \to +0} \int_{-\infty}^{\infty} e^{i\rho\xi_n} (\xi^2 - \tau^2)^{-2} \, d\xi_n = \frac{\pi}{2}(1 + |H|^2)^{-1/2}(m^2 - \tau^2)^{-3/2}$$

for $\operatorname{Im} \tau < 0$, for $x' \in \partial\Omega$, $y' \in \partial\Omega$ we deduce the expression

$$\hat{f}_1(\tau, x', y') = F_1(x', y') \int e^{i\langle Q(x', y')(x' - y'), \xi' \rangle} (\xi'^2 - (\tau - i0)^2)^{-3/2} \, d\xi'$$

with $F_1(x', y') \in C^\infty$. It remains to study two oscillatory integrals, similar to I_1 and I_2. Since the present case is completely analogous to the previous one, we leave the details to the reader.

Finally, combining the action of the pull-back j^* with (5.3), we obtain the following.

Theorem 5.1.1:   *The wave front sets of the distributions $f_k(t, x - y) \in \mathcal{D}'(\mathbf{R}_t \times \partial\Omega \times \partial\Omega)$, $k = 0, 1$, are contained in the set of points*

$$(t, x', y', \tau, \tilde{\xi}, \tilde{\eta}) \in T^*(\mathbf{R}_t \times \partial\Omega \times \partial\Omega) \setminus \{0\}$$

satisfying one of the following conditions:

(i) $t > 0$ and there exists $\xi \in \mathbf{R}^n \setminus \{0\}$ such that $x' = y' \mp t\xi/|\xi|$, $\tau = \pm|\xi|$, $\tilde{\xi} = p_x(\xi)$, $\tilde{\xi} + \tilde{\eta} = 0$;

(ii) $t = 0$, $x' = y'$, $\tilde{\xi} + \tilde{\eta} = 0$. ♠

Here $p_x : T_x^*(\mathbf{R}^n) \to T_x^*(\partial\Omega)$ is the *canonical projection*.

5.2. The distribution $\sigma(\mathbf{t})$

Let $\Omega \subset \mathbf{R}^n$, $n \geq 2$, be a bounded closed domain with smooth boundary $\partial\Omega$. Let $H_0^1(\Omega)$ be the closure of the space $C_0^\infty(\Omega)$ with respect to the norm

$$\|u\|_1^2 = \sum_{|\alpha| \leq 1} \|\partial^\alpha u\|^2,$$

$\|\cdot\|$ being the norm in $L^2(\Omega)$.

Introduce the operator A by $Au = -\Delta u$ for $u \in C_0^\infty(\Omega)$, and extend it in the domain

$$D_A = \{u \in H_0^1(\Omega) : \Delta u \in L^2(\Omega)\},$$

where Δu is interpreted in the sense of distributions. Denoting by $(.,.)$ the inner product in $L^2(\Omega)$, we have

$$(Au, v) = (u, Av) \text{ for } u, v \in D_A.$$

Hence A is a symmetric and closed operator in $L^2(\Omega)$. Moreover,

$$((A+1)u, u) = \|u\|_1^2, \quad u \in D_A,$$

and the form

$$B(u, v) = ((A+1)u, v)$$

is continuous and coercive in $H_0^1(\Omega)$. Consequently, by the Lax–Milgram theorem, for each $f \in L^2(\Omega)$ there exists an unique solution $u \in H_0^1(\Omega)$ of the problem

$$-\Delta u + u = f \text{ in } \mathcal{D}'(\Omega).$$

In this way we see that $A+1$ is a bijection from D_A into $L^2(\Omega)$ and

$$\|(A+1)^{-1}f\|_1 \leq C \|f\|, \quad \forall f \in L^2(\Omega). \tag{5.4}$$

This implies that -1 is not in the spectrum of A, and, by the criteria for self-adjointness, A is a self-adjoint operator in $L^2(\Omega)$, related to the *Laplacian* $-\Delta$ *with Dirichlet boundary conditions on* $\partial\Omega$.

The estimate (5.4) shows that the resolvent $(A+1)^{-1}$ is compact in $L^2(\Omega)$, hence the spectrum of A is formed by an infinite number of eigenvalues

$$0 < \lambda_1^2 \leq \lambda_2^2 \leq \ldots \lambda_m^2 \leq \ldots$$

with finite multiplicities. Let $\{\varphi_j(x)\}_{j=1}^\infty$ be an orthonormal set of eigenfunctions of A so that

$$\begin{cases} -\Delta\varphi_j(x) = \lambda_j^2 \varphi_j(x), & x \in \Omega, \\ \varphi_j(x) = 0, & x \in \partial\Omega. \end{cases}$$

Then $\varphi_j \in C^\infty(\Omega)$, and we can introduce the *spectral function*

$$e(x, y, \lambda) = \sum_{\lambda_j^2 \leq \lambda^2} \varphi_j(x)\varphi_j(y),$$

which is the kernel of the *spectral projection* E_λ of A. Moreover, we have the estimate

$$\sup\{|e(x, y, \lambda)| : x, y \in \Omega\} \leq C\lambda^n, \quad \lambda \geq 1 \tag{5.5}$$

with a constant $C > 0$, independent of λ. We refer to Hörmander [H3, Section. 17.5] for the proof of (5.5).

Let

$$N(\lambda) = \sharp\{j : \lambda_j^2 \leq \lambda^2\}$$

be the *counting function*. Clearly,

$$N(\lambda) = \int_\Omega e(x, x, \lambda^2)\, dx,$$

and (5.5) yields

$$N(\lambda) \leq C_1\lambda^n, \quad \lambda \to +\infty.$$

Introduce the tempered distribution

$$\sigma(t) = \sum_{j=1}^\infty \cos\lambda_j t = \operatorname{Re}\sum_{j=1}^\infty \exp(\lambda_j t) \in S'(\mathbf{R}).$$

Since A is a non-negative operator, by the spectral calculus we may define the operator $\cos(A^{1/2}t)$. Let

$$\mathcal{E}(t, x, y) \in \mathcal{D}'(\mathbf{R} \times \Omega \times \Omega)$$

be the *kernel* of $\cos(tA^{1/2})$. Since

$$(\cos(tA^{1/2})f, f) = \int_0^\infty \cos(\lambda t)\, d\lambda \int_\Omega\int_\Omega e(x, y, \lambda^2)f(x)\overline{f(y)}\, dx dy,$$

a simple calculation implies

$$\mathcal{E}(t, x, y) = \sum_{j=1}^\infty (\cos\lambda_j t)\varphi_j(x)\varphi_j(y).$$

Hence

$$\sigma(t) = \int_\Omega \mathcal{E}(t, x, y) \, dx. \tag{5.6}$$

Thus, $\mathcal{E}(t, x, y)$ is a solution of the problem

$$\begin{cases} (\partial_t^2 - \Delta_x)\mathcal{E}(t, x, y) = 0, \quad t \in \mathbf{R}, \quad x \in \Omega^\circ, \quad y \in \Omega^\circ, \\ \mathcal{E}(t, x, y)_{|x \in \partial\Omega} = \mathcal{E}(t, x, y)_{|y \in \partial\Omega} = 0, \\ \mathcal{E}(0, x, y) = \delta(x - y), \quad \partial_t \mathcal{E}(0, x, y) = 0. \end{cases} \tag{5.7}$$

Following the results for the propagations of singularities in [MS2, H3] (see Theorem 1.3.12), we can describe the singularities of $\mathcal{E}(t, x, y)$. Since the trace defining $\sigma(t)$ is taken over a manifold with boundary, Theorems 1.3.8, 1.3.9, concerning the calculus with wave front sets, cannot be applied immediately. To describe the singularities of $\sigma(t)$, in the next sections we find another representation of $\sigma(t)$, involving integration over the manifold without boundary $\partial\Omega$.

5.3. Poisson relation for convex domains

As before, Ω will be a compact domain in $\mathbf{R}^n$, $n \geq 2$. Our first considerations concern the general case, i.e. we do not assume that Ω is convex.

Let

$$E(t, x, y) = \sum_{j=1}^\infty \lambda_j^{-1} \sin(\lambda_j t) \varphi_j(x) \varphi_j(y)$$

be the *kernel of the operator* $A^{-1/2} \sin(tA^{1/2})$. Clearly,

$$\mathcal{E}(t, x, y) = \partial_t E(t, x, y),$$

and $E(t, x, y)$ is the solution of the problem

$$\begin{cases} (\partial_t^2 - \Delta_x)E(t, x, y) = 0, \quad t \in \mathbf{R}, \quad x \in \Omega^\circ, \quad y \in \Omega^\circ, \\ E(t, x, y)_{|x \in \partial\Omega} = 0, \\ E(0, x, y) = 0, \quad \partial_t E(0, x, y) = \delta(x - y). \end{cases}$$

Next, we study the trace

$$\int_\Omega E(t, x, y) \, dx.$$

For $y \in \partial\Omega$ introduce the distribution $K(t, x, y)$ as the solution of the problem

$$\begin{cases} (\partial_t^2 - \Delta_x)K(t, x, y) = 0, \quad t \in \mathbf{R}, \quad x \in \Omega^\circ, \quad y \in \Omega^\circ, \\ K(t, x, y) = \delta(t) \otimes \delta(x - y) = 0, \quad x \in \partial\Omega, \\ \operatorname{supp} K(t, x, y) \subset \{(t, x, y) \in \mathbf{R}_t \times \Omega \times \partial\Omega : t \geq 0\}. \end{cases} \tag{5.8}$$

Notice that $K(t - s, x, y)$ is the kernel of the continuous operator

$$\mathcal{K} : C_0^\infty(\mathbf{R} \times \partial\Omega) \to \overline{\mathcal{D}}'(\mathbf{R} \times \Omega),$$

$\mathcal{K} : f(s, y) \mapsto (\mathcal{K}f)(x, y)$. Here $(\mathcal{K}f)(t, x)$ is the solution of the problem

$$\begin{cases} (\partial_t^2 - \Delta_x)\mathcal{K}f = 0 & \text{in } \mathbf{R} \times \Omega^\circ, \\ \mathcal{K}f - f = 0 & \text{on } \mathbf{R} \times \partial\Omega, \\ \mathcal{K}f_{|t \ll 0} = 0, \end{cases}$$

while $\overline{\mathcal{D}}'(\mathbf{R} \times \Omega)$ is the space of all distributions u on $\mathbf{R} \times \Omega^\circ$ for which there exists an open neighbourhood $\tilde{\Omega}$ of Ω such that u can be extended to a distribution in $\mathcal{D}'(\mathbf{R} \times \tilde{\Omega})$. We can extend $\mathcal{K}$ as a continuous operator on $\mathcal{E}'(\mathbf{R} \times \partial\Omega)$. To do this, given $f(s, y) \in \mathcal{E}'(\mathbf{R} \times \partial\Omega)$, write it in the form

$$f(s, y) = \sum_{j=1}^N L_j(s, y, D_s, D_y) f_j(s, y),$$

where $f_j(s, y) \in \mathcal{E}'(\mathbf{R} \times \partial\Omega) \cap H_s(\mathbf{R} \times \partial\Omega)$, $s \geq 1$, and $L_j(s, y, D_s, D_y)$ are differential operators, involving only derivatives along directions tangential to $\mathbf{R} \times \partial\Omega$. It follows by [H3, Theorem 24.1.1] that we can define $\mathcal{K}g$ for $g \in H_s(\mathbf{R} \times \partial\Omega)$, $s \geq 1$. Applying this we find $(\mathcal{K}f_j)(t, x, s, y)$ for each $j = 1, \ldots, N$. Now we have

$$(\mathcal{K}f)(t, x) = \sum_{j=1}^N L_j(s, y, D_s, D_y) \mathcal{K}f_j(t, x, s, y).$$

Let $e_0(t, x - y)$ be the fundamental solution of $\square$, defined in Section 5.1. Notice that it coincides with the solution of the Cauchy problem:

$$\begin{cases} (\partial_t^2 - \Delta_x)e_0(t, x - y) = 0, & t \in \mathbf{R}_t^+, \quad x, y \in \mathbf{R}^n, \\ e_0(0, x - y) = 0, \quad \partial_t e_0(0, x - y) = \delta(x - y), \end{cases}$$

extended as 0 for $t < 0$. Next, introduce the distribution

$$\tilde{E}(t, x, y) = \int_{-\infty}^\infty \int_{\partial\Omega} K(s, x, z)e_0(t - s, z - y) \, ds \, dz$$

$$= \int_{-\infty}^\infty \int_{\partial\Omega} K(t - s, x, z)e_0(s, z - y) \, ds \, dz, \qquad (5.9)$$

where the integral is interpreted as the action of the distribution $e_0(t - s, z - y)_{|(z,y) \in \partial\Omega \times \Omega}$ on $K(s, x, z)$, the latter distribution having a compact support with respect to s and z. To justify this action we use Theorem 1.3.9 and the fact that

$$WF(e_0(t-s, z-y)_{|(z,y)\in\partial\Omega\times\Omega}) \cap \{(t, s, y, z, 0, -\sigma, 0, -\zeta)$$
$$\in T^*(\mathbf{R}_t \times \mathbf{R}_s \times \Omega \times \partial\Omega)\backslash\{0\}\} = \emptyset. \tag{5.10}$$

To verify (5.10), mention that $(\partial_t + \partial_s)e_0(t-s, .) = 0$ leads to $\tau + \sigma = 0$ on $WF(e_0(t-s, .))$, while $\eta + \zeta = 0$ on $WF(e_0(t-s, .))$ follows from Theorem 5.1.1.

Since the hyperplane $t = 0$ is not characteristic for $\square$, the traces $\tilde{E}(0, x, y)$ and $(\partial_t\tilde{E})(0, x, y)$ exist. Moreover,

$$\tilde{E}(0, x, y) = \int_{-\infty}^{\infty}\int_{\partial\Omega} K(-s, x, z)e_0(s, z-y)\,ds dz = 0,$$

because $e_0(s, z-y) = 0$ for $s \le 0$. Similarly, we find $(\partial_t\tilde{E})(0, x, y) = 0$. In this way we see that $\tilde{E}(0, x, y)$ is a solution of the problem

$$\begin{cases} (\partial_t^2 - \Delta_x)\tilde{E}(t, x, y) = 0, & t \in \mathbf{R}, \quad x \in \Omega^\circ, \quad y \in \Omega^\circ, \\ \tilde{E}(t, x, y) - e_0(t, x-y) = 0, & (t, x) \in \mathbf{R} \times \partial\Omega, \\ \tilde{E}(0, x, y) = \partial_t\tilde{E}(0, x, y) = 0. \end{cases}$$

Consequently,

$$E(t, x, y) = e_0(t, x-y) - \tilde{E}(t, x, y).$$

In what follows we investigate the singularities of $\sigma(t)$ for $t > 0$. First, note that $e_0(t, 0) = 0$ for $t > 0$, since

$$\operatorname{supp} e_0(t, x-y) \subset \{(t, x, y) : |x-y| \le t\}.$$

Thus, for $t > 0$ we obtain

$$E(t, x, x) = -\tilde{E}(t, x, x) = -\int_{-\infty}^{\infty}\int_{\partial\Omega} K(t-s, x, z)e_0(s, z-x)\,ds dz$$

$$= -\int_{-\infty}^{\infty}\int_{\partial\Omega} K(t-s, x, z)(\partial_s^2 - \Delta_x)h_0(s, z-x)\,ds dz,$$

where $h_0(t, x-y)$ is the fundamental solution of $\square^2$, introduced in Section 5.1. Now, taking the trace of $\tilde{E}(t, x, x)$ and integrating by parts with respect to s and x, we find

$$\int_\Omega \tilde{E}(t, x, x)\,dx = \int_{-\infty}^{\infty} ds \int_{\partial\Omega} dz \int_{\partial\Omega} [\partial_{\nu_x}K(t-s, x, z)h_0(s, z-x)$$
$$- K(t-s, x, z)\partial_{\nu_x}h_0(s, z-x)]\,dx. \tag{5.11}$$

Here ∂_{ν_x} denotes the derivative with respect to the exterior normal ν_x at $x \in \partial\Omega$. Since

$$\text{sing supp } h_0(t, x - y)$$
$$\subset \{(t, x, y) \in \mathbf{R}_t \times \mathbf{R}_x^n \times \mathbf{R}_y^n : |x - y| = t\}, \tag{5.12}$$

the second term in (5.11) becomes

$$\int_{-\infty}^{\infty} ds \int_{\partial\Omega} dz \int_{\partial\Omega} (\partial_{\nu_x} h_0)(t, 0) \, dx \in C^\infty(\mathbf{R}_t^+).$$

The inclusion (5.12) follows from the results on propagation of singularities for the solution of the Cauchy problem

$$\begin{cases} (\partial_t^2 - \Delta_x)^2 h_0(t, x - y) = 0, & t \in \mathbf{R}_t^+, \quad x, y \in \mathbf{R}^n, \\ \partial_t^j h_0(0, x - y) = 0, & j = 0, 1, 2, \\ \partial_t^3 h_0(0, x - y) = \delta(x - y), \end{cases}$$

where $h_0(t, x - y)$ is extended as 0 for $t < 0$.

Since the boundary $\partial\Omega$ is not characteristic for $\square = \partial_t^2 - \Delta_x$, we can apply the partial hypoellipticity of $K(t, x, y)$ with respect to the normal directions ν_x and introduce the trace

$$k(t, x, y) = \partial_{\nu_x} K(t, x, z)_{|(x,z)\in\partial\Omega\times\partial\Omega}.$$

Since $WF(h_0(t - s, z - y)_{|(y,z)\in\partial\Omega\times\partial\Omega})$ satisfies a relation analogous to (5.10), we can define the distribution

$$B(t, x, y) = -\int_{-\infty}^{\infty} \int_{\partial\Omega} k(s, x, z) h_0(t - s, z - y) \, ds dz, \tag{5.13}$$

interpreted as the action of $h_0(t - s, z - y)$ on

$$k(s, x, z) \in \overline{\mathcal{D}}'(\mathbf{R}_s^+ \times \partial\Omega \times \partial\Omega).$$

We are going to study the singularities of $\sigma(t)$ for $t > 0$. To this end consider $B(t, x, y)$ for $t > 0$, $x \in \partial\Omega$, $y \in \partial\Omega$, $|x - y| < t$. A finite speed of propagation argument yields

$$\text{sing supp } k(t, x, z) \subset \{(t, x, z) \in \mathbf{R} \times \partial\Omega \times \partial\Omega : |x - z| \leq t\}.$$

It follows from (5.12) that the integrand in (5.13) is singular in t only if (t, s, x, y, z) satisfies

$$t > |x - y| \geq |z - y| - |x - z| \geq t - 2s.$$

Thus, to describe $WF\left(B(t,x,y)_{|t>0,|x-y|<t}\right)$, we need to know the singularities of $k(t,x,y)$ for $t>0$.

In what follows we use the relation C, introduced in Section 1.2. Recall that C is the set of all

$$(t,x,y,\tau,\xi,\eta) \in T^*(\mathbf{R}\times\Omega\times\Omega)\backslash\{0\},$$

such that $\tau^2 = |\xi|^2$, and (t,x,τ,ξ) and $(0,y,\tau,\eta)$ lie on a generalized bicharacteristic of $\Box$. According to Lemma 1.2.7, the relation C is closed. Denote by C_b the set of those

$$(t,x,y,\tau,\tilde{\xi},\tilde{\eta}) \in T^*(\mathbf{R}\times\partial\Omega\times\partial\Omega)\backslash\{0\}$$

such that there exists $\xi \in T_x^*(\Omega)$, $\eta \in T_y^*(\Omega)$ with $\tilde{\xi} = p_x(\xi)$, $\tilde{\eta} = p_y(\eta)$ and $(t,x,y,\tau,\xi,\eta) \in C$. Here p_x is the projection, introduced at the end of Section 5.1. Repeating the proof of Lemma 1.2.7, we conclude that C_b is closed.

Proposition 5.3.1:   *We have*

$$WF'\left(k(t,x,y)_{|t>0}\right) \subset C_b. \tag{5.14}$$

Proof: It is convenient to introduce the set $\tilde{C}_b$ of those

$$(t,s,x,y,\tau,\sigma,\tilde{\xi},\tilde{\eta}) \in T^*(\mathbf{R}_t\times\mathbf{R}_s\times\partial\Omega\times\partial\Omega)\backslash\{0\}$$

such that $\tau = \sigma$, and there exist $\xi \in T_x^*(\Omega)$, $\eta \in T_y^*(\Omega)$ with $p_x(\xi)=\tilde{\xi}$, $p_y(\eta)=\tilde{\eta}$, $\tau^2 = |\xi|^2 = |\eta|^2$ such that (t,x,τ,ξ) and (s,y,τ,η) lie on a generalized bicharacteristic of $\Box$. We shall prove the inclusion

$$WF'\left(k(t-s,x,y)_{|t>s}\right) \subset \tilde{C}_b. \tag{5.15}$$

Note that (5.14) follows from (5.15), taking the trace on $s=0$.

Consider an arbitrary

$$\rho_0 = (t_0,s_0,x_0,y_0,\tau_0,\sigma_0,\tilde{\xi}_0,\tilde{\eta}_0) \notin \tilde{C}_b$$

with $t_0 > s_0$. Since $\tau = \sigma \neq 0$ on $WF'(k(t-s,.,.))$, we may assume $\tau_0 = \sigma_0 \neq 0$. Since $\tilde{C}_b$ is closed, there exist open conic neighbourhoods Γ_1 of $(s_0,y_0,\tau_0,\tilde{\eta}_0)$ and Γ_2 of $(t_0,x_0,\tau_0,\tilde{\xi}_0)$ so that $(s,y,\sigma,\tilde{\eta}) \in \Gamma_1$ and $(t,x,\tau,\tilde{\xi}) \in \Gamma_2$ imply

$$(t,s,x,y,\tau,\sigma,\tilde{\xi},\tilde{\eta}) \notin \tilde{C}_b.$$

Let $\pi : T^*(\mathbf{R}\times\partial\Omega) \to \mathbf{R}$ be the natural projection. We may choose Γ_1 and Γ_2 in such a way that

$$\pi(\Gamma_1) = (\alpha,\beta), \quad \pi(\Gamma_2) = (\gamma,\delta), \quad \beta < \gamma.$$

Let $A \in L^\circ(\mathbf{R} \times \partial\Omega)$ be a pseudo-differential operator on $\mathbf{R} \times \partial\Omega$ with full symbol, equal to 1 in a small conic neighbourhood of $(s_0, y_0, \tau_0, \tilde{\eta}_0)$, and with wave front set $WF(A) \subset \Gamma_1$. We choose A in such a way that the kernel of A has a compact support, contained in

$$(\alpha, \beta) \times \partial\Omega \times (\alpha, \beta) \times \partial\Omega.$$

In a similar way we choose a pseudo-differential operator $B_0 \in L^\circ(\mathbf{R} \times \partial\Omega)$ with full symbol, equal to 1 in a small conic neighbourhood of $(t_0, x_0, \tau_0, \tilde{\xi}_0)$, such that $WF(B_0) \subset \Gamma_2$ and the kernel of B_0 has a compact support, contained in $\mathbf{R} \times \partial\Omega \times \mathbf{R} \times \partial\Omega$.

Fix $p \geq 2$. Then for each $f \in H_p^{\mathrm{loc}}(\mathbf{R} \times \partial\Omega)$ we have $Af \in H_p(\mathbf{R} \times \partial\Omega)$, $\operatorname{supp} Af \subset (\alpha, \beta) \times \partial\Omega$. By Theorem 24.1.1 in [H3], we can find a solution $(SAf)(t, x)$ of the problem

$$\begin{cases} (\partial_t^2 - \Delta_x)SAf = 0 & \text{in } \mathbf{R} \times \Omega^\circ, \\ SAf - Af = 0 & \text{on } \mathbf{R} \times \partial\Omega, \\ (SAf)_{|t<\alpha} = 0. \end{cases} \tag{5.16}$$

Thus, we obtain a linear map

$$H_p^{\mathrm{loc}}(\mathbf{R} \times \partial\Omega) \ni f \mapsto SAf \in \overline{H}_p^{\mathrm{loc}}(\mathbf{R} \times \Omega^\circ).$$

Define

$$Rf = \frac{\partial}{\partial\nu_x}(SAf)_{|x\in\partial\Omega}.$$

Note that Rf can be obtained by the action of the distribution $k(t-s, x, z)$ to $(Af)(s, z)$. Now for the solution SAf of (5.16) we apply the results on propagation of singularities in [MS2] (see Theorem 1.3.12). Consequently, for $t \in (\gamma, \delta)$, the singularities of $(SAf)(t, x)$ are described by the generalized bicharacteristics of $\square$ issued from points

$$(s, y, \sigma, \eta) \in T^*(\mathbf{R} \times \Omega)$$

such that

$$\sigma^2 = |\eta|^2, \quad (s, y, \sigma, p_y(\eta)) \in WF(Af) \subset \Gamma_1.$$

Our assumptions yield $WF(Rf) \cap \Gamma_2 = \emptyset$. Thus, we obtain a linear map

$$H_p^{\mathrm{loc}}(\mathbf{R} \times \partial\Omega) \ni f \mapsto B_0 Rf \in C^\infty(\mathbf{R} \times \partial\Omega).$$

The energy estimates for (5.16) show that $B_0 R$ is a closed map, defined on $H_p^{\mathrm{loc}}(\mathbf{R} \times \partial\Omega)$, and by the closed graph theorem we conclude that $B_0 R$ is continuous. Therefore, the kernel of $B_0 R$ is C^∞, and

$$B_0(t, x, D_t, D_x)A^*(s, y, D_s, D_y)k(t-s, x, y) \in C^\infty,$$

A^* being the operator formally adjoint to A. Choose a pseudo-differential operator

$$C(t, x, s, y, D_t, D_x, D_s, D_y) \in S^0,$$

which is elliptic at ρ_0 and belongs to $S^{-\infty}$ in a small open neighbourhood of ρ_0. Then $CB_0 A^*$ is an operator in S^0, which is elliptic in ρ_0. Thus, $CB_0 A^* k \in C^\infty$ implies $\rho_0 \notin WF'(k(t-s, x, y))$. This completes the proof of the proposition. ♠

From now on we assume that Ω is convex. We aim to find the composition

$$WF'(k(s, x, z)) \circ WF'\left(h_0(t-s, z-y)_{|(y,z) \in \partial\Omega \times \partial\Omega}\right).$$

By Theorem 5.1.1, if

$$(t, s, z, y, \tau, \sigma, \tilde{\xi}, \tilde{\eta}) \in WF'\left(h_0(t-s, z-y)_{|(y,z) \in \partial\Omega \times \partial\Omega}\right),$$

we have

$$z = y \pm (t-s)\frac{\eta}{|\eta|}, \quad \tilde{\xi} = \tilde{\eta}, \quad z \in \partial\Omega,$$

with $\eta \in T_y^*(\Omega)$, $|\eta|^2 = \tau^2$, $p_y(\eta) = \tilde{\eta}$. Moreover, the intersection

$$\left\{ z = y + \sigma\frac{\eta}{|\eta|} : \sigma \in \mathbf{R} \right\} \cap \Omega$$

is convex. We may consider the generalized bicharacteristic of $\Box$, issued from a point $(z, \zeta) \in T^*(\Omega)$ with $p_z(\zeta) = \tilde{\zeta} = \tilde{\eta}$, $|\zeta|^2 = \tau^2$, as a part of a generalized bicharacteristic issued from (y, η). Then, according to (5.13) and applying Theorem 1.3.9, we obtain

$$WF'\left(B(t, x, y)_{|t>0, |x-y|<t}\right) \subset C_b. \tag{5.17}$$

The advantage of the equality

$$\int_\Omega E(t, x, x)\, dx = \int_{\partial\Omega} B(t, x, x)\, dx \quad \mod C^\infty(\mathbf{R}_t^+)$$

is that $\partial\Omega$ is a manifold without boundary, and we can describe the singularities of the right-hand side integral. Consider the map

$$\kappa : \mathbf{R} \times \partial\Omega \ni (t, x) \mapsto (t, x, x) \in \mathbf{R} \times \partial\Omega \times \partial\Omega.$$

For the normal set N_κ of κ we have $C_b \cap N_\kappa = \emptyset$. Using the pull-back κ^* of κ, one finds by Theorem 1.3.6 that

$$WF(B(t, x, x)_{|t>0})$$

is contained in

$$\{(t,x,\tau,\tilde{\xi}-\tilde{\eta}) \in T^*(\mathbf{R} \times \partial\Omega)\backslash\{0\} : (t,x,x,\tau,\tilde{\xi},\tilde{\eta}) \in C_b\}.$$

Finally, an application of Theorem 1.3.8 for the integral over small open sets ω_j covering $\partial\Omega$ implies that

$$WF\left(\int_{\partial\Omega} B(t,x,x)\,dx_{|t>0}\right)$$

is contained in the set

$$\{(t,\tau) \in T^*(\mathbf{R}_t^+)\backslash\{0\} : \exists (x,\tilde{\xi}) \in T^*(\partial\Omega) \text{ with } (t,x,x,\tau,\tilde{\xi},\tilde{\xi}) \in C_b\}.$$

On the other hand, as we have mentioned in Section 1.2, $(T,x,x,\tau,\xi,\xi) \in C$ means that there exists a periodic generalized bicharacteristic of $\square$ with period T, passing through (x,ξ).

Thus we obtain the following.

Theorem 5.3.2: *For every compact convex domain Ω in $\mathbf{R}^n$, $n \geq 2$, we have*

$$\text{sing supp } \sigma(t) \subset \{0\} \cup \{\pm T_\gamma : \gamma \in \mathcal{L}_\Omega\}, \tag{5.18}$$

where $\mathcal{L}_\Omega$ is the set of all periodic generalized bicharacteristics of $\square$ in Ω. ♠

5.4 Poisson relation for arbitrary domains

Here we treat the same problem as in the previous section but for arbitrary domains Ω, i.e. we do not assume that Ω is convex. If we follow the proof in the convex case, we must establish the inclusion (5.17). Since the intersection of straight lines, issued from y, and Ω could be non-convex, we are going to consider the generalized bicharacteristics passing through all points $(s,z) \in \mathbf{R} \times \partial\Omega$ such that $(t-s,z,y) \in \text{sing supp } h_0(t-s,z-y)_{|y,z\in\partial\Omega}$. This leads to some singularities which are not described by the relation C_b.

Throughout we use the notation from the previous sections of the present chapter. The following assertion can be established repeating the proof of Proposition 5.3.1. We leave the details to the reader.

Proposition 5.4.1: *The intersection*

$$WF'(K(t,x,z)) \cap T^*(\mathbf{R}_t^+ \times \Omega^\circ \times \partial\Omega)$$

is contained in the set of all

$$(t, x, z, \tau, \xi, \tilde{\zeta}) \in T^*(\mathbf{R}_t^+ \times \Omega^\circ \times \partial\Omega)\backslash\{0\}$$

such that there exists $\zeta \in T_z^*(\Omega)$ *with* $p_z(\zeta) = \tilde{\zeta}$, $\tau^2 = |\xi|^2 = |\zeta|^2$ *and* (t, x, τ, ξ) *and* $(0, z, \tau, \zeta)$ *lie on a common generalized bicharacteristic of* $\Box$. ♠

In the following we shall examine the distribution $E(t, x, y)$ for $t > 0$ and $|x - y| < t$. First, we deal with the case $x, y \in \Omega^\circ$.

Proposition 5.4.2:   We have the relation

$$WF'(E(t, x, y)) \cap \{(t, x, y, \tau, \xi, \eta) \in T^*(\mathbf{R} \times \Omega^\circ \times \Omega^\circ) :$$
$$t > 0, \quad |x - y| < t\} \subset C. \tag{5.19}$$

Proof: Assume that $q_0 = (t_0, x_0, y_0, \tau_0, \xi_0, \eta_0) \notin C$ for some $t_0 > 0$, $x_0, y_0 \in \Omega^\circ$, $|x_0 - y_0| < t_0$. Note that for $t > 0$, $x, y \in \Omega^\circ$ we have

$$(\partial_t^2 - \Delta_x)E(t, x, y) = 0, \quad (\partial_t^2 - \Delta_y)E(t, x, y) = 0,$$

consequently $\tau^2 = |\xi|^2 = |\eta|^2$. In particular, $\tau_0^2 = |\xi_0|^2 = |\eta_0|^2$ if $q_0 \in WF'$ $(E(t, x, y))$. We shall assume that $\tau_0 < 0$; the case $\tau_0 > 0$ is treated similarly. Choose small conic neighbourhoods $\Gamma_1 = V_1 \times \Sigma_1$ of (y_0, η_0) and $\Gamma_2 = V_2 \times \Sigma_2$ of (x_0, ξ_0) and $\delta_0 > 0$ such that $(y, \eta) \in \Gamma_1$, $(x, \xi) \in \Gamma_2$, $|t - t_0| < \delta_0$ imply $(t, x, y, \tau, \xi, \eta) \notin C$ for $\tau^2 = |\xi|^2 = |\eta|^2$. We may take V_i so that $\bar{V}_i \subset \Omega^\circ$, $i = 1, 2$.

Let $A_1(y, D_y) \in L^\circ(\Omega^\circ)$ be a pseudo-differential operator, the full symbol $a_1(y, \eta)$ of which equals 1 in some conic neighbourhood of (y_0, η_0), and such that the kernel of A_1 has a compact support in $\Omega^\circ \times \Omega^\circ$ and $WF(A_1) \subset \Gamma_1$. Consider the straight line

$$L_0 = \{z \in \mathbf{R}^n : z = y_0 + \sigma\eta_0, \quad \sigma \in \mathbf{R}\},$$

and let

$$L_0 \cap \Omega = \cup_{j \in P} l_j,$$

where $l_j = [x_j, x_{j+1}]$, $x_j \in \partial\Omega$, $l_j \cap l_k = \emptyset$ for $j \neq k$ and $P \subset \mathbf{Z} \cap [-p, q]$, p, q being non-negative integers or ∞. We may assume that $y_0 \in l_0$. Notice that in general $l_0 \cap \partial\Omega$ could contain non-trivial linear segments lying on $\partial\Omega$.

Let $j \in P$ and $x_j = y_0 + \sigma_j\eta_0/|\eta_0|$. Assume that $-2, 2 \in P$. Then $0 \leq \sigma_1 < \sigma_2$, $\sigma_{-1} < \sigma_0 \leq 0$. Choose $\epsilon' > 0$ with $\epsilon' < (\sigma_2 - \sigma_1)/2$, $\epsilon' < (\sigma_0 - \sigma_{-1})/2$ and two functions $\kappa(t), \chi(t) \in C^\infty(\mathbf{R})$ such that

$$\kappa(t) = \begin{cases} 1 & \text{for } t \geq \sigma_0 - \epsilon', \\ 0 & \text{for } t \leq \sigma_{-1} + \epsilon', \end{cases}$$

$$\chi(t) = \begin{cases} 1 & \text{for } t \leq \sigma_1 + \epsilon', \\ 0 & \text{for } t \geq \sigma_2 - \epsilon'. \end{cases}$$

Now set

$$e_1(t,x,y) = A_1(y,D_y)e_0(t,x-y),$$
$$e_2(t,x,y) = e_0(t,x-y) - e_1(t,x,y),$$
$$\tilde{e}_1(t,x,y) = \kappa(t)\chi(t)A_1(y,D_y)e_0(t,x-y).$$

Notice that $(\partial_t^2 - \Delta_x)(e_1 - \tilde{e}_1)(t,x,y) \neq 0$ implies $\sigma_1 + \epsilon' \leq t \leq \sigma_2 - \epsilon'$ or $\sigma_{-1} + \epsilon' \leq t \leq \sigma_0 - \epsilon'$. For sufficiently small Γ_1 and ϵ' we obtain

$$(\partial_t^2 - \Delta_x)(\tilde{e}_1 + e_2) \in C^\infty(\mathbf{R} \times \Omega \times \Omega),$$
$$(\tilde{e}_1 + e_2)(0,x,y) = 0,$$
$$\partial_t(\tilde{e}_1 + e_2)(0,x,y) - \delta(x-y) \in C^\infty(\Omega \times \Omega).$$

This implies

$$E(t,x,y) - (\tilde{e}_1 + e_2)(t,x,y) + \int_{-\infty}^\infty \int_{\partial\Omega} K(s,x,z)(\tilde{e}_1 + e_2)$$
$$\times (t-s,z,y)\,ds\,dz \in C^\infty(\mathbf{R} \times \Omega \times \Omega). \qquad (5.20)$$

For $|x-y| < t$ the sum $(\tilde{e}_1 + e_2)(t,x,y)$ is smooth, so we have to examine the integral

$$\tilde{E}_1(t,x,y) = \int_{-\infty}^\infty \int_{\partial\Omega} K(s,x,z)\tilde{e}_1(t-s,z,y)\,ds\,dz. \qquad (5.21)$$

Note that by Theorem 1.3.9 we have

$$WF'(\tilde{E}_1(t,x,y)) \cap \{(t,x,y,\tau,\xi,\eta) \in T^*(\mathbf{R}^+ \times V_2 \times V_1) : |x-y| < t\}$$
$$\subset WF'\left(K(s,x,z)|_{x\in V_2, z\in\partial\Omega}\right) \circ WF'\left(\tilde{e}_1(t-s,z,y)|_{y\in V_1, z\in\partial\Omega}\right) (5.22)$$

where the composition is taken with respect to s and z. Now it is easy to see that

$$q_0 \notin WF'(\tilde{E}_1(t,x,y)).$$

Indeed, assume that for some $(\hat{s}, \hat{z}, \hat{\sigma}, \hat{\zeta}) \in T^*(\mathbf{R} \times \partial\Omega)$ we have

$$(t_0, \hat{s}, \hat{z}, y_0, \tau_0, \hat{\sigma}, \hat{\zeta}, \eta_0) \in WF'(\tilde{e}_1(t-s,z,y)),$$
$$(\hat{s}, x_0, \hat{z}, \hat{\sigma}, \xi_0, \hat{\zeta}) \in WF'(K(s,x,z)).$$

Since $t_0 - \hat{s} > 0$ implies

$$\hat{\sigma} = \tau_0, \quad \hat{z} = y_0 + (t_0 - \hat{s})\frac{\eta_0}{|\eta_0|}, \quad \hat{\zeta} = p_{\hat{z}}(\eta_0),$$

the construction of $\tilde{e}_1$ yields $\hat{z} \in l_0$. Thus the generalized bicharacteristic issued from $(t_0 - \hat{s}, \hat{z}, \tau_0, \zeta)$ with $p_{\hat{z}}(\zeta) = \hat{\zeta}$, $|\zeta|^2 = \tau_0^2$ is a part of a generalized bicharacteristic issued from $(0, y_0, \tau_0, \eta_0)$ and passing through $(t_0, x_0, \tau_0, \xi_0)$. This is a contradiction with the choice of q_0. For $t_0 = \hat{s}$ we use a similar argument.

The analysis of the integral involving e_2 is trivial, and we conclude that

$$q_0 \notin WF'(E(t, x, y)).$$

This completes the proof of the proposition. ♠

Now we pass to the analysis of the singularities of $E(t, x, y)$ for x and y close to some point $z_0 \in \partial\Omega$. In some open neighbourhood U of z_0 we choose local coordinates $x = (x_1, \ldots, x_{n-1}, x_n) = (x', x_n)$ such that $\partial\Omega \cap U$ is given by $x_n = 0$, and $\Omega \cap U$ lies in the halfspace $x_n \geq 0$. For $t > 0$ we have $(\partial_t^2 - \Delta_x)E(t, x, y) = 0$ and the boundary $\partial\Omega$ is not characteristic for $\partial_t^2 - \Delta_x$. Therefore the partial hypoellipticity of $E(t, x, y)$ implies that $E(t, x, y)_{|t>0}$ is a C^∞ smooth function of x_n with values in the space of distributions $\mathcal{D}'(\mathbf{R}_t^+ \times \mathbf{R}_{x'}^{n-1} \times \Omega)$. Since the same argument works for the variables $y = (y_1, \ldots, y_{n-1}, y_n) = (y', y_n)$, for sufficiently small $\epsilon > 0$ we obtain a C^∞ function

$$H : [0, \epsilon] \times [0, \epsilon] \to \mathcal{D}'(\mathbf{R}^+ \times U' \times U'), \quad H(x_n, y_n) = E(t, x', x_n, y', y_n),$$

where $U' \subset \mathbf{R}^{n-1}$ is a small neighbourhood of 0. For this purpose we exploit the smoothness of E with respect to y_n and next we use Theorem B.2.9 of [H3], considering y_n as a parameter. Since the proof of this theorem can be trivially extended to cover a smooth dependence on a parameter, we deduce the above assertion.

Let L_Ω be *the set of periods of all periodic generalized bicharacteristics of $\square$ in Ω.* According to Lemma 1.2.10, L_Ω is closed in $\mathbf{R}$.

Proposition 5.4.3: *Let $t_0 > 0$, $t_0 \notin L_\Omega$. Then for all sufficiently small $\epsilon > 0$ we have*

$$(t_0, x', x', \tau, \xi', \xi') \notin WF'(H(x_n, y_n)),$$

whenever $(x_n, y_n) \in [0, \epsilon] \times [0, \epsilon]$, $\tau \in \mathbf{R}$ and $(x', \xi') \in U \times \mathbf{R}^{n-1}$.

The proof of this proposition is rather long, and we begin with some preparations. The main difficulty is to provide uniformity with respect to x_n and y_n. We follow the idea of the proof of Proposition 5.4.2, studying the singularities of a distribution similar to $\tilde{E}_1(t, x, y)$.

Let $q(x, \tau, \xi) = q_2(x, \xi) - \tau^2$ be the principal symbol of $\partial_t^2 - \Delta_x$ in the coordinates (x', x_n). Here $q_2(x, \xi)$ is a positively definite quadratic form in ξ, depending smoothly on x. If

$$(t, x', y', \tau, \xi', \eta') \in WF'(H(x_n, y_n)), \quad t > 0,$$

then the point (t, x', τ, ξ') belongs to the compressed characteristic set Σ_b of the wave operator, introduced in Section 1.2, with respect to the surface $x_n = $ const.

This implies $q_2(x', x_n, \xi', 0) \leq \tau^2$, and $\tau = 0$ yields $\xi' = 0$. Thus, we may assume $\tau \neq 0$.

Consider a point

$$\rho_0 = (t_0, x_0', x_0', \tau_0, \xi_0', \xi_0')$$

with $\tau_0 < 0$, $q_2(x_0', 0, \xi_0', 0) \leq \tau^2$, $(x_0', \xi_0') \in U' \times \mathbf{R}^{n-1}$. The case $\tau_0 > 0$ can be treated by a similar argument.

Below we take $0 < \epsilon < t_0/4$ and obtain

$$(t_0, x_0', x_n, x_0', y_n) \notin \text{sing supp } e_0(t, x - y)$$

for $x_n, y_n \in [0, \epsilon]$. For technical reasons it is more convenient to study

$$\Xi(t - t', x, y) = E(t - t', x, y)Y(t - t'),$$

where $Y(t)$ is the *Heaviside function*, and to take the trace $t' = 0$. Let $A_1(t', y, D_{t'}, D_{y'}) \in L^0(\mathbf{R}_{t'} \times \mathbf{R}_{y'}^{n-1})$ be a pseudo-differential operator depending smoothly on $y_n \in [0, \epsilon]$ such that the full symbol $a_1(t', y, \tau', \eta')$ of A_1 is equal to 1 in some conic neighbourhood $\tilde{\Gamma}_1$ of $(0, x_0', \tau_0, \xi_0')$ for $y_n \in [0, \epsilon]$. Let $\Gamma_1 \supset \tilde{\Gamma}_1$ be another conic neighbourhood of the same point and let

$$WF(A_1(., y_n, .)) \subset \Gamma_1.$$

Moreover, we assume that the kernel of $A_1(., y_n, .)$ has a compact support in $\mathcal{T} \times U' \times \mathcal{T} \times U'$, $\mathcal{T}$ being a small neighbourhood of 0 in $\mathbf{R}$.

Consider the equation

$$q(x_0', 0, \tau_0, \xi_0', \xi_n) = 0 \tag{5.23}$$

with respect to ξ_n. First, we treat the case when this equation has a double real root ξ_n^0. Set $y_0 = (x_0', 0)$, $\eta_0 = (\xi_0', \xi_n^0)$, and consider the line L_0 and the linear segments l_j introduced in the proof of Proposition 5.4.2. Next, define

$$e_1(t - t', x, y) = A_1(t', y, D_{t'}, D_{y'})e_0(t - t', x, y),$$
$$e_2(t - t', x, y) = e_0(t - t', x, y) - e_1(t - t', x, y),$$
$$\tilde{e}_1(t - t', x, y) = \kappa(t - t')\chi(t - t')e_1(t - t', x, y),$$

where $\kappa(t)$ and $\chi(t)$ are the functions used in the proof of Proposition 5.4.2. Thus, for $\Gamma_1, \mathcal{T}$ and ϵ sufficiently small we are going to study

$$\tilde{\Xi}_1(t - t', x, y) = \int_{-\infty}^{\infty} \int_{\partial\Omega} K(s, x, z)\tilde{e}_1(t - t' - s, z, y) \, ds \, dz.$$

Further, we concentrate our attention to the distribution $\tilde{\Xi}_1(t - t', x, y)$. We need an inclusion similar to that in Proposition 5.4.1 for x close to $\partial\Omega$. To this end, using the fact that $\partial\Omega$ is not characteristic for $\square$, for $t > 0$ and small $\epsilon > 0$ we introduce the C^∞ function

$$\mathcal{K} : [0, \epsilon] \to \mathcal{D}'(\mathbf{R}_t^+ \times U' \times \partial\Omega), \quad \mathcal{K}(x_n) = K(t, x', x_n, z).$$

Proposition 5.4.4: *For sufficiently small $\epsilon > 0$ and $x_n \in [0, \epsilon]$, $WF'(\mathcal{K}(x_n))$ is contained in the set of all $(t, x', z, \tau, \xi', \tilde\zeta) \in T^*(\mathbf{R}_t^+ \times U' \times \partial\Omega) \backslash \{0\}$ such that there exist $\zeta \in T_z^*(\Omega)$ and $\xi_n \in \mathbf{R}$ with $p_z(\zeta) = \tilde\zeta$, $\tau^2 = |\zeta|^2$, $q(x', x_n, \tau, \xi', \xi_n) = 0$, and $(0, z, \tau, \zeta)$ and $(t, x', x_n, \tau, \xi', \xi_n)$ lie on a common generalized bicharacteristic of $\square$.*

Proof of Proposition 5.4.4: For $\hat{x}_n > 0$ the result follows from Proposition 5.4.1 taking the trace on $x_n = \hat{x}_n$. For $\hat{x}_n = 0$ the proof is a modification of that of Proposition 5.3.1. By using the notation in this proof, it suffices to apply the results in [MS2] or Section 24 of [H3] concerning the propagation of the generalized wave front set $WF_b(SAf)$ introduced in Section 1.3. Then for some pseudo-differential operator $B_0(t, x', x_n, D_t, D_{x'}) \in L^0(\mathbf{R}_t \times \mathbf{R}_{x'}^{n-1})$, depending smoothly on x_n, we obtain a linear map

$$H_p^{loc}(\mathbf{R} \times \partial\Omega) \ni f \mapsto B_0 SAf \in C^\infty(\mathbf{R}_t \times \mathbf{R}^n),$$

and we repeat the argument of the proof of Proposition 5.3.1. ♠

Consider the inclusion

$$WF'\left(\tilde{\Xi}_1(t - t', x', x_n, y', y_n)\right) \cap T^*(\mathbf{R}_t^+ \times \mathbf{R}_{t'} \times U' \times U')$$
$$\subset WF'\left(K(s, x', x_n, z)_{|x' \in U', z \in \partial\Omega}\right)$$
$$\circ WF'\left(\tilde{e}_1(t - t' - s, z, y', y_n)_{|y' \in U', z \in \partial\Omega}\right), \tag{5.24}$$

where $x_n, y_n \in [0, \epsilon]$ are considered as parameters. Observe that for Γ_1, $\mathcal{T}$ and ϵ sufficiently small the wave front set $WF'(\tilde{e}_1(t - t' - s, z, y', y_n))$ has a projection on $T^*(\mathbf{R}_t \times \mathbf{R}_x^n)$ which is related to the straight lines issued from y with direction η, (y, n) being close to (y_0, η_0). The precise choice of ϵ will be discussed later.

Introduce

$$l_0(\sigma) = \left\{\left(\sigma, y_0 + \sigma\frac{\eta_0}{|\eta|}, \tau_0, \eta_0\right) : 0 \le \sigma \le \sigma_1\right\}.$$

Recall the set $C_t(\mu)$ of those $\nu \in T^*(\mathbf{R} \times \Omega)$ such that there exists a generalized bicharacteristic $\gamma(t)$ of $\square$ with $\gamma(0) = \mu$, $\gamma(t) = \nu$. Consider the metric $D(\rho, \mu)$ defined in Section 1.2. Recall that $D(\rho, \mu) = 0$ implies $\rho = \mu$ or $\rho = (x, \xi)$, $\mu = (x, \eta)$ with $x \in \partial\Omega$, $p_x(\xi) = p_x(\eta)$. We denote by $\gamma(t; \mu)$ one of the generalized bicharacteristics of $\square$ parametrized by the time t and passing through

μ for $t = 0$. Thus we have

$$C_t(\mu) = \cup\gamma(t;\mu),$$

where the union is taken over all bicharacteristics issued from μ. Set $\nu_0 = (0, x_0', 0, \tau_0, \xi_0', \xi_n^0)$.

Lemma 5.4.5: *For each $\delta > 0$ there exists $\epsilon(\delta) > 0$ such that if $D(\mu, l_0(\sigma)) < \epsilon(\delta)$ for some $\sigma \in [0, \sigma_1]$, then for each $\nu \in C_{t_0-\sigma}(\mu)$ we have*

$$D(\nu, C_{t_0}(\nu_0)) = \inf_{\rho \in C_{t_0}(\nu_0)} D(\nu, \rho) < \delta.$$

Remark: This lemma says that if $\bar{\gamma}(\sigma; \nu)$ is a curve which coincides with a linear segment, passing through ν for $0 \leq \sigma \leq \hat{\sigma}$ and with the generalized bicharacteristic issued from $\mu = \bar{\gamma}(\hat{\sigma}, \nu)$ for $\hat{\sigma} \leq \sigma \leq t_0$, then $\bar{\gamma}(t_0; \nu)$ is close to $C_{t_0}(\nu_0)$, provided ν is close to ν_0. We need this property because in general $\bar{\gamma}(\sigma; \nu)$, $0 \leq \sigma \leq t_0$ is not a generalized bicharacteristic of $\Box$.

Proof of Lemma 5.4.5: Assume that there exist sequences $\{\sigma_k\}$, $0 \leq \sigma_k \leq \sigma_1$, and $\{\mu_k\}$, $\nu_k = \gamma(t_0 - \sigma_k; \mu_k)$ so that

$$D(\mu_k, l_0(\sigma_k)) < \frac{1}{k}, \quad D(\nu_k, C_{t_0}(\nu_0)) \geq \delta, \quad \forall k \in \mathbf{N}.$$

Taking subsequences, we can suppose

$$\sigma_k \to \hat{\sigma}, \quad D(\mu_k, l_0(\hat{\sigma})) \to 0 \text{ as } k \to +\infty.$$

A simple argument shows that there exists a subsequence $\{\mu_{k_m}\}$ converging to $\nu_0 = l_0(\hat{\sigma})$ or to v_1 such that $D(v_0, v_1) = 0$. The latter is possible only if $\hat{\sigma} = \sigma_1$ and if l_0 hits transversally $\partial\Omega$ at x_1.

Next, we suppose that $\mu_k \to v$ in the usual sense, where $v = v_0$ or $v = v_1$. Consider the sequence of generalized bicharacteristics

$$\{\gamma(t; \mu_k)\}, \quad 0 \leq t \leq t_0.$$

According to Lemma 1.2.6, there exists a subsequence $\{\mu_{k_m}\}$ and a generalized bicharacteristic $\bar{\gamma}(t; v)$ of $\Box$ so that for all $t \in [0, t_0]$ we have

$$D(\gamma(t, \mu_{k_m}), \bar{\gamma}(t; v)) \to 0 \text{ as } m \to +\infty.$$

Without loss of generality we may assume that $k_m = m$ for each m. Applying the triangle inequality for the metric D, we get

$$D(\nu_k, \tilde{\gamma}(t_0 - \hat{\sigma}; v)) \leq D(\gamma(t_0 - \sigma_k; \mu_k), \gamma(t_0 - \hat{\sigma}; \mu_k))$$
$$+ D(\gamma(t_0 - \hat{\sigma}; \mu_k), \tilde{\gamma}(t_0 - \hat{\sigma}; v)).$$

According to Lemma 1.2.5, for the first term on the right-hand side we apply the estimate

$$D(\gamma(t'; \mu), \gamma(t''; \mu)) \leq C_0 |t' - t''|$$

for $t', t'' \in [0, t_0]$ uniformly on μ in some small neighbourhood of v. Thus for k_0 large enough we get

$$D(\nu_k, \tilde{\gamma}(t_0 - \hat{\sigma}; v)) < \delta, \quad k \geq k_0.$$

Let us define the generalized bicharacteristics

$$\gamma_0(\sigma; \nu_0) = \begin{cases} l_0(\sigma), & 0 \leq \sigma \leq \hat{\sigma}, \\ \tilde{\gamma}(\sigma - \hat{\sigma}; v), & \hat{\sigma} \leq \sigma. \end{cases}$$

Then $\tilde{\gamma}(t_0 - \hat{\sigma}; v) \in C_{t_0}(\nu_0)$ and we obtain a contradiction. This completes the proof. ♠

Proof of Proposition 5.4.3: Consider again the inclusion (5.24). As we have already observed, for $0 \leq t_0 - s \leq \sigma_1$ the set $WF'(\tilde{e}_1(t_0 - s, z, y', y_n))$ has a projection on Ω close to $l_0(\sigma)$. Choosing Γ_1, $\mathcal{T}$ and ϵ sufficiently small, we apply Lemma 5.4.5 and Proposition 5.4.4 and conclude that for $t = t_0$, $t' = 0$ the wave front $WF'(\tilde{\Xi}_1)$, considered as a subset of $T^*(\mathbf{R} \times \Omega)$, is sufficiently close to $C_{t_0}(\nu_0)$. On the other hand, $t_0 \notin L_\Omega$ implies $(t_0, y_0, \tau_0, \eta_0) \notin C_{t_0}(\nu_0)$. Now we take the trace $t' = 0$ and use the fact that $C_{t_0}(\nu_0)$ is closed as a consequence of Lemma 1.2.7. Thus, for small Γ_1 and ϵ we obtain

$$\rho_0 \notin WF'(H(x_n, y_n)) \text{ for } (x_n, y_n) \in [0, \epsilon] \times [0, \epsilon]. \tag{5.25}$$

Here the choice of ϵ depends on that of Γ_1.

Next, we pass to the case when the point $(t_0, x'_0, \tau_0, \xi'_0)$ is hyperbolic for $\Box$, that is equation (5.23) has two distinct real roots $\xi^0_\pm$. Let Γ_1 be a small conic neighbourhood of $(0, x'_0, \tau_0, \xi'_0)$ such that the points $(t, x', \tau, \xi') \in \Gamma_1$ are hyperbolic (see Section 1.2). Let $A_1(t', y, D_{t'}, D_{y'})$ be a pseudo-differential operator as above with $WF(A_1) \subset \Gamma_1$. The singularities of the distribution.

$$e_1(t - t', x, y) = A_1(t', y, D_{t'}, D_{y'}) e_0(t - t', x - y)$$

for small $t - t' > 0$ are propagating along the outgoing and incoming bicharacteristics entering the exterior or interior of Ω, respectively. For $0 < 2\delta < \epsilon$

introduce a function $\kappa_\delta(t) \in C^\infty(\mathbf{R})$ such that

$$\kappa_\delta(t) = \begin{cases} 1, & t \leq \delta \\ 0, & t \geq 2\delta. \end{cases}$$

For δ sufficiently small consider the distributions

$$e_2(t - t', x, y) = (1 - A_1)e_0(t - t', x - y),$$
$$\tilde{e}_1(t - t', x, y) = e_1(t - t', x, y)\kappa_\delta(t - t')\zeta_\Omega(x),$$

$\zeta_\Omega(x)$ being the characteristic function of Ω. For $(t, x) \in \mathbf{R} \times \Omega^\circ$ we have

$$(\partial_t^2 - \Delta_x)\tilde{e}_1(t - t', x, y) = f_\delta(t - t', x, y),$$

and for sufficiently small $\delta > 0$ we deduce

$$\text{sing supp} f_\delta(t - t', x, y) \subset \{(t, x, t', y) : \delta \leq t - t' \leq 2\delta, \ x \in \mathcal{O}_\delta\}, \tag{5.26}$$

where $\bar{\mathcal{O}}_\delta \subset \Omega^\circ$. Hence the singularities of f_δ for t' small enough are bounded away from the boundary $\partial\Omega$. We extend $E(t, x, y)$ as 0 for $t < 0$, and for $t \geq 0$ we write

$$\Xi(t - t', x, y) = (\tilde{e}_1 + e_2)(t - t', x, y)$$
$$- \int_{-\infty}^{+\infty} \int_{\partial\Omega} K(s, x, z)(\tilde{e}_1 + e_2)(t - t' - s, z, y)\, ds dz$$
$$- \int_{-\infty}^{+\infty} \int_{\partial\Omega} E(t - t' - s, x, z) f_\delta(s, z, y)\, ds dz.$$

The integrals are interpreted in the sense of distributions. The smoothness of $e_0(t, x - y)$ with respect to $t \in \mathbf{R}^+$ and (5.26) make this possible.

The analysis of the term involving $\tilde{e}_1$ is easy since the singularities of $\tilde{e}_1$ are concentrated around the bicharacteristics entering Ω°. For the term involving f_δ we apply a similar argument. To this end we prove an assertion similar to Proposition 5.4.4 with $\partial\Omega$ replaced by an open domain $\mathcal{O}$, $\bar{\mathcal{O}} \subset \Omega^\circ$. After this, we establish an analogue of (5.19) for

$$WF'(E(t, x', x_n, y)) \cap T^*(\mathbf{R}_t^+ \times U' \times \mathcal{O})$$

uniformly with respect to $x_n \in [0, \epsilon]$. This can be done applying the arguments from the proofs of Propositions 5.4.2 and 5.4.4 with slight modifications. We leave the details to the reader. Combining these results and taking the trace $t' = 0$, we obtain (5.25).

In the above argument the choice of ϵ has been related to that of the conic neighbourhood Γ_1. To obtain uniformity with respect to U', consider a covering

$$T \times U' \times (\mathbf{R}^n \backslash \{0\}) \subset \cup_{j=1}^M \Gamma_j$$

consisting of conic neighbourhoods Γ_j for which our argument works with $\epsilon = \epsilon_j > 0$. Taking $\epsilon = \min_j \epsilon_j$, we obtain $\epsilon > 0$ which depends on U' and t_0, only. This completes the proof of Proposition 5.4.3, since the relation (5.25) holds for $\epsilon > 0$ chosen above. ♠

Now it is easy to establish the inclusion (5.18). Let

$$\Omega \subset \cup_{k=1}^m U_k \qquad (5.27)$$

be a covering with open sets $U_k \subset \mathbf{R}^n$. Assume that for $k = 1, \dots, m_0$ we have $U_k \subset \Omega^\circ$, while $U_k \cap \partial\Omega \neq \emptyset$ for $k = m_0 + 1, \dots, m$. Let $t_0 \notin L_\Omega, t_0 > 0$.

Fix a neighbourhood $U_k \subset \Omega^\circ$. Applying (5.19), as at the end of Section 5.3, we deduce that

$$WF\left(\int_{U_k} E(t, x, x) \, dx_{|t>0} \right)$$

is contained in the set of all $(t, \tau) \in T^*(\mathbf{R}) \backslash \{0\}$ such that there exists $(x, \xi) \in T^*(U_k)$ with $(t, x, x, \tau, \xi, \xi) \in C$.

Now assume that $U_k \cap \partial\Omega \neq \emptyset$. Introduce in U_k local coordinates (x', x_n) so that $U_k \cap \partial\Omega$ has the form $x_n = 0$, while $U_k \cap \Omega$ is given by $x' \in U'$, $0 \leq x_n \leq \epsilon$. For sufficiently small $\delta_0 > 0$ we have $t_0 - \delta_0 > 0$ and

$$\Delta_0 = (t_0 - \delta_0, t_0 + \delta_0) \cap L_\Omega = \emptyset.$$

Taking δ_0 and ϵ small enough, Proposition 5.4.3 implies

$$(t, x', x', \tau, \xi', \xi') \notin WF(H(x_n, x_n)) \qquad (5.28)$$

for $t \in \Delta_0$, $(x', \xi') \in U' \times \mathbf{R}^{n-1}$, $(x_n, y_n) \in [0, \epsilon] \times [0, \epsilon]$. We consider t, x_n, y_n as parameters and apply the argument in Section 5.3 for the trace $x' = y'$ and the integral over U'. Thus, the relation (5.28) yields

$$\int_{U'} E(t, x', x_n, x', x_n) \, dx'_{|t \in \Delta_0, x_n \in [0, \epsilon]} \in C^\infty,$$

and integrating with respect to x_n, we get

$$\int_0^\epsilon \int_{U'} E(t, x', x_n, x', x_n) \, dx' dx_n \in C^\infty(\Delta_0).$$

Consequently, given $t_0 \notin L_\Omega$, we can find a sufficiently fine covering (5.27) such that

$$t_0 \notin \text{sing supp} \sum_{k=1}^{m} \int_{U_k \cap \Omega} E(t, x, x) \, dx.$$

In this way we obtain the inclusion (5.18) for an arbitrary domain Ω with smooth boundary $\partial\Omega$.

The above argument, with some trivial modifications, can be applied to the analysis of the distribution

$$\sigma_N(t) = \sum_{j=1}^{\infty} \cos \lambda_j t \in S'(\mathbf{R}),$$

where $0 \le \lambda_1^2 \le \lambda_2^2 \le \ldots \le \lambda_m^2 \le \ldots$ are the eigenvalues of the self-adjoint operator A_N in $L^2(\Omega)$ related to the Laplacian in Ω with Neumann or Robin boundary conditions on $\partial\Omega$. The corresponding eigenfunctions $\{\varphi_j(x)\}_{j=1}^{\infty}$ satisfy

$$\begin{cases} -\Delta\varphi_j(x) = \lambda_j^2 \varphi_j(x) \text{ in } \Omega, \\ (\partial_\nu + \alpha(x))\varphi_j(x) = 0 \text{ on } \partial\Omega. \end{cases}$$

Here ∂_ν denotes the derivative with respect to a continuous normal field $\nu(x)$ to $\partial\Omega$ and $\alpha(x) \in C^\infty(\partial\Omega)$. We define $\mathcal{E}(t, x, y)$ and $E(t, x, y)$ in the same way as in Section 5.2.

The proof of (5.18) for arbitrary domains goes without any change, since we may use the results for propagation of singularities in [MS2] concerning the Neumann and Robin boundary problems. For example, we define $K(t, x, y)$ as the solution of the problem

$$\begin{cases} (\partial_t^2 - \Delta_x)K(t, x, y) = 0 \quad \text{in } \mathbf{R} \times \Omega^\circ, \\ (\partial_\nu + \alpha(x))K(t, x, y) - \delta(t) \otimes \delta(x - y) = 0 \quad \text{for } x \in \partial\Omega, \\ \text{supp } K(t, x, y) \subset \{(t, x, y) \in \mathbf{R}_t \times \Omega \times \partial\Omega : t \ge 0\}. \end{cases}$$

The assertions of Propositions 5.4.1 and 5.4.4 are true for K. Summing up the above results, we get the following.

Theorem 5.4.6: *Let Ω be a compact domain in $\mathbf{R}^n$, $n \ge 2$, with C^∞ smooth boundary $\partial\Omega$. Then*

$$\text{sing supp } \sigma(t) \subset \{0\} \cup \{\pm T_\gamma : \gamma \in \mathcal{L}_\Omega\}. \tag{5.29}$$

The same is true for $\text{sing supp } \sigma_N(t)$, where $\sigma_N(t)$ is related to Neumann or Robin boundary problem for the Laplacian. ♠

5.5. Notes

The analysis of the fundamental solutions in Section 5.1 is taken from [BLR]. A detailed investigation of $N(\lambda)$ and the spectral function $e(\lambda, x, y)$ is contained in [H3, H4]. The idea of the proof of the Poisson relation for convex domains was proposed in [BLR]. For strictly convex (concave) domains this relation was obtained previously by Anderson and Melrose [AM]. The argument in [AM] can be generalized for general domains by using the results for propagations of singularities established in [MS1, MS2]. We followed the approach in [AM] exploiting the continuity properties of the generalized bicharacteristics described by Lemmas 1.2.6 and 5.4.5.

6 POISSON SUMMATION FORMULA FOR MANIFOLDS WITH BOUNDARY

In this chapter the leading singularity of $\sigma(t) = \sum_j \cos \lambda_j t$ near the period T_γ of a periodic ordinary reflecting bicharacteristic γ of $\square$ in Ω is examined. We assume that if δ is another periodic bicharacteristic of $\square$ in Ω with the same period T_γ, then the projections of γ and δ on Ω coincide. Moreover, we suppose that the Poincaré map P_γ of γ has no eigenvalues equal to 1.

In Section 6.1 a global parametrix for the mixed problem characterizing $\mathcal{E}(t, x, y)$ is constructed. For this purpose we apply global Fourier integral distributions to express the successive reflections. The principal symbol of the parametrix is investigated in Section 6.2. The singularity of $\sigma(t)$ is studied in Section 6.3 and a Poisson summation formula for manifolds with boundary is obtained in Theorem 6.3.1.

6.1. Global parametrix for mixed problem

In this section we use the notation of Chapter 5. Our aim is to construct a global parametrix for the operator $\mathcal{E}_B = \cos(tA)B(y, D_y)$, where $B(y, D_y)$ is a zero order pseudo-differential operator with $WF(B) \subset \Gamma$, $\Gamma = U \times V$ being a small conic neighbourhood of a fixed point $(y_0, \eta_0) \in T^*(\mathring{\Omega}) \backslash 0$. We assume that $\bar{U} \subset \mathring{\Omega}$ and that the kernel of B has compact support in $U \times U$.

Let $T_0 > 0$ be fixed. We consider the generalized bicharacteristics $\gamma(t; \nu)$ of $\square$ issued from $\nu = (0, y, \pm|\eta|, \eta)$ with $(y, \eta) \in \Gamma$ and parametrized by the time t. In this section we treat the case when $\gamma(t; \nu)$ are reflecting on $\partial\Omega$ and without tangent segments for $|t| \leq T_0$.

Let $F_B(t, x, y)$ be the kernel of $\mathcal{E}_B$. We wish to construct a global Fourier integral distribution $\hat{F}_B(t, x, y)$ so that

$$(F_B(t,x,y) - \hat{F}_B(t,x,y))|_{[0,T_0] \times \Omega \times U} \in C^\infty.$$

The distribution $\hat{F}_B$ will be obtained as a sum of global Fourier integral distributions related to the reflections of $\gamma(t;\nu)$.

We start by the fundamental solution

$$R_0(t, x-y) = \frac{(2\pi)^{-n}}{2} \left(\int e^{it|\eta|+i\langle x-y,\eta\rangle}\, d\eta + \int e^{-it|\eta|+i\langle x-y,\eta\rangle}\, d\eta \right)$$

$$= \frac{1}{2}(R_0^- + R_0^+)$$

so that

$$\begin{cases} (\partial_t^2 - \Delta_x)R_0 = 0, \\ R_0(0, x-y) = \delta(x-y), \quad \partial_t R_0(0, x-y) = 0. \end{cases}$$

We can consider R_0 as a Fourier integral distribution

$$R_0 \in I^{-1/4}(\mathbf{R}_t \times \mathbf{R}_x^n \times \mathbf{R}_y^n, C_0'),$$

where $C_0 = C_0^+ \cup C_0^-$ and

$$C_0^\pm = \left\{ (t,x,y,\tau,\xi,\eta) \in T^*(\mathbf{R} \times \mathbf{R}^n \times \mathbf{R}^n) \backslash 0; \quad x = y \pm t\frac{\eta}{|\eta|}, \right.$$

$$\left. \xi = \eta, \quad \tau = \mp|\eta| \right\}.$$

Below we consider the distribution R_0^+ related to C_0^+ and given by the integral with phase function $\varphi_- = -t|\eta| + \langle x-y,\eta\rangle$. Let $\gamma_+(t;\nu)$ be the generalized bicharacteristic issued from $\nu = (0, y, -|\eta|, \eta)$, $(y,\eta) \in \Gamma$. Denote by $t_1(y,\eta)$ the time of the first reflection of $\gamma_+(t;\nu)$ and set

$$t_1 = \inf_{(y,\eta)\in\Gamma} t_1(y,\eta), \quad T_1 = \sup_{(y,\eta)\in\Gamma} t_1(y,\eta), \quad \mathcal{I}_1 = [t_1, T_1].$$

We need to examine the trace on $\partial\Omega$ of the distribution $(R_B^+)(t,x,y) = (2\pi)^{-n}\int e^{i\varphi_-}\beta(y,\eta)\, d\eta$ for $t \in \mathcal{I}_1$, provided Γ is sufficiently small. Here $\beta(y,\eta)$ is the symbol of $B(y, D_y)$.

Consider the inclusion map

$$\mathbf{R} \times \partial\Omega \ni (t,x) \xrightarrow{i} (t,x) \in \mathbf{R} \times \Omega$$

and denote by i^* the operator of the trace on $\mathbf{R} \times \partial\Omega$. The kernel of i^* is a Fourier integral distribution in $I^{-1/4}(\mathbf{R} \times \partial\Omega \times \mathbf{R} \times \Omega, \mathcal{N}')$, where the canonical relation

$\mathcal{N}$ has the form

$$\mathcal{N} = \left\{ (t, x, \tau, \tilde{\xi}, t, x, \tau, \xi) \in T^*(\mathbf{R} \times \partial\Omega \times \mathbf{R} \times \Omega) \backslash 0 \; : \; x \in \partial\Omega, \tilde{\xi} = \xi|_{T_x(\partial\Omega)} \right\}.$$

We wish to show that for $t \in \mathcal{I}_1$, $i^* R_B^+$ is a Fourier integral distribution. To define the corresponding canonical relation we need some preparation. Set

$$C_\Gamma = \left\{ (t, x, y, \tau, \xi, \eta) \in C_0^+, (y, \eta) \in \Gamma \right\}$$

and introduce

$$Z = T^*(\mathbf{R} \times \partial\Omega) \backslash 0 \times \Delta(T^*(\mathbf{R} \times \Omega) \backslash 0) \times T^*(U) \backslash 0,$$

where $\Delta(\mathcal{U}) = \{(m, m) : m \in \mathcal{U}\}$ denotes the diagonal of $\mathcal{U}$.

Let $\gamma_+(t; \nu)$ hit (transversally) $\partial\Omega$ at $x_1(y, \eta)$ for $t = t_1(y, \eta)$. Assume that $x_1(y, \eta) \in \omega_1 \subset \partial\Omega$ for $(y, \eta) \in \Gamma$. Let ω_1 have the form $x_n = 0$. Suppose that the principal symbol q of $\square$ in the local coordinates becomes

$$q(x, \tau, \xi) = q_2(x, \xi) - \tau^2.$$

A simple calculus yields $\text{codim}(\mathcal{N} \times C_\Gamma) = 4n + 2$, $\text{codim } Z = 2n + 2$. On the other hand, the transversality condition means that

$$\frac{\partial q_2}{\partial \xi_n} \left(y + t \frac{\eta}{|\eta|}, \eta \right) \neq 0, \quad t \in \mathcal{I}_1. \tag{6.1}$$

This implies easily that $\text{codim}((\mathcal{N} \times C_\Gamma) \cap Z) = 6n + 3$. On the other hand, the manifold $\mathcal{N} \times C_\Gamma$ intersects Z cleanly which means that for each $u \in (N \times C_\Gamma) \cap Z$ we have

$$T_u((\mathcal{N} \times C_\Gamma) \cap Z) = T_u(\mathcal{N} \times C_\Gamma) \cap T_u(Z).$$

The excess of this intersection is 1 and the composition $\rho_0 = \mathcal{N} \circ C_\Gamma$ will be a canonical relation (see [H3], Section 21). Since the projection

$$\rho_0 \ni (t_1(y, \eta), x_1(y, \eta), y, \tau, \tilde{\xi}, \eta) \rightarrow (y, \eta) \in \Gamma$$

is a diffeomorphism, the relation ρ_0 is locally the graph of a homogeneous canonical transformation

$$r_0 : \Gamma \rightarrow T^*(\mathbf{R} \times \partial\Omega).$$

By the calculus for Fourier integral operators with clean intersection (see [H4], Section 25) we deduce that $i^* R_B^+$ is a Fourier integral distribution

$$i^* R_B^+(t, x, y) \in I^0(\mathbf{R} \times \partial\Omega \times U, \rho_0'). \tag{6.2}$$

Next we modify $(R_B^+)(t, x, y)$ for $t > T_1 + \varepsilon_1$ and small $\varepsilon_1 > 0$ so that

$$R_B^+ \in C^\infty \quad \text{for } t > T_1 + \varepsilon_1.$$

Thus for $t \geq 0$ the trace $i^* R_B^+$ modulo smooth functions coincides with (6.2). This continuation is possible since the singularities of $(R_B^+)(t, x, y)$ are propagating along the bicharacteristics of $\square$ lying in the exterior of Ω for $T_1 + \varepsilon \leq t \leq T_1 + 2\varepsilon$, provided that ε is small enough.

To construct the parametrix for $t > t_1$, we need to satisfy the boundary condition on $\partial\Omega$. Let ν_x be the exterior unit normal at $x \in \partial\Omega$. Consider the set

$$\Sigma_1 = \left\{ (t, z, \tau, \tilde{\xi}) \in T^*(\mathbf{R} \times \partial\Omega)\backslash 0 : t = t_1(y, \eta), \ z = y + t_1(y, \eta)\frac{\eta}{|\eta|}, \right.$$
$$\left. \tau = -|\eta|, \ \eta|_{T_z(\partial\Omega)} = \tilde{\xi}, \langle v_z, \eta \rangle > 0, \ (y, \eta) \in \Gamma \right\}$$

and introduce the map

$$\Phi : \mathbf{R} \times \Sigma_1 \ni (t, t', z, \tau, \tilde{\xi}) \rightarrow \left(t' + t, z + t\frac{\xi}{|\xi|}, \tau, \xi \right) \in T^*(\mathbf{R} \times \Omega),$$

where

$$|\xi| = -\tau, \quad \langle \xi, \nu_z \rangle = -\langle \eta, \nu_z \rangle,$$

while η is related to $\tilde{\xi}$, as in the definition of Σ_1. It is easy to see that Φ is an immersion. Indeed, for $t = 0, \Phi$ maps Σ_1 diffeomorphically into $T^*(\mathbf{R} \times \Omega)$ since ξ is uniquely determined. Moreover, $d\Phi|_{t=0}$ maps $\frac{\partial}{\partial t}$ into the Hamiltonian field H_q of the principal symbol q of $\square$. Our choice of Σ_1 guarantees that H_q is transversal to $\partial\Omega$. The group property $\Phi(t_1 + t_2, .) = \Phi(t_1, \Phi(t_2, .))$ shows that $d\Phi$ is an immersion for all t.

Introduce the relation

$$C_1 = \{(\Phi(t, \rho), \rho) \in T^*(\mathbf{R} \times \Omega \times \mathbf{R} \times \partial\Omega)\backslash 0; \ t \geq 0, \ \rho \in \Sigma_1\}.$$

To see that C_1 is a canonical relation, consider the symplectic forms $d\alpha$ and $d\alpha_b$ on $T^*(\mathbf{R} \times \Omega)$ and $T^*(\mathbf{R} \times \partial\Omega)$, respectively. Here α and α_b are the canonical one-forms on $T^*(\mathbf{R} \times \Omega)$ and $T^*(\mathbf{R} \times \partial\Omega)$, respectively. For example, if $w \in T_{(s,\sigma)}(T^*(\mathbf{R} \times \partial\Omega))$ with $s \in \mathbf{R} \times \partial\Omega$, $\sigma \in T_s^*(\mathbf{R} \times \partial\Omega)$, we have

$$\langle \alpha_b, w \rangle = \langle \sigma, w' \rangle,$$

$w' \in T_s(\mathbf{R} \times \partial\Omega)$ being the projection of w on $\mathbf{R} \times \partial\Omega$.

By using a trivial lifting, consider α and α_b as one-forms on $W = T^*(\mathbf{R} \times \Omega) \times T^*(\mathbf{R} \times \partial\Omega)$.

To see that C_1 is a canonical relation, consider the set $\Lambda_0 = \{(\Phi(0,\rho),\rho) : \rho \in \Sigma_1\}$. Let

$$d\Phi(0,\rho)w = v \in T_{(z,\zeta)}(T^*(\mathbf{R} \times \Omega)).$$

Then $\zeta|_{T_s(\mathbf{R} \times \partial\Omega)} = \sigma$ and the projection v' of v on $\mathbf{R} \times \partial\Omega$ coincides with w'. Consequently,

$$\langle \alpha, d\Phi(0,\rho)w \rangle = \langle \zeta, w' \rangle = \langle \sigma, w' \rangle$$

and

$$\langle \alpha - \alpha_b, \ (d\Phi(0,\rho)w, w) \rangle = 0.$$

As we have mentioned above, we have

$$d\Phi(0,\rho)\left(\frac{\partial}{\partial t}\right) = H_q,$$

hence

$$\left\langle \alpha - \alpha_b, d\Phi(0,\rho)\left(\frac{\partial}{\partial t}\right) \right\rangle = \langle \alpha, H_q \rangle = q = 0,$$

since the principal symbol q vanishes on $\Phi(0,\rho)$, $\rho \in \Sigma_1$. Thus, Λ_0 is a Lagrangian submanifold of W with respect to the symplectic form $d\alpha - d\alpha_b$. Next

$$\Lambda_t = \{(\Phi(t,\rho),\rho) : t > 0, \quad \rho \in \Sigma_1\}$$

is also a Langrangian submanifold with respect to $d\alpha - d\alpha_b$ because $\Phi(t,.)$ is the Hamiltonian flow of q. Finally, C_1 is a canonical relation in W.

Now, let $\omega_1 = \pi(\Sigma_1)$, where $\pi : (t,x,\tau,\tilde{\xi}) \rightarrow x$ is the projection on $\partial\Omega$. To arrange the Dirichlet boundary condition on ω_1, we introduce a Fourier integral operator

$$R_1^+ \in I^{-1/4}(\mathbf{R} \times \Omega \times \mathbf{R} \times \partial\Omega, C_1')$$

satisfying the conditions

$$\begin{cases} (\partial_t^2 - \Delta_x)R_1^+ \equiv 0, \\ i_{\omega_1}^* R_1^+ Q \equiv Q, \end{cases} \tag{6.3}$$

for each pseudo-differential operator Q such that $WF(Q) \subset \Sigma_1$. Here $\equiv$ means an equality modulo operators with smooth kernels, while $i_{\omega_1}^*$ is the trace on ω_1. Notice that the trace $i^* R_1^+$ does not coincide with the identity on $\partial\Omega$. Moreover, for sufficiently small ω_1, hence for small enough Γ, we have $\Phi(t,\rho) \in \Sigma_1$ for $t = 0$, only. Thus, the operator $i_{\omega_1}^* R_1^+$ is related to the relation $\{(\rho,\rho) : \rho \in \Sigma_1\}$, hence it is a pseudo-differential operator.

To construct R_1^+, we must solve the transport equations for the principal and lower-order symbols of R_1^+. To do this, we need to recall the notion of the principal symbol following for the convenience of the reader the notations in ([H4], Section 25). By using a positive density d on $\mathbf{R} \times \Omega$ and a positive density d_b on

$\mathbf{R} \times \partial\Omega$, we can consider R_1^+ as a linear map

$$R_1^+ : C_0^\infty \left(\mathbf{R} \times \partial\Omega, \Omega_{\mathbf{R} \times \partial\Omega}^{1/2} \right) \;\to\; \mathcal{D}' \left(\mathbf{R} \times \Omega, \Omega_{\mathbf{R} \times \Omega}^{1/2} \right) \qquad (6.4)$$

between the space of C^∞ smooth half-densities and the spaces of distributions with values half-densities. For this introduce the map

$$d^{1/2} R_1^+ \left(u d_b^{-1/2} \right) \qquad \text{for } u \in C_0^\infty \left(\mathbf{R} \times \partial\Omega, \Omega_{\mathbf{R} \times \partial\Omega}^{1/2} \right).$$

For brevity of notation put

$$Y = \mathbf{R} \times \Omega \times \mathbf{R} \times \partial\Omega, \quad \Lambda = C_1'.$$

Then the kernel of (6.4) is a distribution

$$\mathfrak{R}_1^+ \in I^{-1/4} \left(Y, \Lambda; \Omega_Y^{1/2} \right)$$

and its principal symbol a_0 belongs to the class $S^0(Y, M_\Lambda \otimes \Omega_\Lambda^{1/2})$. Here M_Λ is the Maslov bundle of Λ, while $\Omega_\Lambda^{1/2}$ is the bundle of half-densities on Λ. We refer to [H4], Section 25, for the precise definition of M_Λ, $\Omega_\Lambda^{1/2}$ and $S^0(Y, M_\Lambda \otimes \Omega_\Lambda^{1/2})$. We denote by q the principal symbol of $\square$ and consider H_q as a vector field on Y by using a trivial lifting. Since the subprincipal symbol of $\square$ vanishes, we obtain the transport equation

$$\mathcal{L}_{H_q} a_0 = 0, \qquad (6.5)$$

where $\mathcal{L}_{H_q}$ denotes the Lie derivative along H_q. We can solve (6.5) with initial condition on ω_1 so that the principal symbol of the pseudo-differential operator $i_{\omega_1}^* R_1^+$ is 1. Then we get

$$\left(\partial_t^2 - \Delta_x \right) \mathfrak{R}_1^+ \in I^{-1/4} \left(Y, \Lambda; \Omega_Y^{1/2} \right).$$

Next for the lower-order symbols a_j of $\mathfrak{R}_1^+$, we obtain the equations

$$\mathcal{L}_{H_q} a_j = f_j(a_0, \ldots, a_{j-1}), \quad j \geq 1$$

which we solve so that the full symbol of $i_{\omega_1}^* R_1^+$ is equal to 1. Thus we arrange (6.3). The form of a_0 will be discussed in the next section.

Now repeat the above procedure for other reflections. Let $t_k(y, \eta)$ be the time of the kth reflection of $\gamma_+(t; \nu)$. Set

$$t_k = \inf_{(y, \eta) \in \Gamma} t_k(y, \eta), \quad T_k = \sup_{(y, \eta) \in \Gamma} t_k(y, \eta).$$

Since Φ is an immersion, it follows easily that the set $C_1 \times (\mathcal{N} \circ C_\Gamma)$ intersects

$$T^*(\mathbf{R} \times \Omega)\backslash 0 \times \Delta(T^*(\mathbf{R} \times \partial\Omega)\backslash 0) \times T^*(U)\backslash 0$$

transversally. Then the composition $C_1 \circ (\mathcal{N} \circ C_\Gamma)$ is a canonical relation and by the calculus of Fourier integral operators we deduce

$$R_1^+ i^* R_B^+ \in I^{-1/4}\left(\mathbf{R} \times \Omega \times U, (C_1 \circ \mathcal{N} \circ C_\Gamma)'\right).$$

Notice that $C_1 \circ \mathcal{N} \circ C_\Gamma \subset C_+$, where C_+ is the relation introduced in Section 1.2.

Now set

$$V_0^+ = R_B^+, \quad V_1^+ = R_1^+ i^* R_B^+ = R_1^+ i^* V_0^+. \tag{6.6}$$

Then for $0 \le t < t_2$ the trace of V_1^+ on $\partial\Omega$ vanishes modulo operators with smooth kernels. For $t \in [t_2, T_2]$ the generalized bicharacteristics $\gamma_+(t; \nu)$ hit transversally $\partial\Omega$. This implies that

$$\rho_1 = \mathcal{N} \circ C_1 \subset T^*(\mathbf{R} \times \partial\Omega)\backslash 0 \times T^*(\mathbf{R} \times \partial\Omega)\backslash 0$$

is a canonical relation. Moreover, the projection

$$\rho_1 \ni (t, x, \tau, \tilde{\xi}, t', y, \tau, \tilde{\eta}) \to (t', y, \tau, \tilde{\eta}) \in \Sigma_1$$

is a diffeomorphism and we conclude that ρ_1 is locally the graph of a homogeneous canonical transformation

$$r_1 : \Sigma_1 \to T^*(\mathbf{R} \times \partial\Omega).$$

Next we modify $\mathfrak{R}_1^+$ for $t \notin [t_1 - \varepsilon_2, T_2 + \varepsilon_2]$ and small $\varepsilon_2 > 0$ so that $R_1^+ Qf \in C^\infty$ if $t \notin [t_1 - \varepsilon_2, T_2 + \varepsilon_2]$ and Q is a pseudo-differential operator with $WF(Q) \subset \Sigma_1$. The singularities of the kernel $\mathfrak{R}_1^+$ and the form of C_1 make this possible. Setting $\Sigma_2 = r_1(\Sigma_1)$, $\omega_2 = \pi(\Sigma_2) \subset \partial\Omega$, we may pass to the next step of the construction.

Following this procedure, define $\Sigma_k = r_{k-1}(\Sigma_{k-1})$, the canonical relation C_k and a homogeneous canonical transformation

$$r_k : \Sigma_k \to T^*(\mathbf{R} \times \partial\Omega)$$

related to the kth reflection of $\gamma(t; \nu)$. Set $\omega_k = \pi(\Sigma_k)$ and denote by $i_{\omega_k}^*$ the trace on ω_k. Repeating the construction of R_1^+, we can find a Fourier integral operator

$$R_k^+ \in I^{-1/4}\left(\mathbf{R} \times \Omega \times \mathbf{R} \times \partial\Omega, C_k'\right)$$

satisfying the conditions

$$\begin{cases} \left(\partial_t^2 - \Delta_x\right) R_k^+ \equiv 0, \\ i_{\omega_k}^* R_k^+ Q \equiv Q, \end{cases}$$

for each pseudo-differential operator Q with $WF(Q) \subset \Sigma_k$. As above, we modify the kernel $\mathfrak{R}_k^+$ of R_k^+ for $t \notin [t_k - \varepsilon_k, T_{k+1} + \varepsilon_k]$, $\epsilon_k > 0$ being sufficiently small. Our construction shows that the trace $i^* R_k^+$, modulo smoothing operators, coincides with the sum of the traces $i_{\omega_k}^* R_k^+$ and $i_{\omega_{k+1}}^* R_k^+$. To satisfy the boundary conditions, introduce a pseudo-differential operator $M_k \in L^0(\mathbf{R} \times \partial\Omega)$ such that

$$WF(M_k) \cap \Sigma_k = \emptyset, \quad WF(\mathrm{Id} - M_k) \cap \Sigma_{k-1} = \emptyset.$$

After this preparation define

$$V_k^+ = R_k^+ (\mathrm{Id} - M_{k-1}) i^* V_{k-1}^+, \quad k \geq 2$$

and set

$$W_p^+ = \sum_{k=0}^{p} (-1)^k V_k^+.$$

Therefore, for $0 \leq t < t_{p+1}$ we have

$$\begin{cases} (\partial_t^2 - \Delta_x) W_p^+ \in C^\infty, \\ i^* W_p^+ \in C^\infty. \end{cases}$$

The construction of W_p^+ works if $\gamma_+(t; \nu)$ have at least p reflections for $0 \leq t \leq T_0$.

In the same way we treat the relation C_0^- and the generalized bicharacteristics $\gamma_-(t; \nu)$ of $\square$ issued from $\nu = (0, y, |\eta|, \eta)$, $(y, \eta) \in \Gamma$. Let

$$W_p^- = \sum_{k=0}^{p} (-1)^k V_k^-$$

be the corresponding distribution. Therefore,

$$W_p = \frac{1}{2} \sum_{k=0}^{p} (-1)^k (V_k^+ + V_k^-)$$

will be a solution of the problem

$$\begin{cases} (\partial_t^2 - \Delta_x) W_p \in C^\infty, \\ i^* W_p \in C^\infty, \\ W_{p|t=0} = B^*(y, D_y)\delta(x - y), \quad \partial_t W_{p|t=0} = 0 \end{cases}$$

for $0 \leq t \leq \hat{t}_{p+1}$ where $\hat{t}_{p+1}$ depends on the times of the $(p+1)$th reflection of $\gamma_\pm(t; \nu)$ and $B^*(y, D_y)$ is the operator adjoint to $B(y, D_y)$. For large p we obtain the distribution $\hat{F}_B(t, x, y)$ for $0 \leq t \leq T_0$.

Now let γ be a periodic ordinary reflecting bicharacteristic of $\square$ with period $T > 0$ passing through (y_0, η_0). Let $m_\gamma \geq 2$ be the number of reflections of γ. The operator V_k^+ is related to the relation

$$C_k \circ \rho_{k-1} \circ \cdots \circ \rho_1 \circ \rho_0 \subset C_+. \tag{6.7}$$

Let U and $\varepsilon > 0$ be sufficiently small and let $\mathcal{I} = (T - \varepsilon, T + \epsilon)$. Then for $t \in \mathcal{I}$ modulo smooth terms we obtain

$$W_{m_\gamma}^+ \equiv (-1)^{m_\gamma} V_{m_\gamma}^+ \in I^{-1/4}(\mathcal{I} \times U \times U, C_+').$$

A similar result holds for $W_{m_\gamma}^-$ with C_+ replaced by the relation C_- defined in Section 1.2. Thus, we can take

$$\hat{F}_B(t, x, y)|_{\mathcal{I} \times U \times U} = \frac{1}{2}(-1)^{m_\gamma}(V_{m_\gamma}^+ + V_{m_\gamma}^-) = \frac{1}{2}(F_B^+ + F_B^-). \tag{6.8}$$

Now, we shall discuss briefly the case when U is a small open neighbourhood of a point $y_0 \in \partial\Omega$. Assume that $\partial\Omega \cap U$ has the form $x_n = 0$, $y_0 = 0$ and $\Omega \cap U = \{(x', x_n) : x' \in U', \ 0 \leq x_n \leq \alpha\}$. Let $q(x, \tau, \xi)$ be the principal symbol of $\square$ and let $\mu_0 = (0, y_0', \tau_0, \eta_0') \in T^*(\mathbf{R}^n) \backslash 0$ be a hyperbolic point of $\square$. Let $\Gamma = \mathcal{O} \times V$ be an open conic neighbourhood of μ_0 containing hyperbolic points, only. Consider a zero order pseudo-differential operator $B(t', y, D_{t'}, D_{y'})$ depending smoothly on y_n so that $WF(B(., y_n, .)) \subset \Gamma$ and the kernel of $B(., y_n, .)$ has compact support in $\mathcal{O} \times \mathcal{O}$. Set

$$\mathcal{E}_{B'} = \cos((t - t')A)Y(t - t')B(t', y, D_t, D_{y'})$$

and assume that the bicharacteristics $\gamma(t; \nu)$ of $\square$ issued from $\nu = (t, y', y_n, \tau, \eta', \eta_n^\pm)$ with $0 \leq y_n \leq \alpha$, $(t, y', \tau, \eta') \in \Gamma$ are reflecting for $|t| \leq T_0$. Here $\eta_n^\pm$ are the real roots of the equation $q(y', y_n, \tau, \eta', \eta_n) = 0$ with respect to η_n. Then we can repeat the above construction of V_k^+ and construct the distribution

$$W_{m_\gamma} = \frac{1}{2}(\mathcal{F}_B^+ + \mathcal{F}_B^-)$$

satisfying the condition

$$(\mathcal{E}_{B'}(t - t', x, y) - W_{m_\gamma}(t - t', x, y))|_{[0, T_0] \times \mathcal{J} \times \Omega \times \mathcal{U}} \in C^\infty, \tag{6.9}$$

$\mathcal{J} \subset \mathbf{R}$ being a small neighbourhood of 0 and $\mathcal{E}_{B'}(t - t', x, y)$ the kernel of $\mathcal{E}_{B'}$.

6.2. Principal symbol of $\hat{\mathbf{F}}_{\mathbf{B}}$

In this section we shall examine the principal symbol Y of the distribution $\hat{F}_B(t, x, y)$ given by (6.8). To do this we must examine the principal symbol of $V_{m_\gamma}^\pm$ following the rules for composition of Fourier integral operators.

Consider the kernel $\mathfrak{R}_1^+$ of the operator R_1^+. The principal symbol a_0 of $\mathfrak{R}_1^+$ satisfies the equation (6.5); hence a_0 is locally constant along the orbits of the Hamiltonian field H_q. Setting $\Lambda = C_1'$, the symbol a_0 is a section in the bundle $M_\Lambda \otimes \Omega_\Lambda^{1/2}$. If we ignore the Maslov factors in M_Λ, the half-density part of a_0 is constant along the orbits of H_q.

We shall trivialize $\Omega_\Lambda^{1/2}$ by using a half-density d_Λ on Λ which is invariant with respect to the action of H_q. For this purpose consider a diffeomorphism

$$\mu_1 : \Sigma_1 \rightarrow \mu_1(\Sigma_1) \subset T^*(\mathbf{R}_t \times \mathbf{R}_{x'}^{n-1})$$

and introduce local coordinates (t, x', τ, ξ') on $\mu_1(\Sigma_1)$. By the pull-back μ_1^* define a half-density

$$d_1 = \mu_1^*(|dt|^{1/2} \cdot |d\tau|^{1/2} \otimes |dx'|^{1/2} \cdot |d\xi'|^{1/2})$$

on Σ_1 related to the symplectic coordinates (t, x', τ, ξ') in $T^*(\mathbf{R}_t \times \mathbf{R}_{x'}^{n-1})$. Therefore, $\Phi(-t, \rho)^* d_1$ is a half-density d_Λ such that $\mathcal{L}_{H_q} d_\Lambda = 0$. Consequently, the half-density part of a_0 will be $f_0 d_\Lambda$, f_0 being a homogeneous of order 0 function. Since f_0 is constant along the orbits of H_q, the initial condition on ω_1 implies $f_0 = 1$.

Next, consider the distribution $(I - M_1)i^* \mathfrak{R}_1^+$, related to $\rho_1 = $ graph r_1. First, the condition $WF(M_1) \cap \Sigma_2 = \emptyset$ shows that the symbol of M_1 vanishes on Σ_2. Secondly, the projection $\pi_1 : \rho_1 \rightarrow \Sigma_1$ is a diffeomorphism and $d_{\rho_1} = (\pi_1)^* d_1$ becomes a half-density on ρ_1 related to symplectic coordinates (x, ξ). As above, the principal symbol of $(I - M_1)i^* \mathfrak{R}_1^+$, modulo Maslov factors, is d_{ρ_1}. Furthermore, $d_2 = ((r_1)^{-1})^* d_1$ is a half-density on Σ_2 and we can pass to the next step.

Repeating this procedure for $(I - M_k)i^* \mathfrak{R}_{k'}^+$ we apply the rule for the computation of the principal symbol of a product of Fourier integral operators associated to homogeneous canonical transformation (see [H4], Section 25). Setting $\sigma_k = r_k \circ \cdots \circ r_1 : \Sigma_1 \rightarrow \Sigma_{k+1}$, introduce a half-density $(\sigma_k^{-1})^* d_1$ on Σ_{k+1}. Since graph σ_k is diffeomorphic to Σ_{k+1}, we obtain a half-density δ_k on graph σ_k. Then the principal symbol of the distribution

$$(I - M_k)i^* R_k^+ \ldots (I - M_1)i^* \mathfrak{R}_1^+,$$

modulo Maslow factors, is δ_k.

Let $\tilde{\Gamma} \subset \Gamma$ be a small conic neighbourhood of (y_0, η_0) and suppose the symbol $\beta(y, \eta)$ of $B(y, D_y)$ is equal to 1 on $\tilde{\Gamma}$. Put $\tilde{\rho}_0 = \mathcal{N} \circ C_{\tilde{\Gamma}}$. The principal symbol of $i^* R_B^+$ on $\tilde{\rho}_0$ is the half-density d_1. Since C_{m_γ} is related to the Hamiltonian flow H_q, we can construct a half-density d_γ on $C_{m_\gamma} \circ ($graph $\sigma_{m_\gamma - 1}) \circ \rho_0 \subset C_+$ related to symplectic coordinated (x, ξ) used above. Then, the principal symbol of $V_{m_\gamma}^+$ over $\tilde{\Gamma}$, modulo Maslov factors, will be d_γ.

To obtain a local representation of F_B^+, consider a small conic neighbourhood

$Z_0 \subset C_+$ of $\nu_0 = (T, y_0, y_0, -|\eta_0|, \eta_0, \eta_0) \in C_+$. The projection

$$Z_0 \ni (t, x, y, \tau, \xi, \eta) \rightarrow (t, x, \eta) \in \mathcal{I} \times U \times V$$

is a diffeomorphism for Z_0 small enough. This follows easily from the fact that for each t the map $(x, \xi) = \Phi^t(y, \eta)$ is a canonical transformation. Then there exists a phase function $\varphi(t, x, \eta)$, determined in a small conic neighbourhood $\Sigma = \tilde{\mathcal{I}} \times \tilde{U} \times \tilde{V}$ of (T, y_0, η_0) and homogeneous of order 1 in η, so that

$$Z_0 = \{(t, x, \varphi_\eta, \varphi_t, \varphi_x, \eta) : (t, x, \eta) \in \Sigma\}, \ \det \varphi_{x\eta}(t, x, \eta) \neq 0, (t, x, \eta) \in \Sigma.$$

For $\mathcal{I}$, U and V small enough we arrange $\mathcal{I} \times U \times V \subset \Sigma$. Then for $t \in \mathcal{I}$, $x, y \in U$, modulo C^∞ terms, we obtain

$$F_B^+(t, x, y) = (2\pi)^{-n} \int_{|\eta| \geq 1} e^{i\varphi(t,x,\eta) - i\langle y, \eta \rangle} b(t, x, \eta) \, d\eta. \tag{6.10}$$

Here

$$b(t, x, \eta) \sim \sum_{j=0}^\infty b_j(t, x, \eta)$$

and $b_j(t, x, \eta)$ are homogeneous of order $(-j)$ with respect to η. For fixed $t \in \mathcal{I}$ we may consider (x, η) as local coordinates on Z_0 and $|dx|^{1/2} \cdot |d\eta|^{1/2}$ becomes a half-density on Z_0. For this localization the principal symbol of (6.10) modulo Maslov factors has the form $b_0(t, x, \eta) |dx|^{1/2} \cdot |d\eta|^{1/2}$. On the other hand, it is possible to express the principal symbol of F_B^+ by using the half-density d_γ on Z_0, related to symplectic coordinates (x, ξ) on $T^*(U)$. Since $\xi = \varphi_x(x, \eta)$ on Z_0, we obtain

$$d_{\gamma|t=T} = |dx|^{1/2} \cdot |d\xi|^{1/2} = |dx|^{1/2} |\det \varphi_{x\eta}(T, x, \eta)|^{1/2} |d\eta|^{1/2}$$

because d_γ has been constructed so that $d_{\gamma|t=T}$ coincides with $|dx|^{1/2} \cdot |d\xi|^{1/2}$. This explains the choice of the half-density d_1 and the procedure leading to d_γ. The principal symbol of F_B^+ at ν_0, modulo Maslov factors and $d_{\gamma|t=T}$, is equal to $(-1)^{m_\gamma}$. Comparing this with the form of this symbol in the coordinates (x, η), we deduce

$$b_0(T, y_0, \eta_0) = (-1)^{m_\gamma} e^{i\frac{\pi}{2}\sigma} |\det \varphi_{x\eta}(T, y_0, \eta_0)|^{1/2}. \tag{6.11}$$

Here $\sigma \in \mathbf{N}$ and the Maslov factor has the form $e^{i\frac{\pi}{2}\sigma}$, while $(-1)^{m_\gamma}$ appears since $F_B^+ = (-1)^{m_\gamma} V_{m_\gamma}^+$ for $t \in \mathcal{I}$.

The integer σ depends on γ, only. To see this we express σ by the signatures of the matrices $d_{\eta\eta}^2 \varphi_j$. Here φ_j are phase functions parametrizing C_+ in small neighbourhoods Z_j. Let $\gamma = \gamma(t)$ be parametrized by the time t and let $0 = t_0 \leq t_1 \leq \ldots \leq t_L = T$ be a sequence of times along γ with $t_j = t_{j+1}$ only if $\gamma(t_j)$ is a reflection point of γ.

Assume $t_j < t_{j+1}$ and let $\gamma(t)$ have no reflections for $t_j \leq t \leq t_{j+1}$. Suppose $\gamma(t_j) \in Z_j \cap Z_{j+1}$ and let Z_k for $k = j, j+1$ be expressed by φ_k as above Z_0 has been expressed by φ. Then, passing from the representation of F_B^+ by φ_j to that related to φ_{j+1}, we must add the Maslov factor

$$i^{1/2(\operatorname{sgn} d^2_{\eta\eta}\varphi_j - \operatorname{sgn} d^2_{\eta\eta}\varphi_{j+1})}. \tag{6.12}$$

Now suppose $\gamma(t_j) = (t_j, x_0, \tau_0, \xi_0)$ is a reflection point of γ. Let $x_0 \in \omega_k \subset \partial\Omega$ and let ω_k have the form $x_n = 0$, while $x_0 = (x_0', 0)$. Denote by $q(x, \tau, \xi)$ the principal symbol of $\Box$. For (x', τ, ξ') close to (x_0', τ_0, ξ_0') denote by $\xi_n^{\pm}(x, \tau, \xi')$ the roots of the equation $q(x, \tau, \xi', \xi_n) = 0$ with respect to ξ_n. For t close to t_j the distribution F_B^+ has the form

$$F_B^+ = (-1)^k (R_k^+ (I - M_{k-1}) i^* L_{k-1} - L_{k-1}) i^* R_B^+.$$

The operator $i^*_{\omega_k} L_{k-1}$ is related to graph σ_k. Let $\chi_k(t, x', \tau, \xi')$ be the generating function of σ_k, that is

$$\operatorname{graph} \sigma_k^{-1} = \left(\begin{pmatrix} t & x' \\ \chi_t & \chi_{x'} \end{pmatrix}, \begin{pmatrix} \chi_\tau & \chi_{\xi'} \\ \tau & \xi' \end{pmatrix} \right).$$

By convention, assume that $(\xi', \xi_n^{\mp})$ is close to the direction of $\gamma(t)$ for $\mp(t - t_j) > 0$ and $|t - t_j|$ sufficiently small. In other words, ξ_n^- (resp. ξ_n^+) corresponds to incoming (resp. outgoing) segments reflecting on ω_k.

Introduce the phase functions $\varphi^{\mp}(t, x, \tau, \xi')$ as solutions of the Cauchy problems

$$\begin{cases} \dfrac{\partial \varphi^{\mp}}{\partial x_n} = \xi_n^{\mp}(x, \varphi_t^{\mp}, \varphi_{x'}^{\mp}), \\ \varphi^{\mp}|_{x_n=0} = \chi(t, x', \tau, \xi'). \end{cases}$$

Then the kernel of L_{k-1} for t close to t_j admits the representation

$$\int e^{i\varphi^-(t, x, \tau, \xi') - it'\tau - i\langle x', \xi'\rangle} b_k(t, x, \tau, \xi') \, d\tau d\xi', \tag{6.13}$$

while the kernel of $R_k^+ (I - M_{k-1}) i^* L_{k-1}$ has a similar representation by $\varphi^+(t, x, \tau, \xi')$. Hence, putting $\zeta = (\tau, \xi')$, by the initial conditions for $\varphi^{\mp}$, we deduce

$$\operatorname{sgn} d^2_{\zeta\zeta}\varphi^- = \operatorname{sgn} d^2_{\zeta\zeta}\varphi^+. \tag{6.14}$$

Thus, the reflection at $\gamma(t_j)$ does not involve a Maslov factor.

Denote by $\mathfrak{R}$ the set of $j \in \mathbf{R}$ such that $t_j = t_{j+1}$. Then, according to (6.12) and (6.14), the Maslov factor σ in (6.11) has the form

$$\sigma_\gamma = \frac{1}{2} \sum_{\substack{j=0 \\ j \notin \Re}}^{L-1} (\operatorname{sgn} d_{\eta\eta}^2 \varphi_j - \operatorname{sgn} d_{\eta\eta}^2 \varphi_{j+1}). \qquad (6.15)$$

Clearly, σ_γ depends on γ, only, since the choice of $\gamma(0)$ is not important for the sum in (6.15).

For F_B^- we follow a completely similar argument. Thus for $t \in \mathcal{I}$, $x, y \in U$ we get the representation

$$F_B^-(t, x, y) = (2\pi)^{-n} \int_{|\eta| \geq 1} e^{i\Psi(t,x,\eta) + i\langle y, \eta \rangle} c(t, x, \eta) \, d\eta \qquad (6.16)$$

with phase function $\Psi(t, x, \eta)$ representing locally C_- and $c \sim \sum_j c_j(t, x, \eta)$. The principal symbol has the form

$$c_0(T, y_0, \eta_0) = (-1)^{m_\gamma} e^{-i\frac{\pi}{2}\sigma_\gamma} \left| \det \Psi_{x\eta}(T, y_0, \eta_0) \right|^{1/2}. \qquad (6.17)$$

In fact, we repeat the construction of F_B^- following a covering of $\gamma(t)$, where we change the orientation because $\tau > 0$ on C_- and the time t decreases when we move along $\gamma(t)$.

Finally, the results of this section can be applied for the distributions $\mathcal{F}_B^\pm$ introduced at the end of Section 6.1. We leave the details to the reader.

6.3. Poisson summation formula

In this section we use the notation of the previous sections. Let γ be a periodic ordinary reflecting bicharacteristic of $\square$ in Ω with period $T > 0$. Let P_γ be the Poincaré map of γ introduced in Section 2.3. Denote by $\pi : T^*(\mathbf{R} \times \Omega) \to \Omega$ the usual projection. Then $\pi(\gamma) = \tilde{\gamma}$ will be a (generalized) periodic geodesic in Ω. Throughout this section we make the following assumptions:

(i) if δ is a periodic bicharacteristic of $\square$ in Ω with period T, then $\pi(\delta) = \pi(\gamma)$,

(ii) $\det(P_\gamma - I) \neq 0$.

Since by Lemma 1.2.10 the set L_Ω of periods of periodic bicharacteristics in Ω is closed, T is an isolated point in $\operatorname{sing\,supp} \sigma(t)$. Indeed, if there exist bicharacteristics with periods $T_k \to T$, passing over $(x_k, \xi_k) \in T^*(\Omega)$, we can find subsequences $x_{n_k} \to x_0$, $\xi_{n_k} \to \xi_0$, and a bicharacteristic with period T passing over (x_0, ξ_0). This contradicts the assumptions (i) and (ii). Choose $\varepsilon > 0$ and $\mathcal{I} = (T - \varepsilon, T + \varepsilon) \subset \mathbf{R}^+$ so that

$$\operatorname{sing\,supp} \sigma(t) \cap \mathcal{I} = \{T\}.$$

Let $\mathcal{O}_\gamma$ be a sufficiently small open neighbourhood of $\tilde{\gamma}$. Then (i) and the choice of $\mathcal{I}$ yield

$$\{(t, x, x, \tau, \xi, \xi) \in C : t \in \mathcal{I}, \quad x \in \mathring{\Omega} \backslash \mathcal{O}_\gamma\} = \emptyset.$$

Applying the argument at the end of Section 5.4 with $\mathcal{I}$ instead of Δ_0, for $W \subset \mathring{\Omega} \backslash \mathcal{O}_\gamma$ we obtain

$$\int_W \mathcal{E}(t, x, x)\, dx \in C^\infty(\mathcal{I}).$$

For $W \cap \partial\Omega \neq \emptyset$ and $W \cap \mathcal{O}_\gamma = \emptyset$ we obtain the same result by using Proposition 5.4.3.

Thus to study the leading singularity of $\sigma(t)$ for t close to T, we must examine the traces

$$\sum_{j=1}^M \int_{\Omega \cap U_j} \mathcal{E}(t, x, x)\, dx,$$

where

$$\mathcal{O}_\gamma \subset \bigcup_{j=1}^M U_j \tag{6.18}$$

is a covering and $U_j \subset \Omega$ for $j = 1, \ldots, M_0$, $U_j \cap \partial\Omega \neq \emptyset$ for $M_0 + 1 \leq j \leq M$.

First, we study the trace on $U_j \subset \mathring{\Omega}$ and for simplicity we write U instead of U_j. To microlocalize the problem, introduce a covering

$$T^*(U) \backslash 0 \subset \bigcup_{k=1}^N (U \times V_k),$$

V_k being small conic neighbourhoods. As in Section 6.1, suppose γ pass over $(y_0, \eta_0) \in U \times V_{k_0}$. Take the above covering sufficiently fine to arrange $\eta_0 \notin V_k$, whereas $k \neq k_0$. Exploiting the assumption (i) once more, we deduce for $k \neq k_0$

$$\{(t, x, x, \tau, \xi, \xi) \in C : t \in \mathcal{I}, \quad (x, \xi) \in U \times V_k\} = \emptyset. \tag{6.19}$$

Choose pseudo-differential operators $\tilde{B}_j \in L^0(\mathring{\Omega})$ with principal symbols $\tilde{b}_j(y, \eta)$ and full symbols $\tilde{\beta}_j(y, \eta)$ so that

$$\sum_{j=1}^N \tilde{b}_j = 1 \text{ on } T^*(U) \backslash 0,$$

$$\text{conesupp } \tilde{\beta}_j(y, \eta) \subset U \times V_j.$$

The operator $\sum_j \tilde{B}_j$ is elliptic on U and there exists a pseudo-differential operator $P \in L^0(\hat{\Omega})$ such that

$$\sum_j P\tilde{B}_j = \text{Id} + R(y, D_y),$$

$R(y, D_y)$ being an operator with C^∞ kernel. Setting $B_j = P\tilde{B}_j$, we obtain a partition of unity on $T^*(U)\backslash 0$ given by $\sum_j B_j = Id + R$.

The kernel of the operator $\cos(tA)B_k$ has the form $B_k^*(y, D_y)\mathcal{E}(t, x, y)$, where B_k^* is the operator adjoint to B_k. By using Proposition 5.4.2, it is easy to see that

$$WF(B_y^*(y, D_y)\mathcal{E}(t, x, y)|_{\mathcal{I} \times U \times U}) \subset$$
$$\{(t, x, y, \tau, \xi, \eta) \in C : (y, \eta) \in WF(B_k)\}.$$

In fact, assume that $\rho = (t, x, \hat{y}, \tau, \xi, \hat{\eta}) \in C$ with $(\hat{y}, \hat{\eta}) \notin WF(B_k)$. Take a pseudo-differential operator A_k so that $WF(A_k) \cap WF(B_k) = \emptyset$ and the full symbol of A_k is equal to 1 in some small conic neightbourhood $\hat{\Gamma}$ of $(\hat{y}, \hat{\eta})$. Then $\cos(tA)B_kA_ku \in C^\infty$ and the kernel

$$A_k^*(y, D_y)B_k^*(y, D_y)\mathcal{E}(t, x, y) \in C^\infty.$$

As in the proof of Proposition 5.3.1, we can choose a pseudo-differential operator C_k, elliptic at ρ, so that $C_kA_k^*B_k^*\mathcal{E}(t, x, y) \in C^\infty$. Thus

$$\rho \notin WF(B_k^*(y, D_y)\mathcal{E}(t, x, y)|_{\mathcal{I} \times U \times U}).$$

In the case $\rho \notin C$ the result follows immediately. Thus (6.19) implies for $k \neq k_0$

$$\int_U (B_k^*(y, D_y)\mathcal{E})(t, x, x)\,dx \in C^\infty(\mathcal{I}).$$

We are going to study $\exp(\mp itA)B_{k_0}$ and we omit the index k_0 writing B and V instead of B_{k_0} and V_{k_0}. Set $\Gamma = U \times V$ and apply the construction of Section 6.1 for B with $WF(B) \subset \Gamma$. If $F_\pm(t, x, y)$ are the kernels of $\exp(\mp itA)$, we have

$$B^*(y, D_y)F_\pm(t, x, y) - F_B^\pm(t, x, y) \in C^\infty(\mathcal{I} \times U \times U).$$

For the distribution $F_B^+ = (-1)^{m_\gamma}V_{m_\gamma}^+$, given by (6.10), we obtain modulo terms in $C^\infty(\mathcal{I})$

$$\int_U F_B^+(t, x, x)\,dx = (2\pi)^{-n}\int_U dx \int_1^\infty dr \int_{V \cap S^{n-1}} e^{ir(\varphi(t, x, \omega) - \langle x, \omega \rangle)}$$
$$\times \sum_j b_j(t, x, \omega)r^{n-1-j}\,d\omega.$$

Change the coordinates and assume

$$U = \{(x', x_n) \in \mathbf{R}^n : x' \in U'\ \ 0 < \alpha < x_n < \beta\},$$
$$\eta_0 = (0, \ldots, 0, 1), \quad \eta_n > 0 \text{ on } V.$$

Here $U' \subset \mathbf{R}^{n-1}$ is an open neighbourhood of $y_0' = 0$ and $\alpha < y_{0,n} < \beta$. Introduce the integral

$$I_{\alpha,\beta} = (2\pi)^{-n} \int_\alpha^\beta \mathrm{d}x_n \int_1^\infty \mathrm{d}r \int_{U'} \int_{V \cap S^{n-1}} e^{ir(\varphi(t,x,\omega) - \langle x, \omega \rangle)}$$
$$\times \sum_j b_j(t, x, \omega) r^{n-1-j} \, \mathrm{d}x' \mathrm{d}\omega. \tag{6.20}$$

Our aim is to apply a stationary phase method for the integral with respect to x' and ω, considering x_n as a parameter.

The critical points satisfy the equalities

$$\varphi_{x'} = \omega', \quad \varphi_\omega = x.$$

The representation of C_+ by the phase φ, given in Section 6.2, implies

$$\varphi_t(t, x, \omega) = -|\varphi_x(t, x, \omega)| = -1.$$

Moreover, $\omega_n > 0$ and $\varphi_{x_n} > 0$ on $V \cap S^{n-1}$. Thus, at the critical points $\hat{x}'$, $\hat{\omega}$ we get $\varphi_t = -1$, $\varphi_x = \omega$ and the form of C_+ yields

$$(\hat{x}', x_n, \hat{\omega}) = \Phi^t(\hat{x}', x_n, \omega).$$

This is possible for $t = T$, $\hat{x}' = 0$, $\hat{\omega} = \eta_0$, only. Next, introduce local coordinates

$$\mathbf{R}^{n-1} \supset W \ni \eta' \rightarrow (\eta', \sqrt{1 - |\eta'|^2}) \in V \cap S^{n-1}$$

and write the phase in the form

$$\varphi(t, x', x_n, \eta', \sqrt{1 - |\eta'|^2}) - \langle x', \eta' \rangle - x_n \sqrt{1 - |\eta'|^2}.$$

For the stationary phase method we need to examine the matrix

$$\Delta(\rho(x_n)) = \begin{pmatrix} \varphi_{x'x'} & \varphi_{x'\eta'} - I_{n-1} \\ \varphi_{\eta'x'} - I_{n-1} & \varphi_{\eta'\eta'} \end{pmatrix} (\rho(x_n)),$$

where $\rho(x_n) = (T, 0, x_n, \eta_0)$ and I_k denotes the $(k \times k)$ identity matrix. Here we have used the equality

$$\varphi_{\eta_n}(\rho(x_n)) = x_n \quad \text{for } \alpha < x_n < \beta.$$

Consider the generalized Hamiltonian flow $\Phi^T : (\varphi_\eta, \eta) \to (x, \varphi_x)$ defined in Section 1.2. For the differential $d\Phi^T$ we get

$$(d\Phi^T)\begin{pmatrix} \varphi_{\eta x}\delta x + \varphi_{\eta\eta}\delta x \\ \delta\eta \end{pmatrix} = \begin{pmatrix} \delta x \\ \varphi_{xx}\delta x + \varphi_{x\eta}\delta\eta \end{pmatrix},$$

hence

$$d\Phi^T - I_n = Q_\varphi = \begin{pmatrix} \varphi_{\eta x}^{-1} - I_n & -\varphi_{\eta x}^{-1}\varphi_{\eta\eta} \\ \varphi_{xx}\varphi_{\eta x}^{-1} & -\varphi_{xx}\varphi_{\eta x}^{-1}\varphi_{\eta\eta} + \varphi_{x\eta} - I_n \end{pmatrix}$$

$$= \begin{pmatrix} 0 & -I_n \\ I_n & 0 \end{pmatrix}\begin{pmatrix} \varphi_{xx} & \varphi_{x\eta} - I_n \\ \varphi_{\eta x} - I_n & \varphi_{\eta\eta} \end{pmatrix}\begin{pmatrix} \varphi_{\eta x}^{-1} & -\varphi_{\eta x}^{-1}\varphi_{\eta\eta} \\ 0 & I_n \end{pmatrix}.$$

Clearly,

$$\varphi_{x\eta_n}(\rho(x_n)) = \eta_0, \quad \varphi_{\eta\eta_n}(\rho(x_n)) = 0,$$
$$\det\varphi_{x\eta}(\rho(x_n)) = \det\varphi_{x'\eta'}(\rho(x_n)).$$

According to the definition of P_γ in Section 2.3, the Poincaré map P_γ, modulo conjugations, is the restriction of $d\Phi^T$ on the linear space

$$\mathcal{L} = \left\{ (\delta x', 0, \delta\eta', 0) : \delta x' \in \mathbf{R}^{n-1}, \quad \delta\eta' \in \mathbf{R}^{n-1} \right\}.$$

Let Q'_φ be the $(n-1) \times (n-1)$ matrix obtained from Q_φ replacing I_n by I_{n-1} and taking only the derivatives of φ with respect to x' and η'. Therefore, for each $l' = (\delta x', 0, \delta\eta', 0) \in \mathcal{L}$ we have

$$\langle (d\Phi^T - I_n)l', l' \rangle = \left\langle Q'_\varphi\begin{pmatrix} \delta x' \\ \delta\eta' \end{pmatrix}, \begin{pmatrix} \delta x' \\ \delta\eta' \end{pmatrix} \right\rangle.$$

This implies

$$(\det\varphi_{x'\eta'})^{-1}(\det\Delta)(\rho(x_n)) = \det(P_\gamma - I).$$

We may apply the representation (6.11) for the points $\rho(x_n)$, hence

$$b_0(\rho(x_n)) = (-1)^{m_\gamma}\exp\left(i\frac{\pi}{2}\sigma_\gamma\right)\left|\det\varphi_{x'\eta'}\right|^{1/2}(\rho(x_n)),$$

$\sigma_\gamma \in \mathbf{N}$ being the Maslov index of γ. By the Euler equality for the phase φ we deduce

$$\varphi(t, 0, x_n, \eta_0) = T - t + x_n.$$

Now the stationary phase method yields

$$(2\pi)^{-n} \int_{U'} \int_{V \cap S^{n-1}} e^{ir(\varphi(t,x,\omega) - \langle x, \omega \rangle)} \sum_j b_j(t,x,\omega) r^{n-1-j} \, dx' d\omega$$

$$= e^{ir(T-t)} \left(\sum_{k=0}^{N-1} c_k r^{-k} + O(r^{-N}) \right),$$

where

$$c_0 = \frac{i}{2\pi} \exp\left(i\frac{\pi}{2}\beta_\gamma\right) |\det(P_\gamma - I)|^{1/2}$$

and

$$\beta_\gamma = 2m_\gamma + \sigma_\gamma + \frac{\operatorname{sgn}\Delta}{2} - 1. \tag{6.21}$$

The integer β_γ is locally constant since for small $\beta - \alpha$ we have $\operatorname{sgn}\Delta(\rho(x_n)) = $ const and the integers $m_\gamma \in \mathbf{N}$, $\sigma_\gamma \in \mathbf{N}$ depend on γ, only. Thus the constant c_0 is independent of x_n and r.

To complete the computation of the leading term of $I_{\alpha,\beta}$, take the Fourier transform of the Heaviside function $Y(r)$. The integral in r can be interpreted in the sense of distributions, hence

$$\int_1^\infty e^{ir(T-t)} \, dr = -i(t - T - i0)^{-1} + L^1_{\mathrm{loc}}(\mathbf{R}).$$

Finally, for $t \in \mathcal{I}$ we have

$$I_{\alpha,\beta} = \frac{\beta - \alpha}{2\pi} \exp\left(i\frac{\pi}{2}\beta_\gamma\right) |\det(P_\gamma - I)|^{-1/2} (t - T - i0)^{-1} + L^1_{\mathrm{loc}}(\mathbf{R}). \tag{6.22}$$

A similar argument with trivial modifications works for $F_B^-(t,x,y)$. On the other hand, we may use the equality $F_-(t,x,y) = \overline{F_+(t,x,y)}$. Then, modulo $C^\infty(\mathcal{I})$, we get

$$\int_U F_B^-(t,x,x) \, dx = \int_U F^-(t,x,x) \, dx$$

$$= \int_U \overline{F^+(t,x,x)} \, dx = \int_U \overline{F_B^+(t,x,x)} \, dx = \overline{I_{\alpha,\beta}} + L^1_{\mathrm{loc}}\mathbf{R}.$$

Thus

$$\int_U \mathcal{E}(t,x,x) \, dx = \operatorname{Re} \int_U F^+(t,x,x) \, dx = \frac{\beta - \alpha}{2\pi} |\det(P_\gamma - I)|^{-1/2}$$

$$\times \operatorname{Re}\left[\exp\left(i\frac{\pi}{2}\beta_\gamma\right)(t - T - i0)^{-1}\right] + L^1_{\mathrm{loc}}(\mathbf{R}). \tag{6.23}$$

Now we turn to the case $\partial\Omega \cap U \neq \emptyset$. Let

$$\pi(\gamma) = \tilde{\gamma} = \bigcup_{j=1}^{p_\gamma} l_j, \quad l_j = [q_j, q_{j+1}]$$

with $q_j \in \partial\Omega$, $q_{p_{\gamma+1}} = q_1$ and $m_\gamma = mp_\gamma$. Let $q_{j+1} \in U$ and let (x', x_n), $x' \in \mathbf{R}^{n-1}$ be local coordinates chosen at the end of Section 6.1. Choose a small neighbourhood $\mathcal{J}$ of $0 \in \mathbf{R}$ and put $\mathcal{O} = \mathcal{J} \times U'$, U' being the same as in Section 6.1. Introduce a covering

$$T^*(\mathcal{O})\backslash 0 \subset \bigcup_{k=1}^{N} (\mathcal{O} \times V_k),$$

where V_k are small conic neighbourhoods in $\mathbf{R}_t \times \mathbf{R}_{x'}^{n-1}$. Choose zero order pseudo-differential operators $B_k(t', y, D_{t'}, D_{x'})$, depending smoothly on $0 \leq y_n \leq \alpha$ and satisfying

$$WF(B_k(., y_n, .)) \subset \mathcal{O} \times V_k.$$

As above, construct a microlocal partition of unity on $T^*(\mathcal{O})\backslash 0$ given by

$$\sum_j B_j(t', y, D_{t'}, D_{x'}) = \mathrm{Id} + R',$$

R' being an operator with C^∞ smooth kernel.

Recall that $\Omega \cap U = \{(x', x_n) : x' \in U', 0 \leq x_n \leq \alpha\}$ and denote by $q(x, \tau, \xi) = q_2(x, \xi) - \tau^2$ the principal symbol of $\square$ in local coordinates. Assume $q_{j+1} = (y_0', 0)$ and let $\eta_n^{\pm}$ be the roots of the equation $q(y, \tau, \eta', \eta_n) = 0$ with respect to η_n. Assume that γ passes over $(0, y_0', \tau_0, \eta_0')$ and set $\eta_0^{\pm} = \eta^{\pm}(y_0', 0, \tau_0^2, \eta_0')$, $\mu_{\pm} = (0, y_0', \pm\tau_0, \eta_0')$. Without loss of generality, suppose $\tau_0 < 0$. By convention we choose $\eta_n^{\pm}$ so that

$$\pm \frac{\partial q_2}{\partial \xi_n}(y_0', 0, \eta_0', \eta_0^{\pm}) > 0.$$

Then (η_0', η_0^+) (resp. (η_0', η_0^-)) is colinear with the direction of l_{j+1} (resp. l_j) at q_{j+1}. Suppose $\Gamma_{\pm} = \mathcal{O} \times V_{\pm}$ are small open conic neighbourhoods of $\mu_{\pm}$. By using assumptions (i), (ii), we may choose $\Gamma_{\pm}$ small enough so that for $\mu \notin \Gamma_+ \cup \Gamma_-$ there are no periodic bicharacteristics of $\square$ in Ω passing over μ and having periods in $\mathcal{I}$. Taking the partition of unity on $T^*(\mathcal{O})\backslash 0$ fine, assume that for $k \neq k_{\pm}$ and $0 \leq y_n \leq \alpha$ we have

$$WF(B_k(., y_n, .)) \cap \Gamma_{\pm} = \emptyset.$$

Let $\mathcal{F}_{B_k}^{\pm}(t - t', x, y)$ be the kernel of the operator

$$\exp(\mp i(t - t')A)Y(t - t')B_k(t', y, D_{t'}, D_{y'}).$$

Following the results of Section 5.4 for the kernel of $\exp(\mp i(t - t')A)Y(t - t')$, for $t \in \mathcal{F}$, $t' \in \mathcal{I}$ and $k \neq k_{\pm}$ we get

$$(t, t', x', x', \tau, \tau, \xi', \xi') \notin WF(\mathcal{F}_{B_k}^{\pm}(t - t', x', x_n, y', y_n)),$$

whenever $(x_n, y_n) \in [0, \alpha] \times [0, \alpha]$, $\tau \in \mathbf{R}$, $(t', x', \tau, \xi') \in WF(B_k(., x_n, .))$. Then for $k \neq k_{\pm}$ we have

$$\int_{\Omega \cap U} \mathcal{F}_{B_k}^{\pm}(t - t', x, x) \, dx \in C^{\infty}(\mathcal{I} \times \mathcal{J}).$$

Below we treat $\mathcal{F}_{B_{k_{\pm}}}^{\pm}$ and we omit $k_{\pm}$ in the notations of $\mathcal{F}_B^{\pm}$ and B. Assume $\Gamma^{\pm} = \mathcal{O} \times V$ is an open conic neighbourhood of $\mu_{\pm}$ and suppose

$$WF(B(., y_n, .)) \subset \Gamma^+.$$

As we have mentioned in Section 6.1, we can construct Fourier integral distribution $\mathcal{F}_B^{\pm}(t - t', x, y)$ related to the canonical relations

$$\mathcal{M}_{\pm} = \{(t, t', x, y, \tau, \tau', \xi, \eta) \in T^*(\mathcal{I} \times U \times \mathcal{J} \times U) \backslash 0; \quad \tau = \tau',$$
$$\eta_n = \eta^{\pm}(y, \tau^2, \eta'), \ (t, x, \tau, \xi) \text{ and } (t', y, \tau, \eta)$$
$$\text{lie on a generalized bicharacteristic of } \square \text{ and } (t', y', \tau, \eta') \in \Gamma^{\pm}\}.$$

In the following we consider $(x_n, y_n) \in [0, \alpha] \times [0, \alpha]$ as parameters. The projection

$$\mathcal{M}_+ \ni (t, t', x, y, \tau, \tau, \xi, \eta) \rightarrow (t, x', x_n, y_n, \tau, \eta') \in \mathcal{I} \times U' \times [0, \alpha] \times [0, \alpha] \times V$$

is a diffeomorphism. In fact, for fixed $\hat{x}_n$, $\hat{y}_n$ the flow Φ^t induces a homogeneous canonical transformation form $y_n = \hat{y}_n$ into $x_n = \hat{x}_n$. Then, there exists a phase function $\varphi(t, x, y_n, \tau, \eta')$, homogeneous of order 1 in (τ, η'), so that $\mathcal{M}_+$ locally has the form

$$\{(t, \varphi_{\tau}, x, \varphi_{\eta'}, y_n, \varphi_t, \tau, \varphi_x, \eta', \zeta_n^+) : (t, x', \tau, \eta') \in \mathcal{I} \times U' \times V\}.$$

Here $\zeta_n^+(t, x, y_n, \tau, \eta')$ is determined from the equation

$$q(\varphi_{\eta'}, y_n, \tau, \eta', \zeta_n) = 0$$

with respect to ζ_n so that $\zeta_n^+(T, y_0', 0, 0, \tau_0, \eta_0') = \eta_0^+$. Write

$$\mathcal{F}_B^+(t-t', x, y) = (2\pi)^{-n} \int e^{i\varphi(t, x, y_n, \tau, \eta') - it'\tau - i\langle y', \eta'\rangle} \sum_{j=0}^{\infty} b_j(t, x, y_n, \tau, \eta') \, d\tau d\eta'$$

with b_j homogeneous of order $(-j)$ with respect to (τ, η'). Notice that $\tau < 0$ and $|\eta'| \le C_0|\tau|$ on $\mathcal{M}_+$ with C_0 independent of η'. Thus, modulo C^∞ terms, we get

$$I_\alpha = \int_{\Omega \cap U} \mathcal{F}_B^+(t, x, x) \, dx = (2\pi)^{-n} \int_0^\alpha dx_n \int_1^\infty d\tau$$

$$\times \int_{U'} \int_{|\eta'| \le C_0} e^{i\tau \Psi^+(t, x, \eta')} \sum_j b_j(t, x, x_n, -1, \eta') \tau^{n-1-j} \, dx' d\eta'$$

with $\Psi^+(t, x, \eta') = \varphi(t, x', x_n, x_n, -1, \eta') - \langle x', \eta'\rangle$. For the integral with respect to x' and η' we apply a stationary phase argument. The critical points $\hat{x}', \hat{\eta}'$ satisfy

$$\varphi_{x'} = \eta', \quad \varphi_{\eta'} = x',$$
$$q(\hat{x}', x_n, -1, \hat{\eta}', \varphi_{x_n}) = q(\hat{x}', x_n, -1, \hat{\eta}', \zeta_n^+) = 0.$$

Since $\varphi_{x_n}(T, y_0', 0, 0, -1, \eta_0') = \eta_0^+$, we deduce $\varphi_{x_n} = \zeta_n^+$ at the critical points and

$$(\hat{x}', x_n, \hat{\eta}', \zeta_n^+(\ldots)) = \Phi^{t-\varphi_{\tau'}}(\hat{x}', x_n, \hat{\eta}', \zeta_n^+(\ldots)).$$

Consequently,

$$\hat{x}' = y_0', \quad \hat{\eta}' = \eta_0', \quad \varphi_{t'}(T, y_0', x_n, x_n, -1, \eta_0') = 0$$

and by the Euler equality for φ we get

$$\Psi^+(t, y_0', x_n, \eta_0') = T - t.$$

Set

$$\Delta_+ = \begin{pmatrix} \Psi_{x'x'}^+ & \Psi_{x'\eta'}^+ \\ \Psi_{\eta'x'}^+ & \Psi_{\eta'\eta'}^+ \end{pmatrix} (T, y_0', 0, \eta_0').$$

Since l_{j+1} is transversal to $\partial\Omega$, as in the previous case, we conclude that $\det \Delta_+ \neq 0$. According to Morse lemma for the phase $\Psi^+(T, x', x_n, \eta')$ with parameter x_n, there exist smooth functions $x'(x_n), \eta'(x_n), z(x', x_n, \eta') = (z_1, \ldots, z_{2n-2})(x', x_n, \eta')$, determined for $0 \le x_n \le \varepsilon$ and (x', η') close to (y_0', η_0'), so that

$$\text{grad}_{x', \eta'} \Psi^+(T, x'(x_n), x_n, \eta'(x_n)) = 0,$$
$$\Psi^+(T, x'(x_n), x_n, \eta'(x_n)) = \frac{\langle \Delta_+ z, z\rangle}{2}.$$

Assuming α and Γ^+ small enough, we obtain

$$(2\pi)^{-n} \int_0^\alpha dx_n \int_1^\infty d\tau e^{i\tau(T-t)} \int_Z e^{i\tau\langle \Delta_+ z, z\rangle/2} \cdot \sum_j \tilde{b}_j(t, z, x_n) \tau^{n-1-j} dz$$

$$= c_\alpha^+ \exp\left[i\frac{\pi}{2}(2m_\gamma + \sigma_\gamma \right.$$

$$\left. + \frac{\operatorname{sgn}\Delta_+}{2} - 1) \right] |\det(P_\gamma - I)|^{-1/2} (t - T - i0)^{-1} + L_{\text{loc}}^1(\mathbf{R}). \quad (6.24)$$

Here $Z \subset \mathbf{R}^{2n-2}$ and $\tilde{b}_j$ are symbols obtained from b_j by the change of co-ordinates. The precise form of $c_\alpha^+ > 0$ is not important because $c_\alpha^+ \to 0$ as $\alpha \to 0$.

To find the trace of $\mathcal{F}_{B_{k_-}}^-(t - t', x, y)$, we use once more the argument based on the equality

$$\mathcal{F}^-(t - t', x, y) = \overline{\mathcal{F}^+(t - t', x, y)},$$

$\mathcal{F}^-(t - t', x, y)$ being the kernels of the operators

$$\exp(\mp i(t - t')A).$$

Then for $t \in \mathcal{I}$, $t' \in \mathcal{J}$, modulo C^∞ terms, we have

$$\int_{\Omega \cap U} \mathcal{F}_{B_{k_-}}^-(t - t', x, x)\, dx = \int_{\Omega \cap U} \mathcal{F}^-(t - t', x, x)\, dx$$

$$= \int_{\Omega \cap U} \overline{\mathcal{F}^+(t - t', x, x)}\, dx = \int_{\Omega \cap U} \overline{\mathcal{F}_{B_{k_+}}^+(t - t', x, x)}\, dx$$

and taking $t' = 0$ we deduce

$$\int_{\Omega \cap U} \mathcal{F}_{B_{k_-}}^-(t, x, x)\, dx = \bar{I}_\alpha$$

$$= c_\alpha^+ \exp\left[-i\frac{\pi}{2}\left(2m_\gamma + \sigma_\gamma + \frac{\operatorname{sgn}\Delta_+}{2} - 1 \right) \right] |\det(P_\gamma - I)|^{-1/2}$$

$$\times (t - T + i0)^{-1} + L_{\text{loc}}^1(\mathbf{R}). \quad (6.25)$$

On the other hand, we may use a local representation of $\mathcal{M}_-$ by a phase function $\Psi(t, x', x_n, y_n, \tau, \eta')$.

Setting $\Psi^-(t, x, \eta') = \Psi(t, x', x_n, y_n, 1, \eta') - \langle x', \eta' \rangle$,

$$\Delta_- = \begin{pmatrix} \Psi_{x'x'}^- & \Psi_{x'\eta'}^- \\ \Psi_{\eta'x'}^- & \Psi_{\eta'\eta'}^- \end{pmatrix} (T, y_0', 0, \eta_0'),$$

we compute the trace of $\mathcal{F}_{B_{k^-}}^-$ repeating the above argument. Comparing the leading terms for the trace of $\mathcal{F}_{B_{k^-}}^-$, we obtain

$$\operatorname{sgn}\Delta_- = -\operatorname{sgn}\Delta_+ (\operatorname{mod} 4). \tag{6.26}$$

Below we consider β_γ, given by (6.21), as an element of $\mathbf{Z}_4$. Our aim is to show that β_γ depends on γ, only. First, assume that for some segment l_j we have

$$\mathring{\Omega} \supset U_k \cap l_j \neq \emptyset, \quad \mathring{\Omega} \supset U_m \cap l_j \neq \emptyset.$$

Covering $\mathring{l}_j$ by a chain of neighbourhoods connecting U_k and U_m and using the leading term in (6.22), we conclude that β_γ does not depend on the choice of U_k.

To see that β_γ does not depend on l_j, take two sufficiently small neighbourhoods $U_k \subset \mathring{\Omega}$, $k = j, j+1$ so that

$$U_j \cap l_j \neq \emptyset, \quad U_{j+1} \cap l_{j+1} \neq \emptyset$$
$$U_k \subset \{(x'x_n) : x' \in U', \ 0 < \varepsilon \le x_n \le \alpha\}, \quad k = j, j+1.$$

Then from (6.22) and (6.24) we conclude that

$$\beta_\gamma = 2m_\gamma + \sigma_\gamma + \frac{\operatorname{sgn}\Delta_+}{2} - 1.$$

Similarly, taking the trace of F_B^- over U_j, we get

$$\beta_\gamma = 2m_\gamma + \sigma_\gamma - \frac{\operatorname{sgn}\Delta_-}{2} - 1.$$

Then the independence of β_γ follows from (6.26). Finally, $\beta_\gamma \in \mathbf{Z}_4$ depends on γ, only.

Summing up the contributions from the leading terms in (6.23) and letting $\alpha \to 0$ in (6.24), we obtain the following.

Theorem 6.3.1: *Let γ be a periodic ordinary reflecting bicharacteristic of $\square$ in Ω with period $T_\gamma > 0$ and primitive period $T_\gamma^\sharp > 0$. Assume the conditions (i) and (ii) fulfilled. Then the distribution $\sigma(t)$ near T_γ has the form*

$$\sigma(t) = \frac{T_\gamma^\sharp}{2\pi} \operatorname{Re}\left[\exp\left(\mathrm{i}\frac{\pi}{2}\beta_\gamma\right)(t - T_\gamma - \mathrm{i}0)^{-1}\right] |\det(P_\gamma - I)|^{-1/2} + L_{\mathrm{loc}}^1(\mathbf{R}). \tag{6.27}$$

Below we discuss briefly the Neumann and Robin problems. For these problems we repeat the construction from Section 6.1. Let V_k^+ be Fourier integral distributions related to the same canonical relation as in the case of the Dirichlet problem.

We take

$$W_p = \sum_{k=0}^{p} V_k^+. \tag{6.28}$$

We must show that for $0 \le t \le \hat{t}_p$ the boundary conditions

$$i^* \left(\frac{\partial}{\partial \nu} + \alpha \right) (V_k^+ + V_{k-1}^+) \in C^\infty, \quad k = 1, \dots, p, \tag{6.29}$$

hold. Here ν is the unit normal to $\partial\Omega$, pointing into the exterior of Ω, while $\alpha \in C^\infty(\partial\Omega)$.

Using the notation of Section 6.2 consider the term

$$i_{\omega_k} \frac{\partial}{\partial \nu} (R_k^+ (I - M_{k-1}) i^* L_{k-1} + L_{k-1}) i^* R_B^+. \tag{6.30}$$

Let $\partial\Omega$ have locally the form $x_n = 0$ and let

$$q(x, \tau, \xi) = a(x)\xi_n^2 - 2h(x, \xi')\xi_n + r(x, \xi') - \tau^2$$

be the principal symbol of $\square$. Here $h(x, \xi')$ is linear in ξ', $r(x, \xi')$ is homogeneous of order 2 in ξ' and $a(x) \neq 0$. Putting $\mu(x, \tau, \xi') = h^2(x, \xi') - (r(x, \xi') + \tau^2)a(x)$, the roots $\xi_n^\pm$ of the equation $q(x, \tau, \xi', \xi_n) = 0$ with respect to ξ_n become

$$\xi_n^\pm = a^{-1}(x)\big(h(x, \xi') \mp \sqrt{\mu(x, \tau, \xi')}\big).$$

In local coordinates the vector field $\partial/\partial\nu$ is collinear with the vector field $-h(x, \partial/\partial x') + a(x)\partial/\partial x_n$.

Now let $\varphi^\pm(t, x, \tau, \xi')$ be the phase functions introduced in Section 6.2. The kernel of L_{k-1} has the form (6.13) and the principal symbol of $i_{\omega_k} \partial/\partial\nu L_{k-1}$ becomes

$$i[-h(x, \varphi_{x'}^-) + a(x)\xi_n^-(x, \varphi_t^-, \varphi_{x'}^-)]|_{x_n=0} = i\sqrt{\mu(x, \varphi_t^-, \varphi_{x'}^-)}\Big|_{x_n=0}$$

Similarly, the principal symbol of

$$i_{\omega_k} \frac{\partial}{\partial \nu} (R_k^+ (I - M_{k-1}) i^* L_{k-1})$$

is equal to

$$-i\sqrt{\mu(x, \varphi_t^+, \varphi_{x'}^+)}\Big|_{x_n=0}.$$

On the other hand, for $x_n = 0$ we have $\varphi^+ = \varphi^-$, hence the principal symbol of (6.30) vanishes. This explains the choice of the sign $+$ in (6.30). Choosing suitably the lower order symbols of R_k^+, we can arrange (6.29). To find the leading singularity of $\sigma(t)$ near T_γ, we need to know only the principal symbol of $V^+_{m_\gamma}$. Thus, repeating the argument of Section 6.3, we get the following.

Theorem 6.3.2:   *Under the assumptions and notation of Theorem 6.3.1, the distribution $\sigma(t)$, related to the eigenvalues of Neumann and Robin boundary problems in Ω, near T_γ has the form*

$$\sigma(t) = \frac{T^\sharp_\gamma}{2\pi} \operatorname{Re}\left[\exp\left(i\frac{\pi}{2}\delta_\gamma\right)(t - T_\gamma - i0)^{-1}\right]\left|\det(P_\gamma - I)\right|^{-1/2} + L^1_{\mathrm{loc}}(\mathbf{R}) \tag{6.31}$$

where

$$\delta_\gamma = \sigma_\gamma + \frac{\operatorname{sgn}\Delta}{2} - 1 \in \mathbf{N}$$

depends only on γ. ♠

In the special case when $\delta_\gamma = -1$ the formulae (6.27) and (6.31) can be simplified. To do this, notice that

$$2\operatorname{Re}(-i(t - T_\gamma - i0)^{-1}) = i(t - T_\gamma + i0)^{-1} - i(t - T_\gamma - i0)^{-1} = 2\pi\delta(t - T_\gamma).$$

Hence we have the following.

Corollary 6.3.3:   *Under the assumptions of Theorem 6.3.1, let $\beta_\gamma = 2m_\gamma - 1$ (resp. $\delta_\gamma = -1$ for Robin boundary problem). Then for t near T_γ we have*

$$\sigma(t) = \frac{1}{2}(-1)^{m_\gamma} T^\sharp_\gamma \left|\det(P_\gamma - I)\right|^{-1/2}\delta(t - T_\gamma) + L^1_{\mathrm{loc}}(\mathbf{R}),$$

where for Robin problem the factor $(-1)^{m_\gamma}$ is omitted. ♠

This corollary can be applied for ordinary periodic reflecting rays γ in the exterior of a finite disjoint union of strictly convex domains. In this case the Maslov index σ_γ is zero and $\operatorname{sgn}\Delta = 0$ (see [I6]).

Finally, notice that the classical Poisson summation formula has the form

$$1 + 2\sum_{k=1}^{\infty}\cos 2\pi kt = \sum_{k\in Z}e^{-2\pi ikt} = 2\pi\sum_{k\in Z}\delta(t - 2\pi k). \tag{6.32}$$

The Laplace–Beltrami operator $-d^2/ds^2$ on $\mathbf{S}^1$ has eigenvalues $\lambda_k^2 = (2\pi k)^2$ and all periodic geodesics have primitive periods $T^\sharp_\gamma = 2\pi$. Thus, (6.32) describes the singularities of $\sum_{k=1}^{\infty}\cos 2\pi kt$. For this reason the formulae (6.27) and (6.31) are called Poisson summation formula for manifolds with boundary.

6.4. Notes

The construction of the global parametrix in Sections 6.1 and 6.2 follows the work of Guillemin and Melrose [GM1] (see also Chapter 29 in [H4] for a similar construction). Theorems 6.3.1 and 6.3.2 are proved in [GM1]. In Section 6.3 we present a more detailed proof discussing the invariance of the leading term in (6.27) and (6.31). An application of these results is considered in [GM2], where a simple inverse problem for the ellipse is studied.

For manifolds without boundary the singularity of $\sigma(t)$, related to the set of periodic bicharacteristics with periods T_γ, has been examined in [Ch2], [DG].

7 INVERSE SPECTRAL RESULTS FOR GENERIC BOUNDED DOMAINS

In this chapter we study the Poisson relation for generic bounded domains Ω in $\mathbf{R}^n$ with smooth boundaries $\partial\Omega$. We begin with the simplest case $n = 2$ and show that for generic $\Omega \subset \mathbf{R}^2$, without any conditions concerning convexity, the Poisson relation becomes an equality. A similar result for $n \geq 3$ is only known for the space of strictly convex domains. Modulo the results of the previous chapters, the main difficulty here is to deal with the lengths T_γ of the closed geodesics γ on $\partial\Omega$. The central point in this direction is the existence of periodic reflecting rays in Ω approximating a given non-degenerate closed geodesics on $\partial\Omega$. The proof of this approximation result is based on Melrose' interpolating Hamiltonians for the billiard ball map.

7.1. Planar domains

In this section we establish that for generic bounded domains Ω in $\mathbf{R}^2$ with smooth boundaries $X = \partial\Omega$ the Poisson relation becomes an equality, i.e. we have

$$\text{sing supp } \sigma_\Omega = \{0\} \cup \{\pm T_\gamma : \gamma \in \mathcal{L}_\Omega\}. \tag{7.1}$$

To this end we use the results from Chapters 3, 4 and 5. The main point here is to show that for generic Ω, $\mathcal{L}_\Omega$ consists only of periodic reflecting rays and, eventually, multiples of the boundary $\partial\Omega$.

Let X be an arbitrary fixed smooth curve in $\mathbf{R}^2$. A curve γ in $\mathbf{R}^2$ of the form

$$\gamma = \cup_{i=1}^{k-1} l_i,$$

with $x_1, \ldots, x_k \in X$, will be called a *degenerate broken ray* for X if the following conditions are satisfied:

(i) for every $i = 1, \ldots, k, l_i = [x_i, x_{i+1}]$ is a linear segment, the interior of which does not intersect transversally X;

(ii) l_i and l_{i+1} satisfy the law of reflection at x_{i+1} with respect to X for every $i = 1, \ldots, k - 2$;

(iii) the curvature of X vanishes at x_1 and x_k, and l_1 and l_{k-1} are tangent to X at x_1 and x_k respectively.

The points $x_1, \ldots, x_k$ will be called *vertices* of γ.

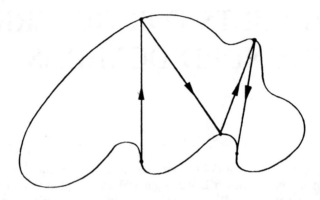

Figure 7.1

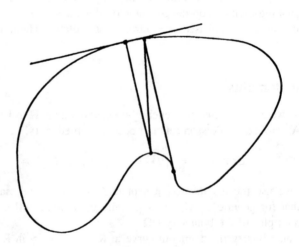

Figure 7.2

Clearly, for $k = 2$ condition (ii) is to be dropped. Examples of degenerate broken rays are drawn on Figures 7.1 and 7.2. If γ contains a segment orthogonal to X at some of its end points, we shall say that γ is *symmetric* (cf. Figure 7.2); otherwise γ will be said to be *non-symmetric*. Note that $l_1 = l_{k-1}$ for symmetric γ.

Proposition 7.1.1: *There exists a residual subset $\mathcal{D}$ of $C(X)$ such that for every $f \in \mathcal{D}$ there are no degenerate broken rays for $f(X)$.*

Proof: We use the technique from Sections 3.2, 3.3 and 3.4, slightly exchanging the definitions of the main objects.

Fix arbitrary integers k and s with $k \geq s \geq 2$, and consider a ns-map

$$\alpha : \{1, \ldots, k\} \to \{1, \ldots, s\} \tag{7.2}$$

(cf. the definition in Section 3.2) such that

$$\alpha(1) = 1, \quad \alpha(2) = 2, \quad \alpha(k) = s. \tag{7.3}$$

Next we shall use the notation $I_i = I_i(\alpha)$, introduced by (3.25). Denote by U_α the set of all $y = (y_1, \ldots, y_s) \in (\mathbf{R}^2)^{(s)}$ such that

$$y_i \notin \text{convex hull } \{y_j : j \in I_i\}$$

for each $i = 1, \ldots, s-1$. Define

$$H = H_\alpha : U_\alpha \to \mathbf{R}$$

by

$$H(y) = \sum_{i=1}^{s-1} \left\| y_{\alpha(i)} - y_{\alpha(i+1)} \right\|.$$

Notice that, given a non-symmetric degenerate broken ray γ for $Y = f(X)$, $f \in C(X)$, there exist integers $k \geq s \geq 2$, a ns-map (7.2) with (7.3) and distinct points $y_1 = f(x_1), \ldots, y_s = f(x_s) \in Y$ such that $y_{\alpha(1)}, \ldots, y_{\alpha(k)}$ are the successive vertices of γ. Then we shall say that γ has type α. In this case we have $x = (x_1, \ldots, x_s) \in X^{(s)}$, $f^s(x) \in U_\alpha$ and

$$\text{grad}_{x'}(H \circ f^s)(x) = 0, \tag{7.4}$$

where

$$x' = (x_2, \ldots, x_{s-1}) \in X^{(s-2)}.$$

The proof of (7.4) is almost the same as that of Proposition 3.2.1. We leave it to the reader. Note also that the curvature of $Y = f(X)$ vanishes at the points $f(x_1)$ and $f(x_s)$, and the segment $[f(x_1), f(x_2)]$ is tangent to $f(X)$ at $f(x_1)$. A condition, similar to the latter one, is satisfied also for $f(x_s)$, however we do not need it here.

Further, introduce the open subset

$$V = \{j^2 f(x) \in J^2(X, \mathbf{R}^2) : \text{ rank } df(x) = 1\}$$

of $J^2(X, \mathbf{R}^2)$. Denote by M the set of all

$$\sigma = (j^2 f_1(x_1), \ldots, j^2 f_s(x_s)) \qquad (7.5)$$

such that $\sigma \in V^s$, $x = (x_1, \ldots, x_s) \in X^{(s)}$, $f^s(x) \in U_\alpha$. Then M is an open subset of the smooth manifold $J_s^2(X, \mathbf{R}^2)$. Let Σ be the set of those elements (7.5) of M such that the curvature of $f_i(X)$ vanishes at $f_i(x_i)$ for $i = 1$ and $i = s$,

$$\text{grad}_{x'}(H \circ (f_1 \times \ldots \times f_s))(x) = 0,$$

and

$$\langle f_2(x_2) - f_1(x_1), N_1 \rangle = 0,$$

N_1 being a non-zero normal vector to $f_1(X)$ at $f_1(x_1)$. It then follows from our reasonings above that for any element f of the set

$$\mathcal{D}_\alpha = \{ f \in C(X) : j_s^2 f(X^{(s)}) \cap \Sigma = \emptyset \}, \qquad (7.6)$$

there are no degenerate broken rays of type α for $f(X)$.

We are going to show that $\mathcal{D}_\alpha$ is residual in $C(X)$. To this end we first prove that Σ is a smooth submanifold of M of codimension $s + 1$.

Consider a coordinate neighbourhood D of an element of Σ in M. We may assume that

$$D = M \cap \prod_{i=1}^s J^2(V_i, \mathbf{R}^2),$$

where $V_1, \ldots, V_s$ are coordinate neighbourhoods of different elements of X such that $V_i \cap V_j = \emptyset$ whenever $i \neq j$. Fix arbitrary smooth charts $\varphi_i : V_i \to \mathbf{R}$, and define the chart

$$\varphi : D \to \mathbf{R}^{(s)} \times (\mathbf{R}^2)^{(s)} \times \mathbf{R}^{2s} \times \mathbf{R}^{2s}$$

by $\varphi(\sigma) = (u; v; a; b)$ for every element σ of D having the form (7.5), where u and v are determined by (3.12) and (3.13),

$$a = (a_i^{(t)})_{1 \leq i \leq s, 1 \leq t \leq 2}, \quad b = (b_i^{(t)})_{1 \leq i \leq s, 1 \leq t \leq 2}, \qquad (7.7)$$

$$a_i^{(t)} = \frac{\partial (f_i^{(t)} \circ \varphi_i^{-1})}{\partial u_i}(u_i), \quad b_i^{(t)} = \frac{\partial^2 (f_i^{(t)} \circ \varphi_i^{-1})}{\partial u_i^2}(u_i)$$

for all $i = 1, \ldots, s, t = 1, 2$. Let us recall that the vector $N_i = (N_i^{(1)}, N_i^{(2)})$, determined by (3.42) for $n = 2$, is orthogonal to $f_i(X)$ at the point $f_i(x_i)$. A straightforward verification shows that the curvature of $f_i(X)$ vanishes at $f_i(x_i)$ if and only if $a_i^{(1)} b_i^{(2)} - a_i^{(2)} b_i^{(1)} = 0$.

To see that Σ is a smooth submanifold of M of codimension $s + 1$, it is sufficient to establish that $\varphi(D \cap \Sigma)$ is a smooth submanifold of $\varphi(D)$ of the same codimension. We use the procedure applied several times in Chapter 3.

Define the map $R : \varphi(D) \rightarrow \mathbf{R}^{s+1}$ by

$$R(\xi) = ((c_i(\xi))_{2 \leq i \leq s-1}; \ L_1(\xi); \ L_s(\xi); \ K(\xi)),$$

where the elements ξ of $\varphi(D)$ are written in the form $\xi = (u; v; a; b)$ with u, v, a and b given by (3.12), (3.13) and (7.6), and the functions C_i, L_1, L_s, K are defined as follows:

$$c_i(\xi) = \frac{\partial H}{\partial y_i^{(1)}}(v)a_i^{(1)} + \frac{\partial H}{\partial y_i^{(2)}}(v)a_i^{(2)}, \quad i = 2, \ldots, s-1,$$

$$L_j(\xi) = a_j^{(1)}b_j^{(2)} - a_j^{(2)}b_j^{(1)}, \quad j = 1, s,$$

$$K(\xi) = (v_2^{(1)} - v_1^{(1)})a_1^{(2)} - (v_2^{(2)} - v_1^{(2)})a_1^{(1)}.$$

It is clear now that

$$\varphi(D \cap \Sigma) = R^{-1}(0).$$

Next, we show that R is a submersion on $\varphi(D)$. This will imply that $R^{-1}(0)$ is a smooth submanifold of $\varphi(D)$ of codimension $s+1$.

Let $\xi \in \varphi(D)$ and assume that

$$\sum_{i=2}^{s-1} C_i \operatorname{grad} c_i(\xi) + A_1 \operatorname{grad} L_1(\xi) + A_s \operatorname{grad} L_s(\xi) + B \operatorname{grad} K(\xi) = 0$$

for some real constants C_i, A_j, B. For a given $j = 1, s$ we have either $a_j^{(1)} \neq 0$ or $a_j^{(2)} \neq 0$. Thus, considering the derivatives with respect to $b_j^{(2)}$ and $b_j^{(1)}$ in the above equality, we find $A_1 = A_s = 0$. In a similar way, using the fact that the functions $c_i(\xi)$ do not depend on $a_1^{(1)}, a_1^{(2)}$ and $v_1 = (v_1^{(1)}, v_1^{(2)}) \neq 0$, one gets $B = 0$. Finally, using an argument very similar to that in the proof of Lemma 3.1.2 and a clear analogue of Lemma 3.2.2 for our present function H, we obtain $C_i = 0$ for all $i = 2, \ldots, s-1$. Consequently, R is a submersion on $\varphi(D)$ which implies that $\varphi(D \cap \Sigma)$ is a smooth submanifold of $\varphi(D)$ of codimension $s+1$.

In this way we have established that Σ is a smooth submanifold of M of codimension $s+1$. It then follows by the definition of $\mathcal{D}_\alpha$ that

$$\mathcal{D}_\alpha = \{f \in \mathbf{C}(X) : j_s^2 f \pitchfork \Sigma\}$$

(cf. for example the end of the proof of Theorem 3.1.1). Now Theorem 1.1.2 yields that $\mathcal{D}_\alpha$ is residual in $\mathbf{C}(X)$.

Set $\mathcal{D}' = \cup_\alpha \mathcal{D}_\alpha$, where α runs over the set of ns-maps (7.2) satisfying (7.3). Then $\mathcal{D}'$ is a residual subset of $\mathbf{C}(X)$. It follows from our considerations above and the definition of the sets $\mathcal{D}_\alpha$ that for every $f \in \mathcal{D}'$ there are no non-symmetric degenerate broken rays for $f(X)$.

As in Chapter 3, the treatment of the symmetric case consists of a slight modification of the argument in the non-symmetric one. We leave to the reader the proof
of the fact that there exists a residual subset $\mathcal{D}''$ of $\mathbf{C}(X)$ such that for $f \in \mathcal{D}''$
there are no symmetric degenerate broken rays for $f(X)$. It is clear now that the
set $\mathcal{D} = \mathcal{D}' \cap \mathcal{D}''$ has the desired properties. ♠

Applying the above proposition, we wish to show that for generic $\Omega \subset \mathbf{R}^2$ every
generalized periodic geodesic in Ω, which is not contained in $\partial\Omega$, is a periodic
reflecting ray. To this end we could apply the results of Melrose and Sjöstrand on
the properties of the generalized bicharacteristics of $\square$ only if we knew that the
curvature of $\partial\Omega$ does not vanish of infinite order. The following proposition shows
that a much stronger condition is satisfied for generic domains. We prove it in a
more general form having in mind another application in Chapter 9.

Proposition 7.1.2: *Let X be a smooth $(n-1)$-dimensional submanifold of $\mathbf{R}^n$,
$n \geq 2$, and let $\mathcal{K}$ be the set of those $f \in \mathbf{C}(X)$ for which there are no points
$y \in f(X)$ and directions $v \in T_y f(X)\backslash\{0\}$ such that the curvature of $f(X)$ at
y in direction v vanishes of order $2n - 3$. Then $\mathcal{K}$ is a residual subset of $\mathbf{C}(X)$.*

In the proof of this proposition we shall use the following lemma which can
be established by the same argument as that in the proof of Thom's transversality
theorem (cf. [GG]).

Lemma 7.1.3: *Let Σ, X and Y be smooth manifolds, $k \geq 1$ be an integer and
let $g : \Sigma \rightarrow J^k(X,Y)$ be a smooth map. Then*

$$R = \{f \in C^\infty(X,Y) : g \pitchfork j^k f\}$$

is a residual subset of $C^\infty(X,Y)$. ♠

Proof of Proposition 7.1.2: Set $k = 2n - 1$ and denote by $P\mathbf{R}^n$ the *projective
space* of all lines through 0 in $\mathbf{R}^n$. Next, denote by M the set of all

$$(w, j^k f(x)) \in P\mathbf{R}^n \times J^k(X,\mathbf{R}^n)$$

such that rank $\mathrm{d}f(x) = n - 1$ and w is tangent to $f(X)$ at $f(x)$. It is easily seen
that M is a smooth submanifold of $P\mathbf{R}^n \times J^k(X,\mathbf{R}^n)$. Let Σ be the set of those
$(w, j^k f(x)) \in M$ such that the curvature of $f(X)$ at $f(x)$ vanishes of order
$2n - 3$ in direction w. Below we describe analytically the latter condition.

We claim that Σ is a smooth submanifold of M of codimension $2n - 2$. To
prove this, consider the open covering $\{\mathcal{O}_j\}_{j=1}^n$ of $P\mathbf{R}^n$, where $\mathcal{O}_j$ is determined
by all vectors $w \in \mathbf{R}^n$ with $w^{(j)} \neq 0$. It is sufficient to show that for every
$j = 1,\ldots,n$,

$$\Sigma_j = \{(w, j^k f(x)) \in \Sigma : w \in \mathcal{O}_j\}$$

is a submanifold of M of codimension $2n - 2$. We shall do this for $j = 1$, the
other cases are the same.

Consider an arbitrary smooth chart $\varphi : U \rightarrow V \subset X$, where U is an open
subset of $\mathbf{R}^{n-1}$, while V is an open neighbourhood of some element of X. Define

the chart

$$\psi \; : \; W \; = \; M \cap (\mathcal{O}_1 \times J^k(X, \mathbf{R}^n)) \; \to \; \mathbf{R}^{n-2} \times \mathbf{R}^{n-1} \times \mathbf{R}^n \times \mathbf{R}^N$$

by $\psi(w, j^k f(x)) \; = \; (\lambda; u; v; a)$. Here $\lambda \; = \; (\lambda_2, \ldots, \lambda_{n-1}) \; \in \; \mathbf{R}^{n-2} \backslash \{0\}$, w is determined by the vector

$$\frac{\partial \varphi}{\partial u_1}(u) + \sum_{j=2}^{n-1} \lambda_j \frac{\partial \varphi}{\partial u_j}(u), \quad \varphi(u) \; = \; x,$$

$v \; = \; f(x)$ and $a \; = \; (a_{\mathbf{i}}^{(t)})$ is given by

$$a_{\mathbf{i}}^{(t)} \; = \; \frac{\partial^{\mathbf{i}}(f^{(t)} \circ \varphi)}{\partial u^{\mathbf{i}}}(u)$$

for every $t \; = \; 1, \ldots, n$ and every multi-index $\mathbf{i} \; = \; (i_1, \ldots, i_{n-1})$ with $|\mathbf{i}| \; = \; i_1 + \ldots + i_k \; \leq \; k$. As usually we use the notation $f \; = \; (f^{(1)}, \ldots, f^{(n)})$.

The assertion will be proved if we show that $\psi(\Sigma \cap W)$ is a smooth submanifold of $\psi(W)$ of codimension $2n - 2$. For each $u \; \in \; U$ choose a unit normal vector $\nu(u)$ to $f(X)$ at $f(\varphi(u))$ so that the map $u \; \mapsto \; \nu(u)$ is continuous. Then the coefficients $b_{ij}(u)$ of the second fundamental form of $f(X)$ at $f(x)$, $x \; = \; \varphi(u)$, are determined by

$$b_{ij}(u) \; = \; \left\langle \frac{\partial^2 \varphi}{\partial u_i \partial u_j}(u), \nu(u) \right\rangle,$$

for $i, j \; = \; 1, \ldots, n - 1$. Now set $\lambda_1 \; = \; 1$, and for $j \; = \; 0, 1, 2, \ldots$ consider the functions

$$K_j(\lambda; u) \; = \; \left(\sum_{i=1}^{n-1} \lambda_i \frac{\partial}{\partial u_i} \right)^j \left(\sum_{p,q=1}^{n-1} \lambda_p \lambda_q b_{pq}(u) \right).$$

The curvature of $f(X)$ at $f(x)$ vanishes of order m in direction

$$w(\lambda) \; = \; \sum_{i=1}^{n-1} \lambda_i \frac{\partial \varphi}{\partial u_i}(u)$$

if and only if $K_j(\lambda; u) \; = \; 0$ for all $j \; = \; 0, 1, \ldots, m$ (cf. [GKM] for example).

It is clear that there exist smooth functions

$$\tilde{\nu}, \tilde{K}_j \; : \; \psi(W) \; \to \; \mathbf{R}$$

such that $\tilde{\nu}(\xi) = \nu(u)$, $\tilde{K}_j(\xi) = K_j(\lambda; u)$ for $\xi \; = \; (\lambda; u; v; a) \; = \; \psi(w; j^k f(x))$. Define the map

$$K \; : \; \psi(W) \; \to \; \mathbf{R}^{2n-2}$$

by $K(\xi) = (\tilde{K}_j(\xi))_{j=0}^{2n-3}$. Then $\psi(W \cap \Sigma) = K^{-1}(0)$, and it is sufficient to show that K is a submersion at any point of $\psi(W \cap \Sigma)$.

Let $\xi = (\lambda; u; v; a) \in \psi(W \cap \Sigma)$ and

$$\sum_{j=0}^{2n-3} A_j \operatorname{grad} \tilde{K}_j(\xi) = 0 \tag{7.8}$$

for some real constants A_j. Since $W \subset M$ and $\xi \in \psi(W)$, there exists $t = 1, \ldots, n$ with $\tilde{\nu}^{(t)}(\xi) \neq 0$. Fix such a t and for the multi-index $\mathbf{i} = (2n-1, 0, \ldots, 0)$ consider in (7.7) the derivatives with respect to $a_{\mathbf{i}}^{(t)}$. We get $A_{2n-3}\tilde{\nu}^{(t)}(\xi) = 0$, which implies $A_{2n-3} = 0$. In the same way one finds $A_{2n-4} = 0$, etc. Thus $A_j = 0$ for all j and K is a submersion. This completes the proof of the fact that Σ is a smooth submanifold of M of codimension $2n - 2$.

Now we apply Lemma 7.1.3 for $\Sigma, X, Y = \mathbf{R}^n$ and $g = \pi \circ i$, where $i : \Sigma \to M$ is the inclusion and $\pi : M \to J^k(X, \mathbf{R}^n)$ is the natural projection. Since

$$\operatorname{codim} \operatorname{Im} T_\sigma \pi \geq (2n-2) - (n-2) = n$$

and $\dim X = n - 1 < n$, the condition $j^k f \pitchfork g$ is equivalent to the relation $j^k f(X) \cap \pi(\Sigma) = \emptyset$. On the other hand, $\mathcal{K}$ coincides with the set of those $f \in \mathbf{C}(X)$ satisfying the latter relation. Therefore by Lemma 7.1.3 $\mathcal{K}$ is a residual subset of $\mathbf{C}(X)$. ♠

Let us turn to the case $X \subset \mathbf{R}^2$. From now on we consider the case when X is *compact and connected*, i.e. X is the boundary of a bounded domain Ω in $\mathbf{R}^2$. Given $f \in \mathbf{C}(X)$ we denote by Ω_f the bounded domain in $\mathbf{R}^2$ with boundary $\partial \Omega_f = f(X)$, and by L_f the length of the curve $f(X)$. In the notation of Propositions 7.1.1 and 7.1.2 we have the following.

Corollary 7.1.4:   *Let $f \in \mathcal{D} \cap \mathcal{K}$. Then every periodic generalized geodesic in Ω_f, which is not contained in $f(X)$, is a periodic reflecting ray in Ω_f.*

Proof: Consider an arbitrary periodic generalized geodesic γ in Ω_f which is not contained in $\partial \Omega_f = f(X)$. Since $f \in \mathcal{K}$ the curvature of $\partial \Omega_f$ can only simply vanish. Therefore γ is the projection of a uniquely extendible bicharacteristic of $\square$ and so γ consists of a finite number linear segments and segments of $\partial \Omega_f$ (see Section 1.2). Clearly, γ has at least one linear segment with points in the interior of Ω_f. If γ contains at least one non-trivial segment of $\partial \Omega_f$, then it would contain a whole degenerate broken ray for $f(X)$, which is a contradiction with $f \in \mathcal{D}$. Thus, γ consists only of linear segments, so it is a periodic reflecting ray in Ω_f. ♠

The next lemma will be useful when we consider generic strictly convex domains $\Omega \subset \mathbf{R}^2$ and try to prove that the length of $\partial \Omega$ is contained in sing supp σ_Ω.

Lemma 7.1.5:   *There exists a residual subset $\mathcal{W}$ of $\mathbf{C}(X)$ such that every $f \in \mathcal{W}$ has the following properties:*

(a) *every periodic reflecting ray in Ω_f is ordinary and non-degenerate;*

(b) $T_\gamma/T_\delta \notin \mathbf{Q}$ *for every two different primitive periodic reflecting rays γ and δ in Ω_f;*

(c) *for every integer $s \geq 2$ there are only finitely many periodic reflecting rays in Ω_f with s reflection points;*

(d) $T_\gamma/L_f \notin \mathbf{Q}$ *for every periodic reflecting ray γ in Ω_f.*

Proof: It follows by Theorems 3.2.3, 3.3.1, 3.4.1 and 3.4.3 that there exists a residual subset $\mathcal{W}'$ such that every f in it has the properties (a), (b) and (c).

Fix arbitrary $p, q, s \in \mathbf{N}$, $s \geq 2$, and denote by $\mathcal{W}(p, q, s)$ the set of those $f \in \mathcal{W}'$ such that $pT_\gamma \neq qL_f$ for every periodic reflecting ray γ in Ω_f having s reflection points. We shall show that $\mathcal{W}(p, q, s)$ is open and dense in $\mathcal{W}'$.

To establish the density we may assume that $\mathrm{id} \in \mathcal{W}'$. Then we have to prove that $\mathcal{W}(p, q, s)$ contains elements arbitrarily close to id in the C^∞ topology. Let $\gamma_1, \dots, \gamma_m$ be all the periodic reflecting rays in Ω_{id} with s reflection points; by $\mathrm{id} \in \mathcal{W}'$ there are only finitely many of them. There exists a non-trivial closed segment Δ of X which does not contain reflection points of γ_i for each $i = 1, \dots, m$. We claim that for every $f \in \mathcal{W}'$ which is sufficiently close to id and $f(x) = x$ for $x \in X \backslash \Delta$ the only periodic reflecting rays in Ω_f with s reflection points are $\gamma_1, \dots, \gamma_m$. If not, there would exists a sequence $\{f_k\} \subset \mathcal{W}'$ converging to id such that for every k there is a periodic reflecting ray δ_k in Ω_{f_k} with s reflection points $y_{1,k}, \dots, y_{s,k}$ and $\delta_k \neq \gamma_i$ for all $i = 1, \dots, m$. The latter implies that at least one reflection point of δ_k is in $f_k(\Delta)$. We may assume $y_{1,k} \in f_k(\Delta)$ for every k. Moreover, using the compactness of X, we may assume also that there exists $\lim_k y_{i,k} = y_i$ for every $i = 1, \dots, s$. Then $y_1 \in \Delta$, and a simple continuity argument shows that $y_1, \dots, y_s$ are the successive reflection points of a periodic reflecting ray γ. Since $y_1 \in \Delta$, we have $\gamma \neq \gamma_i$ for every $i = 1, \dots, m$, which is a contradiction. Thus, for every $f \in \mathcal{W}'$ which is sufficiently close to id and $f(x) = x$ for $x \in X \backslash \Delta$, $\gamma_1, \dots, \gamma_m$ are the only periodic reflecting rays in Ω_f with s reflection points. Moreover, we can choose the maps f so that $pT_{\gamma_i} \neq qL_f$ for every $i = 1, \dots, m$. This shows that $\mathcal{W}(p, q, s)$ is dense in $\mathcal{W}'$.

To prove that $\mathcal{W}(p, q, s)$ is open in $\mathcal{W}'$, consider an arbitrary sequence $\{f_k\} \subset \mathcal{W}' \backslash \mathcal{W}(p, q, s)$ with $f_k \to_k f \in \mathcal{W}'$. We have to show that $f \notin \mathcal{W}(p, q, s)$. Without loss of generality, we may assume $f = \mathrm{id}$. For every k there exists a periodic reflecting ray δ_k in Ω_{f_k} having s reflection points $y_{1,k}, \dots, y_{s,k}$ and such that $pT_{\delta_k} = qL_{f_k}$. Using again the compactness of X, we may assume that there exists $\lim_k y_{i,k} = y_i$ for all i. Then $y_1, \dots, y_s$ are the reflection points of a periodic reflecting ray δ in Ω with $pT_\delta = qL_{\mathrm{id}}$, which implies $\mathrm{id} \notin \mathcal{W}(p, q, s)$. Hence $\mathcal{W}(p, q, s)$ is open in $\mathcal{W}'$.

Finally, setting

$$W = \cap_{p,q,s \in \mathbf{N}, s \geq 2} \mathcal{W}(p, q, s),$$

we obtain a residual subset of $\mathbf{C}(X)$ having the desired properties. ♠

Now we turn to the main point in this section. Denote by Ξ the *family of all bounded domains Ω in $\mathbf{R}^2$ with smooth connected boundaries $\partial\Omega$ which have the following properties:*

(ND) *every periodic reflecting ray in Ω is ordinary and non-degenerate;*

(R) $T_\gamma / T_\delta \notin \mathbf{Q}$ for every two different primitive elements γ and δ of $\mathcal{L}_\Omega$;

(K) the curvature of $\partial \Omega$ does not vanish of order $k \geq 1$ and there are no degenerate broken rays in Ω.

Notice that for strictly convex Ω condition (K) is trivially satisfied. If the curve $X = \partial \Omega = \operatorname{Im} \gamma$ is parametrized by a parameter $t \in \mathbf{R}$ and $\kappa(t)$ is the curvature of X at $\gamma(t)$, then (K) implies that if $\kappa(t) = 0$ for some t, then $\kappa'(t) \neq 0$.

As before, given a domain Ω, we denote by $\{\lambda_j^2\}$ the spectrum of the Laplace operator in Ω, and by $\sigma(t) = \sigma_\Omega(t)$ the corresponding distribution, defined in Section 5.2.

Theorem 7.1.6: (a) *For every bounded domain Ω in $\mathbf{R}^2$ with smooth connected boundary $X = \partial \Omega$ there exists a residual subset $\mathcal{W}$ of $\mathbf{C}(X)$ such that $\Omega_f \in \Xi$ for each $f \in \mathcal{W}$;*

(b) *The equality (7.1) holds for every $\Omega \in \Xi$. Moreover, for each periodic reflecting ray γ in Ω, $\Omega \in \Xi$, spec P_γ can be determined from $\{\lambda_j^2\}$.*

Before proceeding with the proof of the theorem, let us explain that the second part in (b) means the following: If for two domains $\Omega_1, \Omega_2 \in \Xi$ the corresponding Dirichlet problems for the Laplacian have one and the same spectrum $\{\lambda_j^2\}$, then there exists a bijection $\mu : \mathcal{L}_{\Omega_1} \to \mathcal{L}_{\Omega_2}$, $\mu(\gamma) = \gamma'$, such that $T_\gamma = T_{\gamma'}$ and spec $P_\gamma = $ spec $P_{\gamma'}$ for every $\gamma \in \mathcal{L}_{\Omega_1}$.

Proof of Theorem 7.1.6: (a) Given Ω with $X = \partial \Omega$, consider the residual set $\mathcal{W}$ from Lemma 7.1.5. Then each $f \in \mathcal{W}$ has the properties (a)–(d) from Lemma 7.1.5. Hence $\Omega_f \in \Xi$.

(b) Let $\Omega \in \Xi$. Then for every periodic reflecting ray γ in Ω the conditions (ND), (R) are satisfied. Thus, for each $\gamma \in \mathcal{L}_\Omega$ the assumptions of Theorem 6.3.1 are fulfilled, hence $T_\gamma \in$ sing supp σ_Ω.

Suppose that Ω is strictly convex. Parametrizing $\partial \Omega$ by the arc length, we obtain a generalized periodic geodesic δ. It follows from Example 2.1.1 that there exists a sequence $\{\gamma_k\}$ of periodic reflecting rays in Ω such that $T_{\gamma_k} \to_k T_\delta$. Since sing supp σ_Ω is a closed subset of $\mathbf{R}$, we get $T_\delta \in$ sing supp σ_Ω. Now Theorems 5.4.6 and 6.3.1 show that (7.1) holds.

Next, consider the case when Ω is not strictly convex. Using the fact that $\Omega \in \Xi$, we shall show that all elements of $\mathcal{L}_\Omega$ are periodic reflecting rays. Assume that there exists $\delta \in \mathcal{L}_\Omega$ which is not a periodic reflecting ray in Ω. Then the condition (K) and the connectedness of $X = \partial \Omega$ yield that $\operatorname{Im} \delta = X$. On the other hand, since Ω is not strictly convex, there exists a point $x \in X$ at which the curvature of X vanishes. Let $\delta(t_0) = x$, and let $\kappa(t)$ be the curvature of X at $\delta(t)$. Since $\kappa(t_0) = 0$, the condition (K) implies $\kappa'(t_0) \neq 0$, so κ changes its sign at t_0. Then by the properties of the generalized bicharacteristics discussed in Proposition 1.2.3, it follows that there exists $\epsilon > 0$ such that either $\{\delta(t) : t_0 - \epsilon < t \leq t_0\}$ or $\{\delta(t) : t_0 \leq t < t_0 + \epsilon\}$ is a linear segment parallel to $\dot{\delta}(t_0)$ and having end x. Now $\operatorname{Im} \delta = X$ implies that X contains a non-trivial linear segment, which is a contradiction with (K). Thus, in this case $\mathcal{L}_\Omega$ contains only periodic reflecting rays, and as above we conclude that the equality (7.1) holds.

Finally, let $\Omega \in \Xi$ and let γ be a periodic reflecting ray in Ω. We have to show that $\{\lambda_j^2\}$ determines spec P_γ. Since γ is non-degenerate, we have spec $P_\gamma = \{a, 1/a\}$ for some $a \neq \pm 1$. It follows by Theorem 6.3.1 that for t close to T_γ we have

$$(2\pi)\sigma_\Omega(t) = \mathrm{Re}[cT_\gamma^\sharp \left|\det(I - P_\gamma)\right|^{-1/2} (t - T_\gamma - \mathrm{i}0)^{-1}] + L_{\mathrm{loc}}^1,$$

where c is a constant such that $c = \pm 1$ or $c = \pm \mathrm{i}$. Since

$$(t - T_\gamma - \mathrm{i}0)^{-1} - (t - T_\gamma + \mathrm{i}0)^{-1} = 2\pi\mathrm{i}\delta(t - T_\gamma),$$

we find from $\sigma_\Omega(t)$

$$T_\gamma^\sharp \left|\det(I - P_\gamma)\right|^{-1/2}.$$

Therefore $T_\gamma^\sharp \left|\det(I - P_\gamma)\right|^{-1/2}$ can be determined from $\{\lambda_j^2\}$. On the other hand, since $\Omega \in \Xi$, $T_\gamma^\sharp$ can be determined by T_γ. In fact, $T_\gamma^\sharp$ is the smallest positive number $u \in \mathrm{sing\,supp}\,\sigma$ such that T_γ/u is an integer. Hence we can determine

$$d = \left|\det(I - P_\gamma)\right| = 2 - (a + 1/a).$$

Since the elements of spec P_γ are the roots of the equation

$$x^2 - (2 - d)x + 1 = 0,$$

this completes the proof of the assertion. ♠

7.2. Interpolating Hamiltonians

The present and the next sections are devoted to the study of the billiard ball map B in a small neighbourhood of a closed geodesic on the boundary of a strictly convex compact domain in $\mathbf{R}^n$. A very useful tool for this aim are the so-called interpolating Hamiltonians. Their existence was established by Melrose [M1] in a more general situation which we are now going to present briefly.

Let (S, ω) be a smooth symplectic manifold, $\dim S = 2m + 2$, and let F and G be two hypersurfaces of S, i.e. F and G are smooth submanifolds of codimension 1 of S. Let $s_0 \in F \cap G$. There exists an open neighbourhood V of s_0 in S and C^∞ functions $f, g : V \to \mathbf{R}^+$ such that

$$F \cap V = f^{-1}(0), \quad G \cap V = g^{-1}(0), \quad \mathrm{d}f \neq 0, \quad \mathrm{d}g \neq 0 \text{ on } V.$$

Then f and g are called *defining functions* in V for F and G, respectively. Since our next considerations are only local around s_0, we may assume that $S = V$, i.e. f and g are globally defined.

Suppose that F and G have a *transversal intersection* at s_0, i.e. $\mathrm{d}f(s_0)$ and $\mathrm{d}g(s_0)$ are linearly independent. Considering a smaller neighbourhood of s_0, we may assume that F and G have a transversal intersection at any point of

$$J = F \cap G.$$

Then J is a smooth submanifold of S of codimension 2. A point $s \in J$ will be called a *glancing point* of F and G if the *Poisson bracket* $\{f, g\}$ of f and g (cf. [AbM] or [H3]) vanishes at s. Then

$$K = \{s \in J : \{f, g\}(s) = 0\}$$

is the set of all glancing points of F and G. If $s \in K$ and

$$\{f, \{f, g\}\}(s) \neq 0, \quad \{g, \{f, g\}\}(s) \neq 0, \tag{7.9}$$

then s will be called a *non-degenerate glancing point*.

From now on we assume that (7.9) holds for every $s \in K$. Then K is a smooth submanifold of codimension 3 of S. To prove this, first recall that $\{f, g\} = X_f g$, where X_f is the Hamiltonian vector field on S, *determined by the function* f, and that for every $s \in J$, the condition $\{f, g\}(s) = 0$ is equivalent to the fact that the integral curve of X_f (which is contained in F) is tangent to G at s. Given $s \in K$, we have $X_f f(s) = X_f g(s) = 0$, and if $\mathrm{d}\{f, g\}(s)$ is a linear combination of $\mathrm{d}f(s)$ and $\mathrm{d}g(s)$, then

$$\{f, \{f, g\}\}(s) = X_f\{f, g\}(s) = 0,$$

in contradiction with (7.9). Therefore $\mathrm{d}f(s)$, $\mathrm{d}g(s)$ and $\mathrm{d}\{f, g\}(s)$ are linearly independent for each $s \in K$, which shows that K is a smooth submanifold of S of codimension 3.

Next, fix an arbitrary smooth submanifold M_F of F of codimension 1, intersecting transversally ∂S at s_0. For $s \in F$ denote by $\pi_F(s)$ the intersection point of the integral curve of X_f through s with M_F. Assuming $V = S$ sufficiently small, the map $\pi_F : F \to M_F$ is a well-defined smooth submersion. In fact, M_F can be identified with the quotient space $F/\sim$, where $\sim$ is the following equivalence relation on F : $s_1 \sim s_2$ if and only if s_1 and s_2 lie on one and the same integral curve of X_f in F. Then π_F is the canonical projection $F \to F/\sim$. In a similar way one defines $\pi_G : G \to M_G$. Finally, set

$$J_F = \pi_F(J) \subset M_F, \quad J_G = \pi_G(J) \subset M_G.$$

We can now state the result of Melrose [M1] which is crucial for the approximation theorem in the next section.

Theorem 7.2.1: *Under the assumptions and notation above we have the following:*

(a) J_F has a natural structure of a smooth symplectic manifold inherited from (S, ω), such that $\pi_F : J \to J_F$ is a smooth symplectic map and $\partial J_F = \pi_F(K)$. There exist two uniquely determined continuous maps

$$\alpha_\pm : J_F \to J$$

such that

$$\pi_F \circ \alpha_\pm = \text{id}, \quad \text{Im}\, \alpha_+ \cup \text{Im}\, \alpha_- = J,$$

and $\alpha_\pm$ are smooth on $J_F \backslash \partial J_F$. Similar statement is true for J_G. Let $\beta_\pm :$ $J_G \to J$ be the corresponding continuous inverses of π_G. Then the maps

$$\delta_\pm : J_F \to J_F,$$

defined by

$$\delta_\pm = \pi_F \circ \beta_\pm \circ \pi_G \circ \alpha_\pm,$$

are continuous in J_F and smooth and symplectic in $J_F \backslash \partial J_F$, $\delta_\pm(\partial J_F) \subset \partial J_F$, and locally $\delta_\pm \circ \delta_\mp = \text{id}$.

(b) *There exist smooth symplectic coordinates*

$$(x_0, x_1, \ldots, x_m; \ \xi_0, \xi_1, \ldots, \xi_m)$$

in a neighbourhood of s_0 in S such that $s_0 = (0,0)$,

$$F = \{(x; \xi) \in S : x_0 = 0\}, \quad G = \{(x; \xi) \in S : \xi_0^2 - x_0 - \xi_1 = 0\}.$$

These coordinates induce canonical coordinates $(x_1, \ldots, x_m; \xi_1, \ldots, \xi_m)$ in J_F such that $\xi_1 \geq 0$, $\partial J_F = \{\xi_1 = 0\}$, with respect to which the maps $\delta_\pm$ take the form

$$\delta_\pm(x_1, \ldots, x_m; \ \xi_1, \ldots, \xi_m) = (x_1 \pm 2\sqrt{\xi_1}, x_2, \ldots, x_m; \ \xi_1, \ldots, \xi_m).$$

For a proof of this theorem we refer the reader to [M1], see also Section 21.4 in [H3]. ♠

As an immediate consequence of the above result one gets the following.

Corollary 7.2.2: *Let s_0 be a non-degenerate glancing point of F and G. There exist a neighbourhood V of s_0 in S and a smooth function $h : V \to \mathbf{R}^+$, which is a defining function for ∂S in V, such that $\delta_\pm$ take the following form in V*

$$\delta_\pm(s) = \exp(\pm\sqrt{h} X_h)(s). \tag{7.10}$$

If $h_1 : V_1 \to \mathbf{R}^+$ is another smooth function with these properties, then $h - h_1$ vanishes of infinite order on $\partial S \cap V \cap V_1$. ♠

A smooth function $h : V \to \mathbf{R}^+$, which is a defining function for ∂S in V and satisfies (7.10) will be called a *local interpolating Hamiltonian* for the maps $\delta_\pm$.

The following elementary lemma is connected with the canonical form of the maps $\delta_{\pm}$ in Theorem 7.2.1(b) and will be useful later.

Lemma 7.2.3: *Let $a < b$ and $\omega > 0$, $\epsilon > 0$ be real numbers such that $\omega \leq \epsilon^2$. Define the map*

$$u : \mathbf{R} \times \mathbf{R}^+ \to \mathbf{R} \times \mathbf{R}^+$$

by $u(x, \xi) = (x + \sqrt{\xi}, \xi)$. Then for

$$V = [a, a + \epsilon] \times (0, \omega]$$

we have

$$[a, +\infty) \times (0, \omega] \subset \cup_{j=0}^{\infty} u^j(V).$$

Proof: Fix arbitrary $y > a$ and $\eta \in (0, \omega]$. Note that

$$u^j([a, a + \epsilon] \times \{\eta\}) = [a + j\sqrt{\eta}, a + \epsilon + j\sqrt{\eta}] \times \{\eta\}.$$

On the other hand, $a + \epsilon + j\sqrt{\eta} \geq a + (j+1)\sqrt{\eta}$, because $\epsilon \geq \sqrt{\omega} \geq \sqrt{\eta}$. Therefore

$$\cup_{j=0}^{\infty} u^j([a, a + \epsilon] \times \{\eta\}) = [a + \infty) \times \{\eta\},$$

which proves the assertion. ♠

Consider again a non-degenerate glancing point s_0 of F and G, and let U' be a neighbourhood of s_0 in S on which there exist smooth symplectic coordinates $x_1, \ldots, x_m; \xi_1, \ldots, \xi_m$ with the properties listed in Theorem 7.2.1(b). Note that if

$$c : [a, b] \to U = \pi_F(U') \subset J_F, \quad a < b,$$

is the projection under π_F of the integral curve of X_g through s_0, then c has the form

$$c(t) = (A + t, 0, \ldots, 0; \ 0, \ldots, 0),$$

for some real constant $A > 0$. Let ϵ, ω be such that $0 < \epsilon < b - a$, $0 < 4\omega < \epsilon^2$ and let

$$U(a, b, \omega) = \{(x; \xi) : x_1 \in [a, b],$$
$$0 < \xi_1 \leq \omega, \ |x_i| \leq \omega, \ |\xi_i| \leq \omega, i = 2, \ldots, m\} \subset U.$$

For every integer $j \geq 0$ set

$$U_j(a, a + \epsilon, \omega) = U(a, a + \epsilon, \omega) \cap \cap_{i=0}^{j} \delta_+^{-i}(U).$$

It follows by Lemma 7.2.3 and the form of the map δ_+ in the coordinates under considerations that

$$U(a, b, \omega) \subset \cup_{j=0}^{\infty} \delta_+^j(U_j(a, a + \epsilon, \omega)).$$

In particular,

$$\overline{\cup_{j=0}^{\infty} \delta_+^j (U_j(a, a+\epsilon, \omega))}$$

contains an open neighbourhood of $c((a, b))$ in J_F.

We now proceed with the case when $\operatorname{Im} c$ is no longer contained in a small coordinate neighbourhood.

For $\lambda > 0$ denote by V_λ the set of all $x_1, \ldots, x_m; \xi_1, \ldots, \xi_m \in \mathbf{R}^{2m}$ such that $x_1 \in (a - \lambda, b + \lambda), \xi_1 \in [0, \lambda), |x_i| < \lambda, |\xi_i| < \lambda$ for $i = 2, \ldots, m$. Then V_λ as a smooth manifold with boundary $\partial V_\lambda = \{\xi_1 = 0\}$.

Lemma 7.2.4: *Let $a < b$ be real numbers and let*

$$\tilde{c} : [a, b] \to K \subset G$$

be an integral curve of the Hamiltonian vector field X_g such that all points of $\operatorname{Im} \tilde{c}$ are non-degenerate glancing points of F and G. Let $c = \pi_F \circ \tilde{c}$ be its projection on J_F and suppose that δ_+ is defined as a continuous invertible map in a neighbourhood U of $\operatorname{Im} c$ having the properties listed in Theorem 7.2.1(a). Assume that for some $\lambda > 0$,

$$\Phi : V_\lambda \to U$$

is a smooth map, which is a local diffeomorphism and

$$\Phi(t, 0, \ldots, 0; 0, \ldots, 0) = c(t), \quad t \in (a - \lambda, b + \lambda).$$

Then there exists $\mu \in (0, \lambda)$, a continuous invertible map

$$\tilde{\delta}_+ : V_\mu \to V_\lambda \tag{7.11}$$

with

$$\Phi \circ \tilde{\delta}_+ (v) = \delta_+ \circ \Phi(v), \quad v \in V_\mu, \tag{7.12}$$

and a smooth function

$$\tilde{h} : V_\mu \to \mathbf{R}^+$$

which is a defining function of ∂V_μ and

$$\tilde{\delta}_+(v) = (\exp \sqrt{\tilde{h}} X_{\tilde{h}})(v) \tag{7.13}$$

holds for all $v \in V_\mu$, V_μ being considered as a symplectic manifold with respect to the pull-back by Φ of the symplectic structure of J_F inherited from S.

Proof: Clearly, $\operatorname{Im} \tilde{c}$ can be covered with a finite number of coordinate neighbourhoods $U_1', \ldots, U_k'$, each of them possessing smooth symplectic coordinates with the properties listed in Theorem 7.2.1(b). We shall consider only the case $k = 2$; the general case follows in the same way by a simple induction on k.

So we assume that

$$\text{Im}\,\tilde{c} \subset U_1' \cup U_2'.$$

Then there exist open subsets U_1, U_2 of J_F such that

$$\text{Im}\,c \subset U_1 \cup U_2, \quad \bar{U}_i \subset \pi_F(U_i') \subset U, \quad i = 1, 2.$$

We may assume that $c(a) \in U_1$, $c(b) \in U_2$. Set

$$V = U_1 \cap U_2, \quad V_j = V \cap \cap_{i=0}^j \delta_+^{-i}(U), \quad j = 0, 1, 2, \ldots$$

According to the choice of the neighbourhoods U_i', there exist smooth symplectic coordinates $x_1, \ldots, x_m$; $\xi_1, \ldots, \xi_m$ in U_2 having the properties listed in Theorem 7.2.1(b). Then there exists $a' \in (a, b)$ such that

$$c([a, a']) \subset U_1, \quad c([a', b]) \subset U_2.$$

Moreover, with respect to the coordinates in U_2 we have

$$c(t) = (A + t, 0, \ldots, 0; \ 0, \ldots, 0)$$

with some constant A. Exchanging linearly the coordinate x_1, we may assume $A = 0$.

Further, fix an arbitrary $\epsilon > 0$ such that

$$c([a', a' + \epsilon]) \subset U_1 \cap U_2,$$

and take $\omega > 0$ with $4\omega < \epsilon^2$. We shall also assume that ϵ and ω are so small that

$$W = \{(x, \xi) \in U_2 : a' - \omega \le x_1 \le b + \omega, \ 0 \le \xi_1 \le \omega, \ |x_i| \le \omega,$$
$$|\xi_i| \le \omega \text{ for } i = 2, \ldots, m\} \subset U_2 \cap \delta_+^{-1}(U_2),$$

and

$$V = \{(x, \xi) \in W : a' \le x_1 \le a' + \epsilon\} \subset U_1 \cap U_2.$$

It follows now from the reasonings after Lemma 7.2.3 that

$$W \setminus \partial J_F \subset \cup_{j=0}^\infty \delta_+^j(V), \quad W \subset \overline{\cup_{j=0}^\infty \delta_+^j(V)}. \tag{7.14}$$

We now begin the construction of μ, $\tilde{\delta}_+$ and $\tilde{h}$. First, choose $\mu \in (0, \lambda)$ in such a way that every $(x, \xi) = \Phi(v)$ with $v \in V_\mu$ and $x_1 \in [a', b]$ is contained in W. Since Φ is a local diffeomorphism, there is a unique continuous map (7.11) with (7.12). To define $\tilde{h}$, fix an arbitrary local interpolation Hamiltonian h for δ_+ in U_1; its existence is provided by the choice of U_1. Set

$$U_\mu = V_\mu \cap \Phi^{-1}(U_1),$$

and define

$$\tilde{h}(v) = h \circ \Phi(v), \quad v \in U_\mu. \tag{7.15}$$

Then (7.12) and the fact that h is an interpolating Hamiltonian for δ_+ in U_1 imply that (7.13) holds for $v \in U_\mu$. We extend $\tilde{h}$ on V_μ as follows. For $v \in \partial V_\mu$ we simply set $\tilde{h}(v) = 0$; then (7.13) is trivially fulfilled on ∂V_μ. Let now $v \in V_\mu \backslash \partial(V_\mu \cup U_\mu)$. Then $\Phi(v) \in W \backslash \partial J_F$, hence by (7.14) $\Phi(v) = \delta_+^j(s)$ for some $j \geq 0$ and $s \in V$. Set $\tilde{h}(v) = h(s)$. To check the correctness of this definition, assume that $\Phi(v) = \delta_+^l(s')$ for some other $l \geq 0$ and $s' \in V$. Suppose $j \geq l$, the other case is the same. Then $\delta_+^{j-l}(s) = s' \in V \subset U_1$, and since h is constant along the orbits of δ_+, one gets $h(s) = h(s')$. This shows that the definition of $\tilde{h}$ is correct. Moreover, for $\Phi(v) = \delta_+^j(s)$ we can find a small neighbourhood Q of s in $V \backslash J_F$ such that $\delta_+^j(Q)$ is a neighbourhood of $\Phi(v)$ in $U_2 \backslash J_F$, and then we may define $\tilde{h}$ on $\Phi^{-1}(\delta_+^j(Q))$ by $\tilde{h}(\Phi^{-1}(\delta_+^j(s'))) = h(s')$ for all $s' \in Q$. This shows that $\tilde{h}$ is smooth in a neighbourhood of v and (7.13) holds.

Thus, $\tilde{h}$ is smooth in $V_\mu \backslash \partial V_\mu$, and (7.13) holds for all $v \in V_\mu$. It remains to show that $\tilde{h}$ is smooth on ∂V_μ.

Since the function $(x, \xi) \mapsto \xi_1$ is a local interpolating Hamiltonian for δ_+ in U_2, it follows by the second part of Corollary 7.2.3 that the function $h - \xi_1$ vanishes of infinite order on $V \cap \partial J_F$. This implies that for every integer $p > 0$ there exists a constant $C_p > 0$ such that

$$|h(x_1, \ldots, x_m; \xi_1, \ldots, \xi_m) - \xi_1| \leq C_p \xi_1^p \tag{7.16}$$

for all $(x; \xi) \in V$. Let $v \in \partial V_\mu \backslash U_\mu$, then

$$\Phi(v) = s = (x_1, \ldots, x_m; 0, \xi_2, \ldots, \xi_m) \in W.$$

For every $\xi_1 \in (0, \omega)$ we have

$$s' = (x_1, \ldots, x_m; \xi_1, \xi_2, \ldots, \xi_m) \in W \subset U_2,$$

and there exist $y_1 \in [a', a' + \epsilon]$ and $j \in \mathbf{N}$ such that $y_1 + j\sqrt{\xi_1} = x_1$, i.e. $\delta_+^j(s'') = s'$, where

$$s'' = (y_1, x_2, \ldots, x_m; \xi_1, \xi_2, \ldots, \xi_m) \in V.$$

Defining $h_1(s') = h(s'')$ and $h_1 = 0$ on ∂J_F, one gets a function h_1 in a neighbourhood of s in U_2 such that $\tilde{h}(v') = h_1 \circ \Phi(v')$ for v' in a small neighbourhood of v in V_μ. Although the choice of j and y_1 above depends on s', according to (7.16) we always have

$$|h_1(s') - \xi_1| = |h(s'') - \xi_1| \leq C_p \xi_1^p$$

for all $p \geq 0$. This shows that h_1 is smooth in a neighbourhood of s in U_2; in fact all derivatives of h_1 on ∂J_F coincide with the corresponding derivatives of the function ξ_1. Now the smoothness of Φ implies that $\tilde{h}$ is smooth in a neighbourhood of v in V_μ, which concludes the proof of the proposition. ♠

7.3. Approximation of closed geodesics by periodic reflecting rays

Let Ω by a compact strictly convex domain in $\mathbf{R}^n$, $n \geq 2$, with smooth boundary $\partial\Omega$, and let $\gamma : [0, L] \rightarrow \partial\Omega$ be closed geodesics on $\partial\Omega$. Consider the corresponding integral curve $c : [0, L] \rightarrow T^*\partial\Omega$ of the Hamiltonian vector field, determined by the standard Riemannian metric on $\partial\Omega$ (cf. the beginning of Section 4.1). We shall assume that $L > 0$ is the primitive period of γ (resp. c) and that γ (resp. c) is non-degenerate as a curve of period mL for every $m = 1, 2, \ldots$. The latter means that the spectrum of the Poincaré map P_γ does not contain roots of unity.

There is a natural way to define a winding number for any finite sequence of points lying in a small neighbourhood U of Im c in $T^*\Omega$. One can take U such that

$$\text{Im } c \subset \cup_{s \in \text{Im } c} N_s(\epsilon),$$

where $\epsilon > 0$ and $N_s(\epsilon)$ is the ϵ-neighbourhood of s in the orthogonal complement of $T_s^*(\text{Im } c)$ in $T_s^*\Omega$. If ϵ is sufficiently small, then the projection $\mu : U \rightarrow \text{Im } c$, defined by $\mu(s') = s$ for $s' \in N_s(\epsilon)$, is a well defined smooth submersion. As a closed oriented curve without self-intersections, Im c with direction determined by $\dot{c}(0)$, is homeomorphic to the unit circle S^1 with the counterclockwise orientation. Now for every sequence $s_1, \ldots, s_k$ in U define the *winding number* $wn(\{s_i\})$ to be the winding number of the sequence $\mu(s_1), \ldots, \mu(s_k)$ in Im c (cf. Section 2.1). By winding number of a closed billiard trajectory we mean the winding number of the sequence of its successive reflection points.

Let N be an arbitrary natural number. It will stay fixed till the end of the section. Our aim is to prove the following theorem.

Theorem 7.3.1: *Every neighbourhood of* Im γ *in* Ω *contains a closed billiard trajectory in* Ω *with winding number* N.

For the proof of this theorem we need to consider multiples of the closed curves c and γ. To this end it is convenient to extend c and γ in $\mathbf{R}$ periodically with period L, i.e. $c(t + L) = c(t)$ for all t, the same for γ. In fact, we shall need these extensions only on a sufficiently large compact interval I containing $[0, NL]$.

We are now going to use the considerations from the previous section to study the billiard ball map in a neighbourhood of Im γ.

Consider the symplectic manifold $S = T^*\mathbf{R}^n$ endowed with the canonical symplectic form

$$\omega = \sum_{i=1}^{n} \mathrm{d}p_i \wedge \mathrm{d}q_i,$$

$p_1, \ldots p_n, q_1, \ldots, q_n$ being the standard coordinates in $T^* \mathbf{R}^n$. Introduce the hypersurfaces

$$F = T^*_{\partial \Omega} \mathbf{R}^n = \{(p,q) \in T^* \mathbf{R}^n : p \in \partial \Omega\},$$
$$G = S^* \mathbf{R}^n = \{(p,q) \in T^* \mathbf{R}^n : |q| = 1\}.$$

Clearly $g : T^* \mathbf{R}^n \to \mathbf{R}$, $g(p,q) = |q|^2 - 1$, is a defining function for G. To get a similar function for F, fix an arbitrary function φ, defined and smooth in a neighbourhood of $\partial \Omega$ in $\mathbf{R}^n$ such that $\partial \Omega = \varphi^{-1}(0)$ and $d\varphi \neq 0$ on $\partial \Omega$. Then $f(p,q) = \varphi(p)$ gives a defining function f for F. Since f depends on p only and g on q, it is clear that F and G intersect transversally at any point of

$$J = F \cap G = S^*_{\partial \Omega} \mathbf{R}^n.$$

To describe the set K of glancing points, mention that

$$\{f,g\}(p,q) = \sum_{i=1}^{n} \left(\frac{\partial f}{\partial p_i} \frac{\partial g}{\partial q_i} - \frac{\partial f}{\partial q_i} \frac{\partial g}{\partial p_i} \right)$$
$$= 2 \sum_{i=1}^{n} q_i \frac{\partial \varphi}{\partial p_i}(p) = 2\langle q, \nabla \varphi(p) \rangle,$$

where $\nabla \varphi(p)$ is the *gradient* of φ at p. Since $\nabla \varphi(p)$ is a parallel to the *unit normal vector* $\nu(p)$ to $\partial \Omega$ at p, pointing into the interior of Ω, the condition $\{f,g\}(p,q) = 0$ is equivalent to $\langle q, \nu(p) \rangle = 0$. Therefore the set K of glancing points of F and G coincides with $S^* \partial \Omega$.

To apply Theorem 7.2.1, we need to know that the points of K are non-degenerate. For $(p,q) \in K$ we have

$$\{f,\{f,g\}\}(p,q) = 2 \sum_{i=1}^{n} \left(\frac{\partial \varphi}{\partial p_i}(p) \right)^2 \neq 0,$$

since $d\varphi(p) \neq 0$. On the other hand, the strict convexity of $\partial \Omega$ at p implies

$$\{g,\{f,g\}\}(p,q) = 4 \sum_{i,j=1}^{n} q_i q_j \frac{\partial^2 \varphi}{\partial p_i \partial p_j}(p) \neq 0$$

whenever $q \neq 0$. Therefore every point of K is non-degenerate, and so Theorem 7.2.1 is applicable in the situation under consideration.

To describe the maps $\delta_\pm$, we first give a geometric interpretation of the spaces M_F and M_G. Note that

$$X_f(p,q) = (0; -\nabla \varphi), \quad X_g(p,q) = 2(q;0).$$

This shows that every integral curve of X_f has the form

$$p(t) = p(0), \quad q(t) = q(0) - t\nabla\varphi(p(0)).$$

To such a trajectory we assign the point $(p(0), q'(0))$, where $q'(0)$ is the orthogonal projection of $q(0)$ onto $T^*_{p(0)}\partial\Omega$. In other words we shall identify M_F with $T^*\partial\Omega$. Then $\pi_F : F \to M_F = T^*\partial\Omega$ is the projection just defined. Now we have

$$J_F = \pi_F(J) = B^*\partial\Omega = \{(p, q) \in T^*\partial\Omega : |q| \le 1\},$$

and clearly $\partial J_F = S^*\partial\Omega = K$. In fact, $\pi_F = \text{id}$ on K. The maps

$$\alpha_{\pm} : B^*\partial\Omega \to J = S^*_{\partial\Omega}\mathbf{R}^n$$

are now defined as follows. For $(p, q) \in B^*\partial\Omega$ let $q^{\pm} \in S^{n-1}$ be such that $q^{\pm} = q \pm \lambda\nu(p)$ for some $\lambda \ge 0$. Then $\langle q^{\pm}, \nu(p) \rangle = \pm\lambda$, and we set $\alpha_{\pm}(p, q) = (p, q^{\pm})$. It is easy to see that these are exactly the maps from Theorem 7.2.1(a).

To deal with G notice that the integral curves of X_g have the form

$$p(t) = p(0) + 2tq(0), \quad q(t) = q(0).$$

Thus M_G can be naturally identified with the space of all oriented lines in $\mathbf{R}^n$: to the integral curve $(p(t), q(t))$ we assign the line through $p(0)$ with direction $q(0)$. The projection

$$\pi_G : G = S^*\mathbf{R}^n \to M_G$$

is similarly defined: $\pi_G(p, q)$ is the line through p in direction q. In this setting J_G is clearly the subset of M_G consisting of those lines which have a common point with Ω. Then $\partial J_G = \pi_G(K)$ is the set of all oriented lines tangent to $\partial\Omega$, and π_G induces a diffeomorphism between $K = S^*\partial\Omega$ and ∂J_G. It is now easy to find the inverses

$$\beta_{\pm} : J_G \to J = S^*_{\partial\Omega}\mathbf{R}^n.$$

Given an oriented line $l \in J_G$ with direction $q \in S^{n-1}$, denote by p^- and p^+ the intersection points of l and $\partial\Omega$ (which may coincide) in such a way that $p^+ = p^- + \lambda q$, $\lambda \ge 0$ (Figure 7.3(a)). Define $\beta_{\pm}(l) = (p^{\pm}, q)$. These maps clearly have the properties required in Theorem 7.2.1(a).

Let us now describe the map

$$\delta_+ = \pi_F \circ \beta_+ \circ \pi_G \circ \alpha_+ : B^*\partial\Omega \to B^*\partial\Omega.$$

Given $(p, q) \in B^*\partial\Omega$, we have $(p, q^+) \in S^*_{\partial\Omega}\mathbf{R}^n$. Then $l = \pi_G(p, q^+)$ is the line through p with direction q^+. Correspondingly, $\beta_+(l) = (p^+, q^+)$, where p^+ is the other intersection point of l and $\partial\Omega$ (of course, $p^+ = p$ if $q^+ = q$). Finally, $\pi_F(p^+, q^+) = (p^+, r) \in B^*\partial\Omega$, where r is the orthogonal projection of q^+ on $T^*_{p^+}\partial\Omega$, and then $\delta_+(p, q) = (p^+, r)$ (Figure 7.3(b)). Thus δ_+ is globally

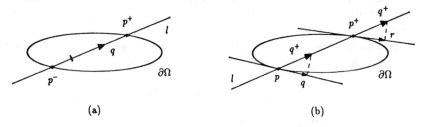

Figure 7.3

defined and is naturally equivalent to the billiard ball map, defined in Section 2.1. In this section, we call δ_+ *the billiard ball map* and denote it by B. Thus

$$B = \pi_F \circ \beta_+ \circ \pi_G \circ \alpha_+ : B^*\partial\Omega \to B^*\partial\Omega$$

and $B^{-1} = \delta_-$.

Set $m = n - 1$, then dim $\partial\Omega = m$, and consider the closed interval

$$I = [-L, NL + L]$$

in **R**. As in Section 4.1, there exists an open neighbourhood $\mathcal{O}$ of Im γ in $\partial\Omega$ and a local diffeomorphism

$$r : \mathcal{O}_u = (-u - L, NL + L + u) \times B_u(0) \to \mathcal{O},$$

with some $u > 0$, $B_u(0)$ being the u-neighbourhood of 0 in $\mathbf{R}^{m-1}$ having the properties:

(i) $\gamma(t) = r(t, 0, \dots, 0)$ for every $t \in (-u - L, NL + L + u)$;
(ii) the 1-jet of g_{11} coincides with the 1-jet of the constant 1 at all points of

$$\gamma_0 = \{(t, 0, \dots, 0) \in \mathbf{R}^m : t \in (-u - L, NL + L + u)\};$$

(iii) $g_{1i} = 0$ on γ_0 for all $i = 2, \dots, m$.

Here $g_{ij}, i, j = 1, \dots, m$ are the components of the standard metric on $\partial\Omega$ with respect to the coordinates $x_1, \dots, x_m$ provided by r. Let $\xi_1, \dots, \xi_m$ be the corresponding dual coordinates in $T^*\partial\Omega$. Then, as before, in a neighbourhood of any point of Im c,

$$x_1, \dots, x_m, \xi_1, \dots, \xi_m$$

can be used as coordinates. With respect to them $\omega = \Sigma_{i=1}^m \, dx_i \wedge d\xi_i$ is the canonical symplectic form on $T^*\partial\Omega$, while

$$H(x; \xi) = \frac{1}{2} \sum_{i,j=1}^m g_{ij}(x)\xi_i\xi_j$$

is the function the Hamiltonian vector field X_{H} of which determines the geodesic flow on $T^*\partial\Omega$, that is, the geodesics on $\partial\Omega$ are exactly the projections of the integral curves of X_{H} in $T^*\partial\Omega$.

For our next considerations it will be convenient to make the correspondence between the coordinates $(x;\xi)$ and points in $T^*\partial\Omega$ more clear. For $u > \lambda > 0$ set

$$\tilde{W}_\lambda = \{(x;\xi) : x \in \mathcal{O}_u, 1-\lambda < 2H(x;\xi) \le 1\}.$$

Then $\tilde{W}$ is a submanifold of $\mathcal{O}_u$ with boundary

$$\partial\tilde{W}_\lambda = \{(x;\xi) \in \mathcal{O}_u : 2H(x;\xi) = 1\}.$$

Let Ψ be the map which assigns to $(x;\xi) \in \tilde{W}_\lambda$ the unique point in $T^*\partial\Omega$ with coordinates $(x;\xi)$ as explained above. If λ is small enough, this map is well defined. Note that

$$W_\lambda = \Psi(\tilde{W}_\lambda)$$

is an open subset of $B^*\partial\Omega$ and

$$\partial W_\lambda = W_\lambda \cap S^*\partial\Omega = \Psi(\partial\tilde{W}_\lambda).$$

Moreover,

$$\Psi : \tilde{W}_\lambda \to W_\lambda$$

is a local diffeomorphism.

In order to apply Lemma 7.2.4, we first slightly exchange Ψ in order to get a map

$$\Phi : V_\lambda \to T^*\partial\Omega$$

satisfying the corresponding requirements. We now denote the points in V_λ (see the text before Lemma 7.2.4 for this notation) by

$$(y,\eta) = (y_1,\ldots,y_m;\eta_1,\ldots,\eta_m).$$

Define the map

$$\psi : \tilde{W}_\lambda \to V_\lambda$$

by $(y;\eta) = \psi(x,\xi)$, where $y = x$, $\eta_i = \xi_i$ for all $i = 2,\ldots,m$, and

$$\eta_1 = 1 - 2H(x;\xi) = 1 - \sum_{i,j=1}^m g_{ij}(x)\xi_i\xi_j.$$

Since $\xi_1 = 1, \xi_2 = \ldots = \xi_m = 0$ along Im c, we have

$$\frac{\partial\eta_1}{\partial\xi_1} = -2\sum_{i=1}^m g_{1i}(x_1,0,\ldots,0)\xi_i = -2$$

at all points of Im c. Therefore, if $\lambda > 0$ is sufficiently small, the map ψ is a diffeomorphism.

From now on we assume that $\lambda > 0$ is fixed small enough to satisfy all the requirements from above. Then $\Phi = \Psi \circ \psi^{-1}$ is a local diffeomorphism between V_λ and $\Phi(V_\lambda) = W_\lambda \subset T^*\partial\Omega$. In order to get the situation in Lemma 7.2.4, we endow V_λ with the symplectic structure induced by the canonical symplectic form ω in $\tilde{W}_\lambda$ via the diffeomorphism ψ. Then Φ becomes a smooth symplectic map. Applying Lemma 7.2.4 and replacing the pair (Φ, V_λ) by $(\Psi, \tilde{W}_\lambda)$ using the diffeomorphism ψ, we get the following.

Lemma 7.3.2: *There exist $\mu \in (0, \lambda)$, a continuous invertible map*

$$\tilde{B} : \tilde{W}_\mu \rightarrow \tilde{W}_\lambda \tag{7.17}$$

with

$$\Psi \circ \tilde{B}(v) = B \circ \Psi(v), \quad v \in \tilde{W}_\mu, \tag{7.18}$$

and a smooth function $\tilde{h} : \tilde{W}_\mu \rightarrow \mathbf{R}_+$ which is a defining function for $\partial \tilde{W}_\mu$ and

$$\tilde{B}(v) = (\exp \sqrt{\tilde{h}} X_{\tilde{h}})(v) \tag{7.19}$$

for all $v \in \tilde{W}_\mu$. ♠

In other words, $\tilde{B}$ is a covering of the billiard ball map B by means of the covering map Ψ, and $\tilde{h}$ is an interpolating Hamiltonian for $\tilde{B}$ on the whole domain $\tilde{W}_\mu$.

Using the canonical local normal fibration of $B^*\partial\Omega$ in the neighbourhood W_λ of Im c, we find a smooth family $\Sigma(t)$, $t \in \mathbf{R}$, of smooth submanifolds of W_λ of codimension one such that for every t, $\Sigma(t)$ contains $c(t)$ being transversal (in fact, orthogonal) to Im c at $c(t)$. Each $\Sigma(t)$ is a manifold with boundary

$$\partial\Sigma(t) = \Sigma(t) \cap S^*\partial\Omega,$$

and

$$\Sigma(t+L) = \Sigma(t) \tag{7.20}$$

for all t. Note that if $(p, q) \in \Sigma(t)$, p_i, q_i being the standard coordinates in $T^*\mathbf{R}^n$, then $p = c(t)$ and $(p, q/|q|) \in \partial\Sigma(t)$.

Next, consider the curve

$$d(t) = (t, 0, \ldots, 0; 0, \ldots, 0) \in \tilde{W}_\lambda,$$

which is the preimage of the integral curve c with respect of Ψ. For each $t \in (-u - L, NL + L + u)$ there exists a smooth submanifold $\tilde{\Sigma}(t)$ of $\tilde{W}_\lambda$, passing through $d(t)$ and transversal to Im d at $d(t)$, such that

$$\Psi(\tilde{\Sigma}(t)) = \Sigma(t).$$

Thus $\tilde{\Sigma}(t)$ is the smooth fibration of $\tilde{W}_\lambda$ corresponding to $\Sigma(t)$ in W_λ. For $t \in I$ define the map

$$\tilde{\mathcal{P}}_t : \tilde{\Sigma}(t) \rightarrow \tilde{\Sigma}(NL+t)$$

locally around $d(t)$ as follows. Given $v \in \tilde{\Sigma}(t)$, consider the integral curve $v(t)$, $v(0) = v$, of the Hamiltonian vector field $X_{\tilde{h}}$ through v, and denote by $\tilde{\mathcal{P}}_t(v)$ the intersection point $v(T)$, $T = T_v$ of this curve with $\tilde{\Sigma}(NL+t)$. Considering the projection $\Psi(v(t))$ of the curve $v(t)$ on W_λ, we get another map

$$\mathcal{P}_t : \Sigma(t) \rightarrow \Sigma(t) = \Sigma(NL+t).$$

Namely, for $w = \Psi(v)$ we set $\mathcal{P}_t(w) = \Psi(v(T))$. Clearly, $\tilde{\mathcal{P}}_t$ is a smooth local symplectic map of a small neighbourhood of $d(t)$ in $\tilde{\Sigma}(t)$ onto a neighbourhood of $d(NL+t)$ in $\tilde{\Sigma}(NL+t)$. $\mathcal{P}_t$ has similar properties and, moreover,

$$\mathcal{P}_t \circ \Psi = \Psi \circ \tilde{\mathcal{P}}_t$$

locally around $d(t)$ in $\tilde{\Sigma}(t)$ for each $t \in I$. Notice that the notation $\mathcal{P}_t$, $\tilde{\mathcal{P}}_t$, $\Sigma(t)$ and $\tilde{\Sigma}(t)$ differ from the corresponding one in Section 4.1. However the restriction of $\mathcal{P}_t$ on $\partial\Sigma(t) = \Sigma(t) \cap S^*\partial\Omega$ is exactly the Poincaré map of the integral curve $c(t)$, $t \in [0, NL]$, of X_H i.e. defined by means of the geodesics flow on $S^*\partial\Omega$.
    Consider the map

$$\tau : W_\lambda \rightarrow \partial W_\lambda = W_\lambda \cap S^*\partial\Omega \tag{7.21}$$

which is the restriction of the orthogonal projection of $B^*\partial\Omega \backslash 0$ on $S^*\partial\Omega$ along the normal fibres. More precisely, using again the standard coordinates p_i, q_i in $T^*\mathbf{R}^n$ from the beginning of this section we have

$$\tau(p, q) = (p, q/|q|).$$

We assume that $\lambda > 0$ is taken so small that $q \neq 0$ for all $(p;q) \in W_\lambda$. Then (7.21) is a well defined smooth submersion. Note that the corresponding map

$$\tilde{\tau} : \tilde{W}_\lambda \rightarrow \partial\tilde{W}_\lambda,$$

for which

$$\Phi \circ \tilde{\tau} = \tau \circ \Phi$$

has the form

$$\tilde{\tau}(x; \xi) = (x; \xi/\sqrt{2H(x; \xi)}),$$

and is also a smooth submersion.
    We are going now to exploit the non-degeneracy of the integral curve c (resp. of the closed geodesics γ).

Lemma 7.3.3:   *For all sufficiently small $\epsilon \in (0, \mu)$ there exists a unique family of smooth maps*

$$c_t : [1 - \epsilon, 1] \to \Sigma(t), \quad t \in I,$$

such that

$$\tau \circ \mathcal{P}_t \circ c_t(s) = \tau \circ c_t(s), \quad s \in [1 - \epsilon, 1], \tag{7.22}$$

and the map

$$I \times (1 - \epsilon, 1] \to W_\lambda, \quad (t, s) \mapsto c_t(s),$$

is smooth.

The uniqueness means the following. If

$$l_t : [1 - \delta, 1] \to \Sigma(t), \quad t \in I$$

is another smooth family with the same properties, then $c_t(s) = l_t(s)$ for all $t \in I, s \in [1 - \min\{\delta, \epsilon\}, 1]$.

Proof of Lemma 7.3.3:   For all sufficiently small $\delta > 0$, Ψ induces a diffeomorphism between

$$\tilde{U}_t = \{(x; \xi) \in \tilde{W}_\delta : t - \delta < x_1 < t + \delta, |x_i| < \delta, |\xi_i| < \delta$$
$$\text{for } i = 2, \ldots, m\} \subset \tilde{W}_\delta$$

and $U_t = \Psi(\tilde{U}_t) \subset W_\delta$ for all $t \in I$. Then we may use $(x; \xi)$ as coordinates in each U_t by means of the chart $(\Psi, \tilde{U})$.

Fix an arbitrary $t_0 \in I$, and introduce new coordinates $(x; \eta)$ in $U = U_{t_0}$ setting $\eta_1 = \xi_1$, $\eta' = \xi'/\sqrt{2H(x; \xi)}$. Here and in what follows we use the notation

$$\eta' = (\eta_2, \ldots, \eta_m), \quad \xi' = (\xi_2, \ldots, \xi_m),$$

etc. Clearly, $(x; \xi) = (x; \eta)$ on $U \cap S^* \partial \Omega$, and in the new coordinates the map τ has the form

$$\tau(x; \eta_1, \eta') = (x; \hat{\eta}_1, \eta').$$

Fix an arbitrary $t \in \Delta = (t_0 - \delta, t_0 + \delta)$. Then the points in $\Sigma(t)$ have the form $(t, x'; s, \eta')$, where $s \in (1 - \delta, 1]$, and in these coordinates the map

$$\mathcal{P}_t : \Sigma(t) \to \Sigma(t) \subset U$$

takes the form

$$\mathcal{P}_t(t, x'; s, \eta') = (t, z'; \chi, \zeta'). \tag{7.23}$$

Fix $s \in [1 - \delta, 1]$, and consider the map

$$Q_s : \mathbf{R}^{2m-2} \to \mathbf{R}^{2m-2},$$

defined in a small neighbourhood of 0 by

$$Q_s(x'; \eta') = (z'; \zeta'),$$

where z' and ζ' are determined by (7.23). As we have already mentioned above, for $s = 1$, $\mathcal{P}_t$ coincides with the Poincaré map of c, therefore the non-degeneracy of c implies that $\mathrm{id} - dQ_1$ is invertible at 0. Since dQ_s depends smoothly on s, this yields that $\mathrm{id} - dQ_s(0)$ is an invertible map for all $s \leq 1$ sufficiently close to 1. Applying the implicit function theorem to the system

$$\begin{cases} x' = z'(x'; s, \eta') \\ \eta' = \zeta'(x'; s, \eta'), \end{cases}$$

we find $\epsilon > 0$, $\epsilon < \mu < \lambda$, and maps

$$s \mapsto x'(s), \quad s \mapsto \eta'(s),$$

defined and smooth for $s \in [1 - \epsilon, 1]$ such that

$$\mathcal{P}_t(t, x'(s); s, \eta'(s)) = (t, x'(s); \chi(s), \eta'(s)), \quad s \in [1 - \epsilon, 1]. \quad (7.24)$$

Now we determine the curve

$$c_t : [1 - \epsilon, 1] \to \Sigma(t)$$

by

$$c_t(s) = (t, x'(s); s, \eta'(s)). \quad (7.25)$$

The smoothness and the uniqueness of the map $(t, s) \mapsto c_t(s)$ for $t \in \Delta$, $s \in [1 - \epsilon, 1]$ follow by the implicit function theorem, while (7.22) for $t \in \Delta$ is a consequence of the definition of $c_t(s)$, (7.24) and the form of the map τ in the coordinates $(x; \eta)$.

Finally, take $t_1, \ldots, t_k \in I$ such that $\Delta_i = (t_i - \delta, t_i + \delta)$, $i = 1, \ldots, k$, cover I, and choose $\epsilon > 0$ so small that the maps $c_t^{(i)}(s)$, determined as above replacing Δ by Δ_i, are all defined for $s \in [1 - \epsilon, 1]$. By the uniqueness of these functions it follows that $c_t^{(i)} = c_t^{(j)}$ for all $t \in \Delta_i \cap \Delta_j$. Therefore setting $c_t(s) = c_t^{(i)}(s)$ whenever $t \in \Delta_i$, $s \in [1 - \epsilon, 1]$, we obtain a smooth family of maps having the desired properties. ♠

For $t \in I$ there exists an unique smooth map

$$\tilde{c}_t : [1 - \epsilon, 1] \to \tilde{\Sigma}(t) \subset \tilde{U}$$

such that

$$c_t = \Psi \circ \tilde{c}_t.$$

Denote by $T_t(s)$ the time required for the point $\tilde{c}_t(s) \in \tilde{\Sigma}(t)$ to flow along the corresponding integral curve of the Hamiltonian $X_{\tilde{h}}$ to the point $\tilde{\mathcal{P}}_t(\tilde{c}_t(s)) \in$

$\tilde{\Sigma}(NL+t)$. Then assuming that $\epsilon > 0$ is sufficiently small, we have that $T_t(s)$ is a continuous bounded function of (t, s) with

$$T_t(1) = NL > 0, \quad t \in I.$$

Lemma 7.3.4: *For all sufficiently small $\epsilon \in (0, \mu)$ there exists an integer $k_0 > 0$ such that if $0 < \epsilon \le \epsilon_0$, then for every $k \ge k_0$ there is a unique smooth function*

$$s_k : I \rightarrow [1 - \epsilon, 1]$$

with

$$k\sqrt{\tilde{h}(\tilde{c}_t(s_k(t)))} = T_t(s_k(t)) \tag{7.26}$$

for all $t \in I$.

Proof: Recall that $\tilde{h}$ is a defining function for

$$\partial \tilde{W}_\mu = \tilde{W}_\mu \cap S^* \partial \Omega$$

in $\tilde{W}_\mu$ (cf. Lemma 7.3.2). On the other hand, from the proof of Lemma 7.3.3 we have that in the coordinates $(x; \eta)$, the curve $c_t(s)$ has the form (7.25). Since the map $(x; \eta) \mapsto \eta_1$ is also a defining function for $\tilde{W}_\mu$ in $\tilde{W}_\mu$, we may choose $\epsilon \in (0, \mu)$ with the properties from Lemma 7.3.3 and such that

$$\frac{\partial \tilde{h}}{\partial \eta_1}(x; \eta) \ne 0$$

for all $(x; \eta) \in W_\mu$ with $\eta_1 \in [1 - \epsilon, 1]$.

Next, mention that all derivatives of $\tilde{h}$, except that with respect to η_1, are zero for $\eta_1 = 1$. Consequently, the absolute values of these derivatives will be arbitrarily small for all $\eta_1 \in [1 - \epsilon, 1]$, provided ϵ is chosen sufficiently small. Now, according again to (7.25), we see that for such ϵ, we have

$$\frac{d}{ds}\tilde{h}(\tilde{c}_t(s)) \ne 0, \quad s \in [1 - \epsilon, 1].$$

This clearly implies that for any sufficiently large integer $k > 0$ the function

$$g_t(s) = k\sqrt{\tilde{h}(\tilde{c}_t(s))} - T_t(s)$$

is strictly monotone in $[1 - \epsilon, 1]$ and takes values with distinct signs at $1 - \epsilon$ and 1. Therefore there exists a unique $s_k(t) \in (1 - \epsilon, 1]$ such that $g_t(s_k(t)) = 0$, i.e. (7.26) holds. It follows by the implicit function theorem that $s_k(t)$ depends smoothly on t. ♠

As a remark which will be useful later, let us mention that if ϵ is chosen sufficiently small as in the proof above, there exist constants $C_2 > C_1 > 0$ such that

$$C_1(1 - \eta_1) \leq \left| \tilde{h}(x; \eta) \right| \leq C_2(1 - \eta_1), \quad \eta_1 \in [1 - \epsilon, 1]. \tag{7.27}$$

From now on we assume that $\epsilon \in (0, \mu)$ is chosen so small that the corresponding requirements of Lemmas 7.3.3 and 7.3.4 are satisfied and (7.27) holds, provided $(x; \eta) \in \tilde{W}_\mu$.

For $k \geq k_0$ set

$$\rho_k(t) = c_t(s_k(t)), \quad t \in I.$$

Thus, we obtain a smooth function

$$\rho_k : I \to W_\mu \subset B^* \partial \Omega$$

such that

$$\rho_k(t) \in \Sigma(t), \quad t \in I.$$

We claim that

$$B^k(\rho_k(t)) \in \Sigma(t) \tag{7.28}$$

for all $t \in I$. To prove this we shall use (7.18) and (7.19) for the covering map (7.17) of B. For $v = \tilde{c}_t(s_k(t))$ (7.19) implies

$$\tilde{B}^k(v) = (\exp k \sqrt{\tilde{h}} X_{\tilde{h}})(v).$$

Combining this with (7.26) and the definition of $T_t(s)$, we find

$$\tilde{B}^k(v) = \tilde{\mathcal{P}}_t(v).$$

Applying the map Ψ and taking into account (7.18), one gets

$$B^k(\Psi(v)) = \mathcal{P}_t(\Psi(v)).$$

Finally, $c_t = \Psi \circ \tilde{c}_t$ and the definition of $\rho_k(t)$ show that (7.28) holds.

Further, considering the functions $\rho_k(t - L)$ (defined in a subinterval of I) and using the uniqueness from Lemmas 7.3.3 and 7.3.4 (and decreasing ϵ once again if necessary), we see that $\rho_k(t)$ is periodic in t with period L, i.e.

$$\rho_k(t + L) = \rho_k(t) \tag{7.29}$$

for all $t \in I$ with $t + L \in I$ and all $k \geq k_0$. Thus each ρ_k can be considered as a smooth map

$$\rho_k : S^1 \to B^* \partial \Omega.$$

Lemma 7.3.5 *For every $k \geq k_0$ and every $t \in [0, L]$ the sequence $\{B^j(\rho_k(t))\}_{j=0}^{k-1}$ has winding number N, and*

$$\rho_k(t) \to_{k \to \infty} c(t)$$

uniformly on $t \in [0, L]$.

Proof: The first assertion follows immediately from the construction of $\rho_k(t)$. To establish the second one it is sufficient to show that

$$s_k(t) \to_{k \to \infty} 1$$

uniformly on $t \in [0, L]$, which will trivially follow form the inequalities

$$\frac{K_1^2}{k^2 C_2} < 1 - s_k(t) < \frac{K_2^2}{k^2 C_1} \tag{7.30}$$

for $t \in [0, L]$, $k \geq k_0$. Here C_1 and C_2 are the constants from (7.27), while $K_i > 0$ can be chosen such that

$$K_1 \leq T_t(s) \leq K_2 \tag{7.31}$$

for all $t \in I$, $s \in (1 - \epsilon, 1]$.

To prove (7.30), first combine (7.26) and (7.31) to get

$$\frac{K_1^2}{k^2} \leq \tilde{h}(\tilde{c}_t(s_k(t))) \leq \frac{K_2^2}{k^2}. \tag{7.32}$$

Recall that using the $(x; \eta)$ coordinates from Lemma 7.3.2, the η_1 component of $\tilde{c}_t(s)$ is exactly s. Thus, applying (7.27), we find

$$C_1(1 - s_k(t)) \leq \tilde{h}(\tilde{c}_t(s_k(t))) \leq C_2(1 - s_k(t)). \tag{7.33}$$

Combining (7.32) and (7.33), one gets immediately (7.30), which proves the assertion. ♠

We now turn to the final step in the proof of the main theorem.

Proof of Theorem 7.3.1: According to Lemma 7.3.5, the theorem will be proved if we show that for every $k \geq k_0$ there exists $t \in [0, L]$ with

$$B^k(\rho_k(t)) = \rho_k(t). \tag{7.34}$$

Fix an arbitrary $k \geq k_0$. Let

$$\pi : T^* \mathbf{R}^n \to \mathbf{R}^n$$

be the natural projection on the first component, i.e. $\pi(x, \xi) = x$, and let $\| \cdot \|$ be the standard norm in $\mathbf{R}^n$. Define the function

$$F_k : B^* \partial \Omega \to \mathbf{R}$$

by

$$F_k(v) = \sum_{j=0}^{k-1} \left\| \pi(B^j(v)) - \pi(B^{j+1}(v)) \right\|.$$

Clearly, F_k is continuous. It is convenient to write it as a composition $F_k = R \circ Q$, where $Q : B^*\partial\Omega \to (\partial\Omega)^k$, $R : (\partial\Omega)^k \to \mathbf{R}$ are given by

$$Q(v)=(\pi(v), \pi(B(v)),\ldots,\pi(B^{k-1}(v))), R(x_1,\ldots,x_k)=\sum_{j=1}^{k} \left\| x_j - x_{j+1} \right\|.$$

The latter maps are also continuous, Q is smooth for $v \in B^*\partial\Omega\backslash S^*\partial\Omega$ with $dQ(v) \neq 0$ and R is smooth in $(\partial\Omega)^{(k)}$.

Since the map $\rho_k(t)$ is periodic, the function $F_k \circ \rho_k(t)$ has a minimum and a maximum on $[0, L]$. Therefore there exists at least one $t \in [0, L]$ with

$$\frac{d(F_k \circ \rho_k)}{dt}(t) = 0. \tag{7.35}$$

Fix an arbitrary t with this property. We are going to prove that (7.34) holds for this choice of t.

It follows by $\rho_k(t) \in \Sigma(t)$ that $\rho_k(t) = (\gamma(t); \xi)$, γ being the initial geodesic on $\partial\Omega$ (see the beginning of this section). Then by (7.28), $B^k(\rho_k(t)) = (\gamma(t); \zeta) \in \Sigma(t)$ which implies

$$\zeta = C\xi, \quad C > 0.$$

It is now sufficient to show that $C = 1$, this would clearly imply (7.34).

Set $Q(\rho_k(t)) = (x_1,\ldots,x_k)$ and

$$v_1 = \frac{x_2 - x_1}{\|x_2 - x_1\|}, \quad v_k = \frac{x_1 - x_k}{\|x_1 - x_k\|}.$$

Then $x_1 = \gamma(t)$ and, identifying $T_x\partial\Omega$ with $T_x^*\partial\Omega$ via the natural duality, we have that the orthogonal projections of v_1 and v_k on $T_{\gamma(t)}\partial\Omega$ coincide with ξ and ζ, respectively. As in the proof of Proposition 2.1.3, take arbitrary smooth charts

$$\varphi_j : \mathbf{R}^{n-1} \to U_j \subset \partial\Omega, \quad j = 1,\ldots,k,$$

with $\varphi_j(0) = x_j$. Consider the function

$$G : (\mathbf{R}^{n-1})^k \to \mathbf{R},$$

defined by

$$G(u_1,\ldots,u_k) = R(\varphi_1(u_1),\ldots,\varphi_k(u_k)).$$

Since the segments $[x_{j-1}, x_j]$ and $[x_j, x_{j+1}]$ satisfy the law of reflection with respect to $\partial\Omega$ for every $j = 2, 3,\ldots, k$, using the calculations from the proof of

Proposition 2.1.3, we find

$$\frac{\partial G}{\partial u_j^{(i)}}(0) = 0$$

for all $j = 2, \ldots, k$, $i = 1, \ldots, n-1$. Moreover,

$$\frac{\partial G}{\partial u_1^{(i)}}(0) = \left\langle v_1 + v_k, \frac{\partial \varphi_1}{\partial u_1^{(i)}}(0) \right\rangle = \left\langle \zeta - \xi, \frac{\partial \varphi_1}{\partial u_1^{(i)}}(0) \right\rangle$$

$$= (C-1)\left\langle \xi, \frac{\partial \varphi_1}{\partial u_1^{(i)}}(0) \right\rangle, \tag{7.36}$$

according to the above remark on v_1 and v_k and taking into account the fact that $\frac{\partial \varphi_1}{\partial u_1^{(i)}}(0)$ is tangent to $\partial\Omega$ at x. There exists a smooth map

$$\chi = (\chi_1, \ldots, \chi_k) : \Delta \to (\mathbf{R}^{n-1})^k,$$

defined in a small open interval Δ around t in $\mathbf{R}$ such that $\chi(0) = 0$ and $G(\chi(s)) = F_k(\rho_k(s))$. Now by (7.35) and (7.36) we have

$$0 = d(G \circ \chi)(t) = \sum_{j=1}^{k}\sum_{i=1}^{n-1} \frac{\partial G}{\partial u_j^{(i)}} d\chi_j^{(i)}(t) = \sum_{i=1}^{n-1} \frac{\partial G}{\partial u_1^{(i)}} d\chi_1^{(i)}(t)$$

$$= (C-1)\left\langle \xi, \sum_{i=1}^{n-1} \frac{\partial \varphi_1}{\partial u_1^{(i)}} d\chi_1^{(i)}(t) \right\rangle = (C-1)\langle \xi, d(\varphi_1 \circ \chi_1)(t)\rangle.$$

Therefore, to establish the equality $C = 1$, it is sufficient to show that $\langle \xi, d(\varphi_1 \circ \chi_1)(t)\rangle \neq 0$. To this end mention that $\varphi_1 \circ \chi_1(s)$ coincides with the first component of $Q \circ \rho_k(s)$ which is in fact $\gamma(s)$. Thus, identifying $\dot\gamma(t)$ with $d\gamma(t)$ according to the natural identification of $T\partial\Omega$ with $T^*\partial\Omega$, we have

$$\langle \xi, d(\varphi_1 \circ \chi_1)(t)\rangle = \langle \xi, \dot\gamma(t)\rangle,$$

which is not zero for sufficiently large k, because

$$\rho_k(t) \to_{k\to\infty} c(t) = (\gamma(t), b\dot\gamma(t))$$

with some $b > 0$. This completes the proof of the theorem. ♠

7.4. Poisson relation for generic strictly convex domains

Here we prove that the equality (7.1) holds for generic strictly convex domains Ω in $\mathbf{R}^n$, $n \geq 3$ with smooth boundaries $\partial\Omega$.

As we have already mentioned, for a strictly convex $\Omega \subset \mathbf{R}^n$ every $\gamma \in \mathcal{L}_\Omega$ is either a closed geodesic on $\partial\Omega$ or a periodic reflecting ray in Ω. If γ is not a multiple of another element δ of $\mathcal{L}_\Omega$, then γ will be called *primitive*. We shall say that γ is a *non-degenerate* element of $\mathcal{L}_\Omega$ if the Poincaré map P_γ of γ has no eigenvalues which are roots of unity.

Consider the family $\Xi = \Xi(n)$ of all strictly convex compact domains Ω in $\mathbf{R}^n$ with C^∞ smooth boundaries $\partial\Omega$ satisfying the following two conditions:

(R) $T_\gamma / T_\delta \notin \mathbf{Q}$ for every two different primitive elements γ and δ of $\mathcal{L}_\Omega$;
(ND) Each element of $\mathcal{L}_\Omega$ is non-degenerate.

Given a strictly convex domain Ω (we always consider domains with smooth boundaries), denote by $\mathcal{O}_\Omega$ the set of all $F \in \mathbf{C}(\partial\Omega)$ such that Ω_F is strictly convex. Recall that Ω_F is the bounded domain in $\mathbf{R}^n$ with $\partial\Omega_F = F(\partial\Omega)$. Clearly $\mathcal{O}_\Omega$ is an open subset of $\mathbf{C}(\partial\Omega)$ containing id. In particular, $\mathcal{O}_\Omega$ is a Baire topolocial space with respect to the topology inherited by $\mathbf{C}(\partial\Omega)$, therefore every residual subset of $\mathcal{O}_\Omega$ is dense in it.

The first main result in this section shows that the family Ξ under considerations is 'very big' in some topological sense. In particular, for every strictly convex domain Ω there exist smooth perturbations of Ω, arbitrarily close to id with respect to the C^∞ topology, such that the perturbed domain is in Ξ. Moreover, almost all perturbations in $\mathcal{O}_\Omega$ have this property.

Theorem 7.4.1: *Let Ω be an arbitrary strictly convex compact domain in $\mathbf{R}^n$ with smooth boundary $\partial\Omega$. There exists a residual subset $R(\Omega)$ of $\mathcal{O}_\Omega$ such that $\Omega_F \in \Xi$ for every $F \in R(\Omega)$.*

The proof of this theorem is postponed to the end of the section. We now proceed to prove that the Poisson relation becomes an equality for all elements of Ξ.

Define the *length spectrum* L_Ω of Ω by

$$L_\Omega = \{T_\gamma : \gamma \in \mathcal{L}_\Omega\}.$$

Applying the approximation theorem from the previous section, we find the following characterization of the isolated points in L_Ω for $\Omega \in \Xi$.

Proposition 7.4.2: *Let $\Omega \in \Xi$ and $\gamma \in \mathcal{L}_\Omega$. Then γ is a periodic reflecting ray in Ω if and only if T_γ is an isolated point in L_Ω.*

Proof: Let γ be a closed geodesic on $\partial\Omega$. Since Ω is strictly convex and γ is non-degenerate by (ND), it follows by Theorem 7.3.1 that γ can be approximated by periodic reflecting rays in Ω. In particular, there exists a sequence $\{\gamma_k\}$ of periodic reflecting rays in Ω such that $T_{\gamma_k} \to T_\gamma$. Condition (R) implies $T_{\gamma_k} \neq T_\gamma$ for each k. Therefore T_γ is not isolated in L_Ω.

Suppose now that γ is a periodic reflecting ray in Ω. We are going to show that T_γ is isolated in L_Ω. Assume that there exists a sequence $\{\gamma_k\} \subset \mathcal{L}_\Omega \backslash \{\gamma\}$ with $T_{\gamma_k} \to T_\gamma$. For any k choose an arbitrary point $x_k \in \gamma \cap \partial\Omega$ and denote by v_k, $v_k \in S^{n-1}$, the outgoing direction of γ at x_k. We may assume $x_k \to x \in \partial\Omega$ and $v_k \to v \in S^{n-1}$. According to the continuity of the generalized

Hamiltonian flow (cf. Section 1.2), there exists $\delta \in \mathcal{L}_\Omega$ passing through x in direction v and such that $T_\delta = \lim_k T_{\gamma_k} = T_\gamma$. Then (R) implies $\delta = \gamma$, i.e. δ is a periodic reflecting ray in Ω. Consequently, v is transversal to $\partial\Omega$ at x, and therefore v_k is transversal to $\partial\Omega$ at x_k for all sufficiently large k. Thus, γ_k is a periodic reflecting ray with the same number of reflection points as $\delta = \gamma$ for all sufficiently large k. This implies however $1 \in \operatorname{spec} P_\gamma$ which is a contradiction with the non-degeneracy of γ. Hence T_γ is an isolated point in L_Ω. ♠

The central moment in this section is the following.

Theorem 7.4.3: *For every $\Omega \in \Xi$ we have*

$$\operatorname{sing supp} \sigma_\Omega(t) = \{0\} \cup \{\pm T_\gamma : \gamma \in \mathcal{L}_\Omega\}. \tag{7.37}$$

Proof: The inclusion $\subset$ follows from Theorem 5.4.6. To check the converse inclusion, consider an arbitrary periodic reflecting ray γ in Ω. Since (R) and (ND) are satisfied, it follows by Theorem 6.3.1 that $T_\gamma \in \operatorname{sing supp} \sigma_\Omega$. Finally, let γ be a closed geodesics on $\partial\Omega$. Since there exists a sequence $\{\gamma_k\}$ of periodic reflecting rays in Ω with $T_{\gamma_k} \to T_\gamma$ (see the first part of the proof of Proposition 7.4.2) and $T_{\gamma_k} \in \operatorname{sing supp} \sigma_\Omega$ for all k, we get $T_\gamma \in \operatorname{sing supp} \sigma_\Omega$. This proves (7.37). ♠

The above theorem shows that knowing the point spectrum of the Laplacian for an $\Omega \in \Xi$, we can recover the length spectrum of Ω, which clearly contains some geometric information about Ω. Going a little bit further, one can recover some part of $\operatorname{spec} P_\gamma$ for every $\gamma \in \mathcal{L}_\Omega$. To do this, the following result of Stark is helpful.

Lemma 7.4.4: *Let P and Q be $2n \times 2n$ symplectic matrices such that*

$$\left|\det(I - P^k)\right| = \left|\det(I - Q^k)\right|$$

for all $k = 1, 2, \ldots$. Then $(\operatorname{spec} P) \backslash S^1 = (\operatorname{spec} Q) \backslash S^1$ and there exists an integer $N = N(P, Q) > 0$ such that $\operatorname{spec} P^N = \operatorname{spec} Q^N$.

For a proof of this lemma we refer the reader to the Appendix in [DG]. Combining it with Theorem 7.4.3 and the main result of Chapter 6, we obtain the following.

Corollary 7.4.5: *Let $\Omega_1, \Omega_2 \in \Xi$ be such that the Dirichlet problem for the Laplacian has one and the same spectrum in Ω_1 and Ω_2. Then there exists a bijection*

$$\mathcal{L}_{\Omega_1} \to \mathcal{L}_{\Omega_2}, \quad \gamma \mapsto \gamma',$$

such that for every $\gamma \in \mathcal{L}_{\Omega_1}$ we have: $T_{\gamma'} = T_\gamma$; γ is a periodic reflecting ray in Ω_1 if and only if γ' is a periodic reflecting ray in Ω_2; $(\operatorname{spec} P_{\gamma'}) \backslash S^1 = (\operatorname{spec} P_\gamma) \backslash S^1$ and there exists an integer $N = N(\gamma) > 0$ with $\operatorname{spec} P_{\gamma'}^N = \operatorname{spec} P_\gamma^N$.

Proof: It follows by our assumption and (7.37) for $\Omega = \Omega_1$ and $\Omega = \Omega_2$, that

$$\{T_\gamma : \gamma \in \mathcal{L}_{\Omega_1}\} = \{T_{\gamma'} : \gamma' \in \mathcal{L}_{\Omega_2}\}.$$

Since (R) is satisfied for both Ω_1 and Ω_2, for every $\gamma \in \mathcal{L}_{\Omega_1}$ there exists a unique $\gamma' \in \mathcal{L}_{\Omega_2}$ with $T_{\gamma'} = T_\gamma$. Moreover, the map $\gamma \mapsto \gamma'$ is a bijection.

Consider a periodic reflecting ray γ in Ω_1. Then by Proposition 7.4.2, T_γ is isolated in L_{Ω_1}. Consequently, $T_{\gamma'}$ is isolated in L_{Ω_2}, so γ' is a periodic reflecting ray in Ω_2. Moreover, applying Theorem 6.3.1 to the k-multiples of γ and γ', we see that

$$\left|\det(I - P_\gamma^k)\right| = \left|\det I - P_{\gamma'}^k)\right|$$

for all $k = 1, 2, \ldots$. It follows now from Lemma 7.4.4 that the third property in the assertion is fulfilled. To prove this when γ is a closed geodesics on $\partial\Omega_1$ we use the classical result of Duistermaat and Guillemin [DG] concerning the Laplace–Beltrami operator on manifolds without boundary, the condition (ND) and Lemma 7.4.4. ♠

The following lemma will be used in the proof of Theorem 7.4.1 below.

Lemma 7.4.6:   *Let M be a compact smooth $(n-1)$-dimensional submanifold of $\mathbf{R}^n$ and let γ be a primitive closed geodesics on M and K be a finite subset of $\operatorname{Im}\gamma$. There exists $q_0 \in \operatorname{Im}\gamma \backslash K$ such that for every neighbourhood U of q_0 in M there is $\lambda > 0$ and a continuous family $F_\mu = \operatorname{id} + f_\mu$, $\mu \in (-\lambda, \lambda)$ of elements of $\mathbf{C}(M)$ such that $f_0 = 0$ and for each μ,*

$$\operatorname{supp} f_\mu \subset U, \tag{7.38}$$

the curve $\gamma_\mu = F_\mu \circ \gamma$ is a closed geodesics on $F_\mu(M)$ with period $\theta(\mu)$ depending smoothly on μ and $\dot{\theta}(0) \neq 0$.

Proof: Let $\gamma : [0, \theta] \to M, \theta > 0$ being the minimal period of γ. Clearly, there exists $t_0 \in (0, \theta)$ such that $\ddot{\gamma}(t_0) \neq 0$ and $q_0 = \gamma(t_0) \notin K$. Fix t_0 and q_0 with these properties, and consider an arbitrary neighbourhood U of q_0 in M with $U \cap K = \emptyset$. We may assume that U is so small that there exist semi-geodesics coordinates along $\operatorname{Im}\gamma$ in U. Namely, U is the image of a chart

$$r : V = (t_0 - a, \; t_0 + a) \times B_a(0) \to U \subset M,$$

where $a > 0$ and $B_a(0)$ is the open ball with radius a and centre 0 in $\mathbf{R}^m$, $m = n - 2$, and in the coordinates $x_0, x_1, \ldots, x_m$ in U, provided by r, for all $y = (y_1, \ldots, y_m) \in B_a(0)$ the curves $\{(t; y) : t \in (t_0 - a, t_0 + a)\}$ are geodesic lines orthogonal to any surface $\{(s; y) : y \in B_a(0)\}$, $s \in (t_0 - a, t_0 + a)$. Then for $x \in V$ we have $g_{00}(x) = 1$, $g_{0i}(x) = 0$ for all $i \geq 1$. Here g is the standard metric on M. We shall assume that $a > 0$ is chosen in such a way that $w(t) = \frac{\partial r}{\partial x_0}(t; 0) \neq 0$ for all $t \in [t_0 - a, t_0 + a]$.

Fix arbitrary smooth functions

$$\rho : \mathbf{R}^m \rightarrow [0,1]$$

with compact supp $\rho \subset B_a(0)$, $\rho(y) = 1$ for all $y \in B_{a/2}(0)$, and

$$\varphi : \mathbf{R} \rightarrow [0,1]$$

such that $\varphi(t_0) > 0$ and supp $\varphi \subset [t_0 - a/2, t_0 + a/2]$. Define the map $v : V \rightarrow \mathbf{R}^n$ by

$$v(t;y) = -\frac{\varphi(t)\rho(y)}{\|\dot{w}(t)\|^2}\dot{w}(t).$$

Then v is smooth with compact supp $v \subset [t_0 - a/2, t_0 + a/2] \times B_a(0)$, and

$$\left\langle w(t), \frac{\partial v}{\partial x_0}(t;0) \right\rangle = \varphi(t)$$

for all $t \in (t_0 - a, t_0 + a)$. For μ close to 0 define $f = f_\mu : M \rightarrow \mathbf{R}^n$ by

$$f(z) = \begin{cases} z, & z \notin U, \\ r(x) + \mu v(x), & z = r(x) \in U. \end{cases}$$

Clearly, there exists a sufficiently small $\lambda > 0$ such that for all $\mu \in (-\lambda, \lambda)$ we have $F_\mu = \mathrm{id} + f_\mu \in \mathbf{C}(M)$.

Fix an arbitrary $\mu \in (-\lambda, \lambda)$. The map

$$\tilde{r} : V \rightarrow \mathbf{R}^n,$$

defined by $\tilde{r} = r + \mu v$, provides coordinates $x_0, \ldots, x_m$ around $\mathrm{Im}\,\gamma_\mu$, $\gamma_\mu = F_\mu \circ \gamma$, $\tilde{M} = F_\mu(M)$. Denote by $\tilde{g}$ the standard metric on $\tilde{M}$. Then for $s \in (t_0 - a, t_0 + a)$, $y \in B_{a/2}(0)$ we clearly have

$$\tilde{g}_{00}(s;y) = 1 + 2\mu\varphi(s) + O(\mu^2), \tag{7.39}$$

the last term being dependent only on s. Moreover, $\tilde{g}_{0i}(s;y) = 0$ for all $i = 1, \ldots, m$. It is now clear that the curve $\gamma_\mu(t) = \tilde{r}(t;0)$, $t \in [0, \theta]$, is a closed geodesic on $\tilde{M}$ (although t is a not a natural parameter for it). Denote by $\theta(\mu)$ the minimal period of γ_μ. It follows from the construction of F_μ that $\theta(\mu)$ is smooth.

It remains to show that $\dot{\theta}(0) \neq 0$. By (7.39) we have

$$\theta(\mu) = \int_0^\theta \sqrt{\tilde{g}_{00}(t;0)}\, dt$$

$$= \int_0^\theta \sqrt{1 + 2\mu\varphi(t) + O(\mu^2)}\, dt.$$

Differentiating this equality with respect to μ and evaluating at $\mu = 0$, we find

$$\theta'(0) = \int_0^\theta \varphi(t)\, dt > 0,$$

which proves the assertion. ♠

Proof of Theorem 7.4.1: Set $M = \partial\Omega$ and fix an arbitrary $q > 0$. Denote by $\mathcal{L}_\Omega(q)$ the set of all $\gamma \in \mathcal{L}_\Omega$ such that $T_\gamma \leq q$, and if γ is a periodic reflecting ray, then it has not more than q reflection points. It follows by the results of Chapters 3 and 4 that there exists a residual subset R'_q of $\mathcal{O}_\Omega$ such that every $F \in R'_q$ has the following properties: each element of $\mathcal{L}_{\Omega_F}(q)$ is non-degenerate; $T_\gamma \neq T_\delta$ for any two different periodic reflecting rays $\gamma, \delta \in \mathcal{L}_{\Omega_F}(q)$. Note that $\mathcal{L}_{\Omega_F}(q)$ is finite provided $F \in R'_q$ (cf. Theorem 3.4.3).

Let R_q be the set of those $F \in R'_q$ such that $T_\gamma \neq T_\delta$ for any two different elements γ and δ of $\mathcal{L}_{\Omega_F}(q)$. We are going to show that R_q is open and dense in R'_q.

To establish the openess, we shall show that $R'_q \backslash R_q$ is closed in R'_q. Let

$$\{F_k\} \subset R'_q \backslash R_q, \quad F_k \to F \in R'_q$$

in the C^∞ topology. We have to check that $F \notin R_q$. Without loss of generality, we may assume $F = \mathrm{id}$. Exploiting the fact that $F_k \notin R_q$, we find two elements $\gamma_k \neq \delta_k$ of $\mathcal{L}_{\Omega_k}(q)$, $\Omega_k = \Omega_{F_k}$, with $T_{\gamma_k} = T_{\delta_k}$, γ_k being a closed geodesics, passing through some point $F_k(x_k)$, $x_k \in M$, with direction $dF_k(u_k)$, $u_k \in S_{x_k}M$. Considering an appropriate subsequence, we may assume that either δ_k is a periodic reflecting ray with at most q reflection points for all k, or δ_k is a closed geodesic, passing through some point $F_k(y_k)$, $y_k \in M$, with direction $dF_k(v_k)$, $v_k \in S_{y_k}M$ for all k. In the first case, using a standard continuity argument and taking subsequences, one finds a closed geodesic γ on M and a periodic reflecting ray δ in Ω with at most q reflection points such that $T_\gamma = T_\delta \leq q$, which implies $\mathrm{id} \notin R_q$. We leave the details in this case to the reader.

Consider the second case when all δ_k are closed geodesics. We may assume $x_k \to x$, $y_k \to y$ in M, $u_k \to u$, $v_k \to v$ in S^{n-1}, $T_{\gamma_k} = T_{\delta_k} \to T$. Then there are closed geodesics γ and δ on M, determined by (x, u) and (y, v), respectively, and such that $T_\gamma = T_\delta = T$. We claim that $\gamma \neq \delta$. Assume the opposite. Let g be the standard metric on M and g_k be the Riemannian metric on M so that $F_k : (M, g_k) \to F_k(M)$ is an isometry. If $t' \in [0, T]$ is the time for which the geodesics flow on (M, g) shifts (y, v) to (x, u), then for the same time t' the geodesic flow on (M, g_k) shifts (y_k, v_k) along δ_k to a $(y'_k, v'_k) \neq (x_k, u_k)$ such that $(y'_k, v'_k) \to (x, u)$. Therefore without loss of generality we may assume $y = x$, $v = u$.

Denote by Σ the hyperplane in $\mathbf{R}^n$ passing through x and orthogonal to u. Shifting x_k and y_k along the geodesic γ_k, we may also assume $x_k, y_k \in \Sigma$ for all sufficiently large k. Using the natural identification of T^*M with TM, we can consider the Poincaré map $\mathcal{P}_\gamma$ as a local symplectic map

$$\mathcal{P}_\gamma : S_{\Sigma \cap M}M \to S_{\Sigma \cap M}M,$$

defined in a small neighbourhood U of (x, u). For (z, w) consider the geodesic on (M, g_k), passing through z in direction w. After time close to T this curve intersects Σ at some point z' in direction $w' \in S^{n-1}$. Define $\mathcal{P}_k(z, w) = (z', w')$. Since $g_k \to g$, there exists a neighbourhood $V \subset U$ of (x, u) and $k_0 > 0$ such that for $k \geq k_0$ the map $\mathcal{P}_k$ is well defined and smooth on V, and $(x_k, u_k) \in V$, $(y_k, v_k) \in V$. So, $\mathcal{P}_k$ can be viewed as a Poincaré map for γ_k in the neighbourhood V of both (x_k, u_k) and (y_k, v_k). Since $(x_k, u_k) \neq (y_k, v_k)$ are fixed points of $\mathcal{P}_k$, for large k there exists $(z_k, w_k) \in V$ such that $(d\mathcal{P}_k - \mathrm{id})(z_k, w_k)$ is not invertible. Now $\mathcal{P}_k \to \mathcal{P}_\gamma$, $z_k \to x$, $w_k \to u$ imply that $(d\mathcal{P}_\gamma - \mathrm{id})(x, u)$ is not invertible. Thus, $1 \in \mathrm{spec}\,P_\gamma$, in contradiction with the non-degeneracy of γ. In this way, we have shown that $\gamma \neq \delta$, which clearly implies $\mathrm{id} \notin R_q$. Hence R_q is open in R'_q.

To establish the density, it is sufficient to assume $\mathrm{id} \in R'_q$ and to show that there are elements of R_q arbitrarily close to id. So, assume $\mathrm{id} \in R'_q$, and let $\mathcal{W}$ be an arbitrary neighbourhood of id in $\mathbf{C}(M)$. By $\mathrm{id} \in R'_q$, there are only finitely many closed geodesics $\gamma_1, \ldots, \gamma_s$ on M with periods $\leq q$, and finitely many periodic reflecting rays $\delta_1, \ldots, \delta_k$ with periods $\leq q$ and not more than q reflection points. Applying Lemma 7.4.6 to $\gamma = \gamma_1$ and

$$K = \mathrm{Im}\,\gamma_1 \cap (\cup_{i=2}^s \mathrm{Im}\,\gamma_i \cup \cup_{j=1}^k \mathrm{Im}\,\delta_j),$$

we find $F = \mathrm{id} + f \in \mathcal{W}$ such that $\mathrm{supp}\,f$ is contained in a small neighbourhood U of a point of $\mathrm{Im}\,\gamma_1$ with

$$U \cap (\cup_{i=2}^s \mathrm{Im}\,\gamma_i \cup \cup_{j=1}^k \mathrm{Im}\,\delta_j) = \emptyset,$$

$\tilde{\gamma}_1 = F \circ \gamma_1$ is a closed geodesic on $F(M)$, and

$$T_{\tilde{\gamma}_1} \notin \mathbf{Q}\{T_{\delta_1}, \ldots, T_{\delta_k}\}.$$

We choose F in such a way that $T_{\tilde{\gamma}_1} < q$ if $T_{\gamma_1} < q$, and $T_{\tilde{\gamma}_1} > q$ if $T_{\gamma_1} = q$. Notice that when U is sufficiently small and f is sufficiently close to 0 in the C^∞ topology, then $\delta_1, \ldots, \delta_k$ are the only periodic reflecting rays in Ω_F with periods $\leq q$ and not more than q reflection points. Moreover, the closed geodesics on $F(M)$ with periods $\leq q$ are $\gamma_2, \ldots, \gamma_s$ and eventually $\tilde{\gamma}_1$. Thus, choosing U small enough and f sufficiently close to 0, we have $F \in R'_q$.

Repeating this procedure $s - 1$ times, we find $G \in \mathcal{W} \cap R_q$, which proves the density of R_q in R'_q. Therefore R_q is a residual subset of $\mathcal{O}_\Omega$. Finally, setting

$$R(\Omega) = \cap_{q=1}^\infty R_q,$$

we obtain a residual subset of $\mathcal{O}_\Omega$ which has the desired properties. This proves the theorem. ♠

7.5. Notes

A slightly different version of Proposition 7.1.2 was proved by Landis [Lan]. Lemma 7.1.3 is also contained in [Lan]. The rest of Section 7.1 is taken from [PS2], see also [PS1]. Theorem 7.2.1 and Corollary 7.2.2 are due to Melrose [M1]. The other material in Sections 7.2 and 7.3 is a modification of a part of Magnuson's thesis [Mag]. The results in Section 7.4 have been proved in [S3]. As it is clear from the text, their proofs rely very heavily on previous results of Anderson and Melrose [AM], Duistermaat and Guillemin [DG], Guillemin and Melrose [GM1], Magnuson [Mag], Melrose [M1], Petkov and Stojanov [PS2, PS3, PS4], Stojanov [S1, S3] and others.

8 POISSON RELATION FOR THE SCATTERING KERNEL

This chapter is devoted to the proof of a relation analogous to that obtained in Section 5.4. We introduce the scattering kernel $s(t, \theta, \omega)$ in Section 8.1 making a brief discussion on the link between $s(t, \theta, \omega)$ and the scattering amplitude $a(\lambda, \theta, \omega)$ used in the physical literature. In Section 8.2 the contributions of the rays incoming with direction ω are localized. The Poisson relation for the scattering kernel has the form

$$\text{sing supp } s(t, \theta, \omega) \subset \{-T_\gamma : \gamma \in \mathcal{L}_{\omega,\theta}(\Omega)\}.$$

Here $\mathcal{L}_{\omega,\theta}(\Omega)$ is the set of all (ω, θ)-rays in Ω and T_γ is the sojourn time of γ. This relation is established in Section 8.3 under the assumption that each (ω, θ)-ray is the projection of a uniquely extendible generalized bicharacteristic. The results for propagation of singularities in [MS2] are not sufficient to eliminate all contributions which must be cancelled from a physical point of view. To overcome this difficultly we use an argument based on the $(i\lambda)$-outgoing solutions of the reduced wave equation $(\Delta + \lambda^2)u = 0$.

8.1. Representation of the scattering kernel

Let K be a compact subset of $\mathbf{R}^n$, $n \geq 3$, n odd, with smooth boundary ∂K such that

$$\Omega = \overline{\mathbf{R}^n \backslash K}$$

is connected. Clearly, $\partial \Omega = \partial K$. In this section we study the scattering kernel related to the scattering operator for the wave operator $\Box = \partial_t^2 - \Delta_x$ in the exterior of K.

Fix $\rho_0 > 0$ such that

$$K \subset \{x \in \mathbf{R}^n : |x| \leq \rho_0\}.$$

Consider the Dirichlet problem

$$\begin{cases} (\partial_t^2 - \Delta_x)u = 0 & \text{in } \mathbf{R} \times \Omega^\circ, \\ u = 0 & \text{on } \mathbf{R} \times \partial\Omega, \\ u_{|t=0} = f_1, \quad u_{t|t=0} = f_2. \end{cases} \qquad (8.1)$$

Denote by $H_D(\Omega)$ the completion of the space $C_0^\infty(\Omega^\circ)$ with respect to the norm

$$\|\varphi\|_D = \left(\int_\Omega |\partial_x \varphi|^2 \, dx \right)^{1/2}$$

and introduce the *energy space*

$$H = H_D(\Omega) \oplus L^2(\Omega).$$

Then there exists a one-parameter group $U(t) = e^{itG}$ of unitary transformations of H such that for $f = (f_1, f_2) \in H$ we have

$$U(t)f = (u(t,.), u_t(t,.)),$$

$u(t,x)$ being the solution of (8.1) in the sense of distributions.

Similarly, define $H_D(\mathbf{R}^n)$ as above, replacing Ω by $\mathbf{R}^n$, and consider the energy space

$$H_0 = H_D(\mathbf{R}^n) \oplus L^2(\mathbf{R}^n).$$

The solution $u_0(t,x)$ of the Cauchy problem for $\square$ in $\mathbf{R}_t \times \mathbf{R}_x^n$ with initial data $f = (f_1, f_2) \in H_0$ is expressed by

$$U_0(t)f = (u_0(t,x), \partial_t u_0(t,x)),$$

where $U_0(t) = e^{itG_0}$ is a unitary group in H_0.

The space H can be considered as a subspace of H_0, extending $f \in H$ as 0 in K. Let

$$J : H_0 \to H$$

be the orthogonal projection, and consider the wave operators

$$W_\pm f = \lim_{t \to \mp\infty} U(t) J U_0(-t) f, \quad f \in H_0.$$

These operators exist for each $f \in H_0$ and moreover they are isometries from H_0 onto H. The operator W_+ (resp. W_-) is related to the evolution when the time $t \to +\infty$ (resp. $t \to -\infty$).

The *scattering operator*

$$S = (W_+)^{-1} \circ W_-$$

is a unitary operator from H_0 onto H_0. We refer the reader to [LPh] for the existence of the operators $W_\pm$ and the main properties of S. Note that for all $t \in \mathbf{R}$ we have

$$SU_0(t) = U_0(t)S.$$

By using the Radon transform, we can construct an isometric isomorphism

$$\mathcal{R} : H_0 \to L^2(\mathbf{R} \times S^{n-1})$$

so that

$$\mathcal{R}U_0(t) = T_t\mathcal{R},$$

where T_t is the *translation operator* in $L^2(\mathbf{R} \times S^{n-1})$ having the form

$$T_t f(\sigma, \omega) = f(\sigma - t, \omega), \quad f \in L^2(\mathbf{R} \times S^{n-1}).$$

Therefore

$$\tilde{S} = \mathcal{R} \circ S \circ \mathcal{R}^{-1} : L^2(\mathbf{R}_t \times S^{n-1}) \to L^2(\mathbf{R}_t \times S^{n-1})$$

becomes a unitary operator commuting with the translations in t. Clearly, $\tilde{S} - \mathrm{Id}$ is a linear continuous map from $C_0^\infty(\mathbf{R} \times S^{n-1})$ into $\mathcal{D}'(\mathbf{R} \times S^{n-1})$. Hence by the Schwartz theorem the operator $\tilde{S} - \mathrm{Id}$ has a kernel

$$s(t - t', \theta, \omega) \in \mathcal{D}'(\mathbf{R}_t \times S^{n-1} \times \mathbf{R}_{t'} \times S^{n-1}),$$

where $\theta, \omega \in S^{n-1}$. Since for fixed ω and θ, s depends on $t - t'$ only, we can consider the distribution

$$s(t, \theta, \omega) \in \mathcal{D}'(\mathbf{R} \times S^{n-1} \times S^{n-1}),$$

called the *scattering kernel*.

To obtain a representation for $s(t, \theta, \omega)$, consider the solution $w(t, x; \omega)$ of the problem

$$\begin{cases} (\partial_t^2 - \Delta x)w(t, x; \omega) = 0 & \text{in } \mathbf{R} \times \Omega^\circ, \\ w = 0 & \text{on } \mathbf{R} \times \partial\Omega, \\ w_{|t < -\rho_0} = \delta(t - \langle x, \omega \rangle). \end{cases} \tag{8.2}$$

Then we have the following representation

$$s(\sigma, \theta, \omega) = C_n \int_{\partial\Omega} \partial_t^{n-2} \partial_\nu w(\langle x, \theta \rangle - \sigma, x; \omega) \, dS_x, \tag{8.3}$$

where ν is the unit normal to $\partial\Omega$ pointing into Ω, dS_x is the measure induced on $\partial\Omega$ and

$$C_n = (-1)^{(n+1)/2} 2^{-n} \pi^{1-n}.$$

The integral (8.3) is interpretted in the sense of distributions, and for $\rho \in C_0^\infty(\mathbf{R})$ we have

$$\langle s(t, \theta, \omega), \rho(t)\rangle = C_n \int_{-\infty}^{\infty} \int_{\partial \Omega} \partial_\nu w(t, x; \omega) \frac{d^{n-2}\rho}{dt^{n-2}}(\langle x, \theta\rangle - t)\, dt dS_x,$$

(8.4)

and $\frac{d^{n-2}\rho}{dt^{n-2}}(\langle x, \theta\rangle - t)$ has a compact support in $\mathbf{R} \times \partial\Omega$. Moreover, it follows from (8.4) that $s(t, \theta, \omega)$ depends smoothly on (θ, ω) and takes values in the space of tempered ditributions in t. We refer the reader to [Ma2, P5] for a proof of the representation (8.3).

In the physical literature the function $a(\lambda, \theta, \omega)$, given by

$$\overline{a(\lambda, \theta, \omega)} = \left(\frac{2\pi}{i\lambda}\right)^{(n-1)/2} \mathcal{F}_{t\to\lambda} s(t, \theta, \omega),$$

is called the *scattering amplitude*. Here $\mathcal{F}_{t\to\lambda}$ is the *Fourier transform* with respect to t. It is easy to see that

$$\mathcal{F}_{t\to\lambda}((\partial_t^{n-2}\partial_\nu)w(\langle x, \theta\rangle - t, x; \omega))$$
$$= (-i\lambda)^{n-2} e^{-i\lambda\langle x, \theta\rangle} \partial_\nu (e^{i\lambda\langle x, \omega\rangle} + \overline{v_{\text{sc}}(\lambda, x; \omega)}),$$

where

$$v_{\text{sc}}(\lambda, x; \omega) = \mathcal{F}_{t\to\lambda}(w(t, x; \omega) - \delta(t - \langle x, \omega\rangle)).$$

Consequently,

$$a(\lambda, \theta, \omega)$$
$$= -\frac{(i\lambda)^{(n-3)/2}}{2(2\pi)^{(n-1)/2}} \int_{\partial\Omega} \left[e^{i\lambda\langle x, \theta\rangle} \partial_\nu v_{\text{sc}}(\lambda, x; \omega) - e^{i\lambda\langle x, \theta-\omega\rangle} i\lambda\langle \nu, \omega\rangle\right] dS_x.$$

The function $v_{\text{sc}}(\lambda, x; \omega)$ is a solution of the problem

$$\begin{cases} (\Delta_x + \lambda^2)v_{\text{sc}}(\lambda, x; \omega) = 0 & \text{in } \Omega^\circ, \\ v_{\text{sc}}(\lambda, x; \omega) + e^{-i\lambda\langle x, \omega\rangle} = 0 & \text{on } \partial\Omega. \end{cases} \quad (8.5)$$

In general (8.5) does not have a unique solution. The uniqueness depends on the behaviour of $v_{\text{sc}}(\lambda, x; \omega)$ as $|x| \to \infty$.

To study this question, set $w^+ = w - \delta(t - \langle x, \omega\rangle)$ and introduce the function

$$H_1(t) = \begin{cases} t, & t \geq 0, \\ 0, & t < 0. \end{cases}$$

Consider the solution w_1 of the problem

$$\begin{cases} \Box w_1 = 0 & \text{in } \mathbf{R} \times \Omega^\circ \\ w_1 + H_1(t - \langle x, \omega \rangle) = 0 & \text{on } \mathbf{R} \times \partial\Omega, \\ w_{1|t<-\rho_0} = 0. \end{cases}$$

Therefore, $w^+ = (\mathrm{d}^2/\mathrm{d}t^2)w_1$, and it suffices to examine $\mathcal{F}_{t\to\lambda}w_1$. Let $\chi \in C_0^\infty(\mathbf{R}^n)$ be a function such that $\chi(x) = 1$ for $|x| \le \rho_0 + 1$, $\chi(x) = 0$ for $|x| \ge \rho_0 + 2$. Then we have

$$\Box(\chi w_1) = -2\langle \nabla\chi, \nabla w_1 \rangle - (\Delta\chi)w_1 = h_1(t,x),$$

and for each $t \in \mathbf{R}$ we obtain

$$h(t,.) = (0, h_1(t,.)) \in H_0.$$

The function χ_{w_1} is an outgoing solution of the wave equation, vanishing for $t < -\rho_0$. Hence

$$\chi(x)w_1(t,x;\omega) = \int_{-\infty}^t (U_0(t-\tau)h(\tau,x))_1 \, \mathrm{d}\tau = \int_0^\infty (U_0(\sigma)h(t-\sigma,x))_1 \, \mathrm{d}\sigma.$$

Here $(g)_1$ denotes the first component of $g = (g_1, g_2) \in H_0$.

Denote by $v_1(\lambda, x; \omega)$ the Fourier transform of $\chi w_1(t, x; \omega)$ with respect to t. For each $\varphi(\lambda) \in S(\mathbf{R})$ we have

$$\langle v_1(\lambda, x; \omega), \varphi(\lambda) \rangle = \langle \chi w_1(t, x; \omega), \int e^{-i\lambda t}\varphi(x)\,\mathrm{d}\lambda \rangle$$

$$= \int_{-\infty}^\infty \int_0^\infty (U_0(\sigma)(0, h_1(t-\sigma, x)))_1 \mathrm{d}\sigma \left(\int e^{-i\lambda t}\varphi(\lambda)\mathrm{d}\lambda \right) \mathrm{d}t$$

$$= \left\langle \int_0^\infty e^{-i\lambda\sigma}(U_0(\sigma)(0, g_1(\lambda, x)))_1 \, \mathrm{d}\sigma, \varphi(\lambda) \right\rangle,$$

where $g_1(\lambda, x) = \mathcal{F}_{t\to\lambda}h_1(t, x)$. Consequently,

$$v_1(\lambda, x; \omega) = \int e^{-i\lambda\sigma}(U_0(\sigma)(0, g_1(\lambda, x)))_1 \, \mathrm{d}\sigma.$$

The last integral can be transformed, and we obtain the representation

$$v_1(\lambda, x; \omega) = \int G_{i\lambda}^+(x-y)g_1(\lambda, y) \, \mathrm{d}y, \tag{8.6}$$

where $G_{i\lambda}^+$ is the $(i\lambda)$-*outgoing Green function* having the form

$$G_{i\lambda}^+(x) = \frac{(-1)^{(n-1)/2}}{2(2\pi)^{(n-1)/2}} \left(\frac{1}{r}\partial_r\right)^{(n-3)/2} \left(\frac{e^{-i\lambda r}}{r}\right) \qquad (8.7)$$

with $r = |x|$. The reader may consult [LPh] or Chapter 2 in [P5] for the justification of the above representation. Notice that

$$G_{i\lambda}^+(x) = -\frac{(i\lambda)^{(n-3)/2}}{2(2\pi)^{(n-1)/2}} \frac{e^{-i\lambda r}}{r^{(n-1)/2}} + O\left(\frac{e^{-i\lambda r}}{r^{(n+1)/2}}\right)$$

as $r \to \infty$. According to (8.6) and (8.7) it is not hard to see that $v_1(\lambda, x; \omega)$ and $v_{sc}(\lambda, x; \omega)$ satisfy the so called $(i\lambda)$-outgoing Sommerfeld radiation condition. More precisely, for $r \to \infty$ we have the asymptotics

$$\begin{cases} v_{sc}(\lambda, x; \omega) = \dfrac{e^{-i\lambda r}}{r^{(n-1)/2}} b\left(\lambda, \dfrac{x}{|x|}, \omega\right) + O\left(\dfrac{e^{-i\lambda r}}{r^{(n+1)/2}}\right), \\[2ex] \dfrac{\partial v_{sc}}{\partial r}(\lambda, x; \omega) + i\lambda v_{sc}(\lambda, x; \omega) = O\left(\dfrac{e^{-i\lambda r}}{r^{(n+1)/2}}\right). \end{cases} \qquad (8.8)$$

For the sake of brevity, a solution $u(x)$ of the reduced wave equation $(\Delta + \lambda^2)u(x) = g(x)$ with $g \in L^2(\mathbf{R}^n)$ and supp $g \subset \{x : |x| \leq R\}$ will be called *outgoing* if for $|x| \to \infty$, $u(x)$ satisfies the condition (8.8) with a suitable $b(\lambda, x/|x|)$ instead of $b(\lambda, x/|x|, \omega)$. The outgoing solution of the problem (8.5) is unique (see [LPh]).

Now we are going to show that $b(\lambda, \theta, \omega)$ coincides with the scattering amplitude. To this end, exploiting (8.8) and applying the Green formula, we obtain

$$v_{sc}(\lambda, x; \omega) = \int_{\partial\Omega} \left[G_{i\lambda}^+(x-y)\frac{\partial v_{sc}}{\partial \nu}(\lambda, y; \omega) - \frac{\partial G_{i\lambda}^+}{\partial \nu}(x-y)v_{sc}(\lambda, y; \omega)\right] dS_y.$$

Multiplying both sides of this equality by $e^{i\lambda r}r^{(n-1)/2}$ and setting $x = r\theta$, $r = |x|$, we find

$$\begin{aligned} b(\lambda, \theta, \omega) &= \lim_{r\to\infty} e^{i\lambda r}r^{(n-1)/2}v_{sc}(\lambda, r\theta; \omega) \\[1ex] &= -\frac{(i\lambda)^{(n-3)/2}}{2(2\pi)^{(n-1)/2}} \int_{\partial\Omega} e^{i\lambda\langle x,\theta\rangle}\left[\frac{\partial v_{sc}}{\partial \nu}(\lambda, x; \omega) - i\lambda\langle\nu, \theta\rangle v_{sc}(\lambda, x; \omega)\right] dS_x \\[1ex] &= -\frac{(i\lambda)^{(n-3)/2}}{2(2\pi)^{(n-1)/2}} \int_{\partial\Omega} \left(e^{i\lambda\langle x,\theta\rangle} \cdot \frac{\partial v_{sc}}{\partial \nu}(\lambda, x; \omega) - i\lambda\langle\nu, \omega\rangle e^{i\lambda\langle x,\theta-\omega\rangle}\right) dS_x. \end{aligned}$$

Here we have used the equality

$$\int_{\partial\Omega} e^{i\langle x,\theta-\omega\rangle} \langle \nu, \theta+\omega\rangle \, dS_x = \int_K \frac{\partial}{\partial(\theta+\omega)} (e^{i\langle y,\theta-\omega\rangle}) \, dy = 0.$$

Thus, $a(\lambda, \theta, \omega) = b(\lambda, \theta, \omega)$, and the scattering amplitude can be considered as the asymptotic profile of the outgoing solution $v_{sc}(\lambda, x; \omega)$ of the problem (8.5).

The scattering amplitude determines uniquely the obstacle K (see [LPh]). On the other hand, in the applications it is possible to measure only the singularities of $s(t, \theta, \omega)$ and their leading terms. Hence in general we cannot measure the Fourier transform of $s(t, \theta, \omega)$ and for this reason it is more important to investigate inverse scattering problems related to the singularities of $s(t, \theta, \omega)$.

8.2. Localization of the singularities of $s(t, \theta, \omega)$

In this section we begin the analysis of the singularities of $s(t, \theta, \omega)$. Let $\theta \neq \omega$ be fixed. Recall that $\mathcal{L}_{\omega,\theta}(\Omega)$ denotes the set of all (ω, θ)-rays in Ω. As usual π is the composition of the natural projections

$$T^*(\mathbf{R} \times \Omega) \rightarrow \mathbf{R} \times \Omega \rightarrow \Omega.$$

We fix $\rho_0 > 0$ as in the previous section. Recall that γ is a (ω, θ)-ray if $\gamma = \pi \circ \tilde{\gamma}$, where

$$\tilde{\gamma}(t) = (t, x(t), \pm 1, \xi(t)) \in T^*(\mathbf{R} \times \Omega)$$

is a generalized bicharacteristic of $\square$ in Ω such that there exist real numbers $t_1 < t_2$ with

$$\xi(t) = -\omega \quad \text{for } t \leq t_1, \quad \xi(t) = -\theta \quad \text{for } t \geq t_2. \tag{8.9}$$

This means that the curve $\gamma(t) = x(t)$ has direction ω for $t \leq t_1$ and direction θ for $t \geq t_2$. We assume that the time t increases when we move along $\tilde{\gamma}(t)$. Denote by T_γ the *sojourn time* of γ introduced in Section 2.4.

Below we consider a fixed t_0 such that

$$-t_0 \notin \{-T_\gamma : \gamma \in \mathcal{L}_{\omega,\theta}(\Omega)\}.$$

Take $T > 0$ with $|t_0| < T$ and consider the set

$$\Gamma_T = \{T_\gamma : |T_\gamma| \leq T, \quad \gamma \in \mathcal{L}_{\omega,\theta}(\Omega)\}.$$

To check that this set is closed, take a sequence $\{\gamma_k\} \subset \mathcal{L}_{\omega,\theta}(\Omega)$ with $T_{\gamma_k} \rightarrow T_0$. There exists a compact set M such that $\gamma_k(t) \in M$ for $|t| \leq T + 3\rho_0$ and all k. Let $\tilde{\gamma}_k$ be a generalized bicharacteristic of $\square$ in Ω such that $\gamma_k = \pi \circ \tilde{\gamma}_k$. There exist $t_1 < t_2$ such that $|t_i| < T + 3\rho_0, i = 1, 2$, and for each k (8.9) hold replacing ξ by ξ_k. It follows by Lemma 1.2.6 that there exists a generalized bicharacteristic $\tilde{\gamma}$ of $\square$ such that $\pi \circ \tilde{\gamma}$ is a (ω, θ)-ray with sojourn time T_0. Hence Γ_T is closed.

Choose $\epsilon_0 > 0$ so that

$$T_\gamma \notin [t_0 - \epsilon_0, t_0 + \epsilon_0], \quad \gamma \in \mathcal{L}_{\omega,\theta}(\Omega). \tag{8.10}$$

Let $\rho(t) \in C_0^\infty(\mathbf{R})$, $\rho(t) = 1$ for $|t| \leq \frac{1}{2}$, $\rho(t) = 0$ for $|t| \geq 1$. Set $\rho_\delta(t) = \rho(t/\delta)$ for $0 < \delta \leq \epsilon_0/2$, and consider the integral

$$J(\lambda) = \langle s(t, \theta, \omega), \rho_\delta(t + t_0)e^{-i\lambda t} \rangle$$

$$= \sum_{k=0}^{n-2} c_k(-i\lambda)^{n-2-k} \int_{\mathbf{R}} \int_{\partial\Omega} e^{i\lambda(t - \langle x, \theta \rangle)} \rho_\delta^{(k)}(\langle x, \theta \rangle - t + t_0)\frac{\partial w}{\partial \nu}(t, x; \omega)\, dt dS_x.$$

Here $w(t, x; \omega)$ is the solution of (8.2), $c_k = $ const, $c_0 = C_n$ and

$$\rho_\delta^{(k)} = \frac{d^k \rho_\delta}{dt^k}.$$

Our aim is to show that for δ sufficiently small the integral $J(\lambda)$ is rapidly decreasing with respect to λ. Below we study the term with $k = 0$. The analysis of the other terms is completely analogous.

Without loss of generality we may assume that

$$\omega = (0, \ldots, 0, 1).$$

Consider the hyperplane

$$Z(\tau) = \{x \in \mathbf{R}^n : x_n = \tau\},$$

where $\tau < -\rho_0$ will be fixed below. Let $\mathbf{R}_\tau^+ = \{t \in \mathbf{R} : t > \tau\}$ and let $\varphi_j(x') \in C_0^\infty(\mathbf{R}^{(n-1)})$, $x' = (x_1, \ldots, x_{n-1})$. Consider the problems:

$$\begin{cases} \square v_j = 0 & \text{in } \mathbf{R}_\tau^+ \times \mathbf{R}_x^n, \\ v_j(\tau, x') = \varphi_j(x')\delta(\tau - x_n), \\ \dfrac{\partial v_j}{\partial t}(\tau, x) = \varphi_j(x')\delta'(\tau - x_n), \end{cases} \tag{8.11}$$

$$\begin{cases} \square W_j = 0 & \text{in } \mathbf{R} \times \Omega^\circ, \\ W_j = 0 & \text{on } \mathbf{R} \times \partial\Omega, \\ W_j(\tau, x) = \varphi_j(x')\delta(\tau - x_n), \\ \dfrac{\partial W_j}{\partial t}(\tau, x) = \varphi_j(x')\delta'(\tau - x_n). \end{cases} \tag{8.12}$$

There exists a compact set $F_0' \subset \mathbf{R}^{n-1}$ such that if $\operatorname{supp}\varphi_j \cap F_0' = \emptyset$, then the straight lines issued from (x', τ), $x' \in \operatorname{supp}\varphi_j$, with direction ω do not meet $\partial\Omega$.

Hence for such j we obtain

$$WF\left(\left(\frac{\partial W_j}{\partial \nu}\right)_{|\mathbf{R} \times \partial \Omega}\right) \cap \left\{\left(t, x, 1, -\theta_{|T_x(\partial \Omega)}\right) : |t| \leq T + \rho_0 + 1, \quad x \in \partial \Omega\right\}.$$

(8.13)

Covering $\partial \Omega$ by small open neighbourhoods ω_k, we can apply Theorem 1.3.4 to the integrals over $\mathbf{R} \times \omega_k$. Then we obtain

$$\int_{\mathbf{R}} \int_{\partial \Omega} e^{i\lambda(t - \langle x, \theta \rangle)} \rho_\delta(\langle x, \theta \rangle - t + t_0) \frac{\partial W_j}{\partial \nu} \, dt dS_x = O(|\lambda|^{-m}), \quad m \in \mathbf{N}.$$

(8.14)

Set

$$F_0 = \{x \in \mathbf{R}^n : x' \in F_0', \ x_n = \tau\}$$

and denote by $l(u_0)$ the straight line passing through $u_0 \in F_0$ with direction ω. First, consider the case

$$\emptyset \neq l(u_0) \cap \bar{K} \subset \partial \Omega,$$

that is $l(u_0)$ could meet $\partial \Omega$ only at points, where it is tangent to $\partial \Omega$. Let γ_0 be a generalized bicharacteristic of $\square$ in Ω with $\text{Im}(\pi \circ \gamma_0) = l(u_0)$. Then γ_0 is uniquely extendible in the sense of Definition 1.2.2. To prove this, assume that $\partial \Omega$ is locally given by $\varphi(x) = 0$ and Ω by $\varphi(x) \geq 0$. If $\hat{x} \in l(u_0) \cap \partial \Omega$, the above geometric assumption implies $\varphi_{x_n x_n}(\hat{x}) \geq 0$, $\varphi_{x_n}(\hat{x}) = 0$. Now we can apply the argument from the proof of Corollary 1.2.4 based on condition (c) to conclude that γ_0 is uniquely extendible.

Next, we use the sets $C_t(\mu)$ and the metric $D(\rho, \mu)$ introduced in Section 1.2. Set $\mu_u = (\tau, u, 1, -\omega)$. Then we have $C_t(\mu_{u_0}) = \gamma_0(t)$. Applying the argument of the proof of Lemma 1.2.6 for fixed $\epsilon > 0$, we find a small neighbourhood $\mathcal{O}(u_0) \subset F_0$ of u_0 such that for $|t| \leq T + \rho_0 + 1$ and all $u \in \mathcal{O}(u_0)$ we have

$$D(C_t(\mu_u), \gamma_0(t)) = \inf_{\nu \in C_t(\mu_u)} D(\nu, \gamma_0(t)) < \epsilon.$$

Now, take φ_j with $\text{supp}\,\varphi_j \subset \mathcal{O}(u_0)$ and consider the solution W_j of (8.12). The singularities of W_j are contained in

$$\{C_t(\mu_u) : u \in F_0 \cap \text{supp}\,\varphi_j\}.$$

Using the above inequality with ϵ small enough, we arrange (8.13) and hence (8.14) holds.

Secondly, consider the case when $l(u_0)$ has common points with the interior of K. Choose $x_1(u_0) \in l(u_0)$ such that the linear segment $[u_0, x_1(u_0)]$ is the maximal one which has no common points with the interior of K. There are two possibilities:

(i) $l(u_0)$ meets transversally $\partial \Omega$ at $x_1(u_0)$;

(ii) $l(u_0)$ is tangent to $\partial\Omega$ at $x_1(u_0)$ and ω is an asymptotic direction for $\partial\Omega$ at $x_1(u_0)$.

Notice that in the case (ii) for each neighbourhood V of $x_1(u_0)$ we have $l(u_0) \cap V \cap K^\circ \neq \emptyset$. Set $t_1(u_0) = |u_0 - x_1(u_0)|$. As in Section 5.1 it is easy to write the solution v_j of (8.11) as an oscillatory integral and to find that $WF(v_j)$ is contained in the set of all $(t, x, \pm\sigma, \mp\sigma\omega) \in T^*(\mathbf{R}^{n+1})\{0\}$ such that $\sigma > 0$ and there exist $\hat{x} \in Z(1)$, $\hat{x}' \in \operatorname{supp}\varphi_j$, $s \geq 0$ with $t = \tau \pm s$, $x = \hat{x} \pm s\omega$. In the case (i) we modify v_j on the intersection of the interior of K with a small neighbourhood of $x_1(u_0)$ so that $\tilde{v}_j = v_j$ for $t < t_1 + \epsilon$, $\tilde{v}_j = 0$ for $t > t_1 + 2\epsilon$. Here $t_1 = \max\{t_1(u) : u \in \mathcal{O}(u_0)\}$, $\mathcal{O}(u_0)$ and $\epsilon > 0$ are chosen sufficiently small and $\operatorname{supp}\varphi_j \subset \mathcal{O}(u_0)$. Thus, we preserve the condition

$$\Box\tilde{v}_j = 0 \quad \text{in } \mathbf{R}_\tau^+ \times \Omega. \tag{8.15}$$

In the case (ii) we repeat the same procedure, modifying v_j in the interior of K so that $\tilde{v}_j = 0$ for $t > t_1 + 2\epsilon$. For this choose $z \in l(u_0) \cap K^\circ$ sufficiently close to $x_1(u_0)$ and modify v_j in a small neighbourhood $W \subset K^\circ$ of z so that (8.15) holds.

Set $h_j = (\tilde{v}_j)_{|\mathbf{R}_\tau^+ \times \partial\Omega}$ and notice that $h_j = 0$ for t sufficiently close to τ. Extending h_j as 0 for $t < \tau$, consider the solution w_j of the problem

$$\begin{cases} \Box w_j = 0 & \text{in } \mathbf{R} \times \Omega^\circ, \\ w_j + h_j = 0 & \text{on } \mathbf{R} \times \partial\Omega, \\ w_{j|t<\tau} = 0. \end{cases} \tag{8.16}$$

Since $\frac{\partial}{\partial t}(w_j + \tilde{v}_j)_{|\mathbf{R}_\tau^+ \times \partial\Omega} = 0$, we are going to study the integrals

$$I_{j,\delta}(\lambda) = \int_{\mathbf{R}}\int_{\partial\Omega} e^{i\lambda(t-\langle x,\theta\rangle)}\rho_\delta(\langle x,\theta\rangle - t + t_0)\left(\frac{\partial}{\partial\nu} - \langle\nu,\theta\rangle\frac{\partial}{\partial t}\right) w_j \, dt dS_x,$$

$$J_{j,\delta}(\lambda) = \int_{\mathbf{R}}\int_{\partial\Omega} e^{i\lambda(t-\langle x,\theta\rangle)}\rho_\delta(\langle x,\theta\rangle - t + t_0)\left(\frac{\partial}{\partial\nu} - \langle\nu,\theta\rangle\frac{\partial}{\partial t}\right) \tilde{v}_j \, dt dS_x.$$

We shall examine these integrals in the next section.

8.3. Poisson relation for the scattering kernel

In this section we use the notation from the previous sections. We begin with the analysis of the integral

$$I(\lambda) = \int_{\mathbf{R}}\int_{\partial\Omega} e^{i\lambda(t-\langle x,\theta\rangle)}\rho_\delta(\langle x,\theta\rangle - t + t_0)\left(\frac{\partial}{\partial\nu} - \langle\nu,\theta\rangle\frac{\partial}{\partial t}\right) v \, dt dS_x.$$

Here $v \in \bar{D}'(\mathbf{R} \times \Omega)$ is the solution of the problem

$$
\begin{cases}
\Box v = F & \text{in } \mathbf{R} \times \Omega^\circ, \\
v = h & \text{on } \mathbf{R} \times \partial\Omega, \\
v_{|t<\tau} = 0,
\end{cases}
\tag{8.17}
$$

where $\tau < -\rho_0$ is fixed, $F \in \mathcal{E}'(\mathbf{R} \times \Omega^\circ)$, $h \in H^s_{\text{loc}}(\mathbf{R} \times \partial\Omega)$ and $F = 0$, $h = 0$ for $t < \tau$. By $\bar{D}'(\mathbf{R} \times \Omega)$ we denote the space of distributions in $\mathbf{R} \times \Omega^\circ$ admitting extensions as distributions on $\mathbf{R}_t \times \mathbf{R}^n_x$. Since $\mathbf{R} \times \partial\Omega$ is not characteristic for $\Box$, the traces

$$
\left. \left(\frac{\partial^j v}{\partial \nu^j} \right) \right|_{\mathbf{R} \times \partial\Omega} \in \mathcal{D}'(\mathbf{R} \times \partial\Omega),
$$

$j = 0, 1$, exist (see [H3]), and we interpret $I(\lambda)$ in the sense of distributions. Set $T_1 = \rho_0 + |t_0| + 1$.

Next, we use the generalized wave front set

$$
WF_b(v) \subset T^*(\mathbf{R} \times \Omega^\circ) \cup T^*(\mathbf{R} \times \partial\Omega) = \tilde{T}^*(\mathbf{R} \times \Omega)
$$

and the map $\sim$ introduced in Section 1.3. Recall that for $x \in \partial\Omega$ we have

$$
\sim : T^*(\mathbf{R} \times \Omega) \ni (t, x, \tau, \xi) \rightarrow (t, x, \tau, \xi_{|T_x(\partial\Omega)}) \in T^*(\mathbf{R} \times \partial\Omega).
$$

To examine $I_{j,\delta}(\lambda)$ and $J_{j,\delta}(\lambda)$ we need the following.

Proposition 8.3.1: *Assume that there exists $\eta > 0$ such that*

$$
\begin{aligned}
WF_b(v) \cap \{\mu \in \tilde{T}^*(\mathbf{R} \times \Omega) : \mu = \overbrace{(t, x, 1, -\theta)}, \\
T_1 + \eta \le t \le T_1 + 2\eta\} = \emptyset, \\
WF(F) \cap \{\mu \in T^*(\mathbf{R} \times \Omega^\circ) : \mu = (t, x, 1, -\theta)\} = \emptyset.
\end{aligned}
\begin{aligned}
\quad (8.18) \\
\\
(8.19)
\end{aligned}
$$

Then $I(\lambda) = O(|\lambda|^{-m})$ for all $m \in \mathbf{N}$.

Proof: Suppose that $F = 0$ for $|x| \ge R$ and choose two functions $\alpha(t) \in C^\infty_0(\mathbf{R})$, $\beta(x) \in C^\infty_0(\mathbf{R}^n)$ such that

$$
\alpha(t) = \begin{cases} 1 & \text{for } t \le T_1 + \eta, \\ 0 & \text{for } t \ge T_1 + 2\eta, \end{cases}
$$

$$
\beta(x) = \begin{cases} 1 & \text{for } |x| \le \tau_1 + R + T_1 + 2\eta, \\ 0 & \text{for } |x| \ge \tau_1 + R + T_1 + 3\eta \end{cases}
$$

with $\tau_1 = \rho_0 - \tau$. Setting $\tilde{v}(t, x) = \alpha(t)\beta(x)v(t, x)$, we get

$$
\begin{cases}
\Box \tilde{v} = \tilde{F} & \text{in } \mathbf{R} \times \Omega^\circ, \\
\tilde{v} = \alpha\beta h & \text{on } \mathbf{R} \times \partial\Omega, \\
\tilde{v}_{|t<\tau} = 0
\end{cases}
$$

with

$$\tilde{F} = 2\alpha_t \beta v_t + \alpha_{tt}\beta v - 2\alpha\langle \nabla\beta, \nabla v\rangle - \alpha(\Delta\beta)v + \alpha\beta F.$$

A finite speed of propagation argument implies $v \in C^\infty$ for $|t| \leq T_1 + 2\eta$, $|x| \geq \tau_1 + R + T_1 + 2\eta$. Consequently, the terms involving derivatives of β are smooth and (8.18) and (8.19) yield

$$WF_b(\tilde{F}) \cap \{\mu \in \tilde{T}^*(\mathbf{R} \times \Omega) : \mu = \widetilde{(t, x, 1, -\theta)}\} = \emptyset, \qquad (8.20)$$

Since $\tilde{v}$ and $\tilde{F}$ have compact supports, we can take the partial Fourier transform with respect to t. Setting

$$V(\lambda, x) = \langle \tilde{v}(t, x), e^{-i\lambda t}\rangle,$$
$$f(\lambda, x) = \langle \tilde{F}(t, x), e^{-i\lambda t}\rangle,$$
$$g(\lambda, x) = \langle \alpha\beta h(t, x), e^{-i\lambda t}\rangle,$$

we obtain

$$\begin{cases} (\Delta + \lambda^2)V(\lambda, x) = -f(\lambda, x) & \text{in } \Omega^\circ, \\ V = g & \text{on } \partial\Omega, \\ V \text{ is } (i\lambda)\text{-outgoing.} \end{cases}$$

The $(i\lambda)$-outgoing condition follows from $\tilde{v}_{|t<\tau} = 0$ by using the argument from Section 8.1. Thus, for $|x| \to \infty$ we have the representation

$$V(\lambda, x) = \int_{\partial\Omega} \left[\frac{\partial V}{\partial\nu}(\lambda, y)G_{i\lambda}^+(x-y) - V(\lambda, y)\frac{\partial G_{i\lambda}^+}{\partial\nu}(x-y) \right] \mathrm{d}S_y$$

$$- \int_\Omega G_{i\lambda}^+(x-y)f(\lambda, y)\,\mathrm{d}y,$$

$G_{i\lambda}^+$ being the Green function given by (8.7). Let $x = r\theta$, $r = |x|$, and multiply the above equality by $e^{i\lambda r}r^{(n-1)/2}$. As in Section 8.1, letting $r \to \infty$, we deduce

$$\int_{\partial\Omega} e^{i\lambda\langle x,\theta\rangle} \left[\frac{\partial V}{\partial\nu}(\lambda, x) - i\lambda\langle\nu, \theta\rangle V(\lambda, x) \right] \mathrm{d}S_x$$

$$= \int_{\mathbf{R}}\int_\Omega e^{-i\lambda(t-\langle x,\theta\rangle)} \tilde{F}(t, x)\,\mathrm{d}t\mathrm{d}x, \qquad (8.21)$$

the integrals being interpreted in the sense of distributions.

Let $\{U_k\}_{k=1}^N$ be a covering of

$$\{x \in \Omega : |x| \leq \tau_1 + R + T_1 + 3\eta\}$$

consisting of small open sets $U_k \subset \mathbf{R}^n$. For $U_k \subset \Omega^\circ$ we apply Theorem 1.3.4

and (8.20) to get

$$\int_{\mathbf{R}} \int_{U_k} e^{-i\lambda(t-\langle x,\theta\rangle)} \tilde{F}(t,x)\,dtdx = O(|\lambda|^{-m}), \quad m \in \mathbf{N}.$$

In the case $U_k \cap \partial\Omega \neq \emptyset$, let $\partial\Omega$ be locally given by $x_n = \psi(x')$ and let $d\psi(\hat{x}') = 0$. Changing the variables, we may assume

$$\Omega \cap U_k = \{x \in \mathbf{R}^n : x' \in U', \quad 0 \le x_n \le c\}.$$

Since $\mathbf{R} \times \partial\Omega$ is not characteristic for $\square$, the distribution v depends smoothly on $x_n \in [0,c]$. The integral over $\mathbf{R} \times (\Omega \cap U_k)$ becomes

$$\int_{\mathbf{R}} \int_0^c dx_n e^{i\lambda x_n \theta_n} \int_{U'} e^{-i\lambda(t-\psi(x')\theta_n - \langle x',\theta'\rangle)} F_k(t,x',x_n)\,dtdx',$$

where $F_k(.,x_n) \in \mathcal{E}'(\mathbf{R}_t \times \mathbf{R}_{x'}^{n-1})$ depends smoothly on the parameter $x_n \in [0,c]$. For U' and c small enough by (8.20) we obtain

$$(t,x',1,-d\psi(x')\theta_n - \theta') \notin WF(F_k(.,x_n)).$$

Thus, we may apply Theorem 1.3.4 for the integral with respect to t, x'. Consequently, the right-hand side of (8.21) can be estimated by $O\,|\lambda|^{-m})$ for all $m \in \mathbf{N}$. Next, we have

$$(2\pi)^{-1} \int_{\mathbf{R}} d\lambda e^{i\lambda t} \int_{\partial\Omega} e^{i\lambda\langle x,\theta\rangle} \left[\frac{\partial V}{\partial \nu}(\lambda,x) - i\lambda\langle \nu,\theta\rangle V(\lambda,x) \right] dS_x$$

$$= \int_{\partial\Omega} \left(\frac{\partial \tilde{v}}{\partial \nu} - \langle \nu,\theta\rangle \frac{\partial \tilde{v}}{\partial t} \right)(t+\langle x,\theta\rangle,x)\,dS_x \in C_0^\infty(\mathbf{R})$$

and

$$I(\lambda) = \int_{\mathbf{R}} \left(\int_{\partial\Omega} \left(\frac{\partial \tilde{v}}{\partial \nu} - \langle \nu,\theta\rangle \frac{\partial \tilde{v}}{\partial t} \right)(t+\langle x,\theta\rangle,x)\,dS_x \right)$$
$$\times e^{i\lambda t} \rho_\delta(-t+t_0)\,dt = O(|\lambda|^{-m}), \quad m \in \mathbf{N}.$$

This proves the assertion. ♠

Now consider the integral $J_{j,\delta}(\lambda)$ defined in Section 8.2. Notice that $\tilde{v}_j = v_j$ for $t < \tau$. Choose a function $\alpha_1(t) \in C_0^\infty(\mathbf{R})$ such that

$$\alpha_1(t) = \begin{cases} 0 & \text{for } t \le \tau - T_1 - 2\epsilon, \\ 1 & \text{for } t \ge \tau - T_1 - \epsilon \end{cases}$$

with $\epsilon > 0$. Then $\theta \neq \omega$ implies

$$WF(\tilde{v}_j) \cap \{(t, x, 1, -\theta) \in T^*(\mathbf{R}^{n+1}) : t < \tau\} = \emptyset.$$

Setting $v(t, x) = \alpha_1(t)\tilde{v}_j(t, x)$, we can apply Proposition 8.3.1 with η large enough. The integral $J_{j,\delta}(\lambda)$ does not change when we replace $\tilde{v}_j$ by v, hence

$$J_{j,\delta}(\lambda) = O(|\lambda|^{-m}), \quad m \in \mathbf{N}. \tag{8.22}$$

We turn to the analysis of $I_{j,\delta}(\lambda)$. To this end we make an additional assumption.

$(U_{\omega,\theta})$ $\left\{\begin{array}{l} \text{Each } (\omega, \theta)\text{-ray } \gamma \text{ in } \Omega \text{ is the projection of a uniquely extendible} \\ \text{generalized bicharacteristic } \tilde{\gamma} \text{ of } \square. \end{array}\right.$

Notice that $(U_{\omega,\theta})$ concerns only the (ω, θ)-rays. In particular, there could exist generalized bicharacteristics of $\square$ which are not uniquelly extendible the projections on Ω of which are not (ω, θ)-rays.

For each $u_0 \in F_0$ satisfying the conditions (i), (ii) from Section 8.2 choose a sufficiently small neighbourhood $\mathcal{O}(u_0) \subset Z(\tau)$ and take φ_j with $\mathrm{supp}\,\varphi_j \subset \mathcal{O}(u_0)$. The singularities of the solution w_j of (8.16) are contained in the set of generalized bicharacteristics of $\square$ issued from $\mathcal{O}(u_0)$ with direction ω. For the sake of brevity set $C_t(u) = C_t(\mu_u)$ with $\mu_u = (\tau, u, 1, -\omega)$. There are two cases.

Case A. For all $\sigma > T_1$ we have

$$C_\sigma(u_0) \cap \{(\sigma, x, 1, -\theta) \in T^*(\mathbf{R} \times \Omega) : \rho_0 \leq |x| \leq \tau_1 + \sigma + 1\} = \emptyset. \tag{8.23}$$

Then for all $t \geq \tau$

$$C_t(u_0) \cap \{(t, x, 1, -\theta) \in T^*(\mathbf{R} \times \Omega) : \rho_0 \leq |x|\} = \emptyset.$$

Indeed, suppose that for some $\hat{t} \in [\tau, T_1]$ there exists a generalized bicharacteristic γ such that $\gamma(t) \in C_t(u_0)$ and $(\hat{t}, \hat{x}, 1, -\theta) \in \gamma(\hat{t})$ with $|\hat{x}| \geq \rho_0$. Then $\gamma(\sigma)$ would have direction θ for all $\sigma \geq \hat{t}$ which is a contradiction with (8.23).

Exploiting Lemma 1.2.6, we find a small neighbourhood $\mathcal{O}(u_0)$ such that for all $u \in \mathcal{O}(u_0)$ and all $t \in [\tau, T_1]$ we have

$$C_t(u) \cap \{(t, x, 1, -\theta) \in T^*(\mathbf{R} \times \Omega) : \rho_0 \leq |x| \leq \rho_0 + 2\} = \emptyset.$$

Now choose a function $\beta \in C_0^\infty(\mathbf{R})$ such that

$$\beta(x) = \begin{cases} 1 & \text{for } |x| \leq \rho_0, \\ 0 & \text{for } |x| \geq \rho_0 + 1. \end{cases}$$

Take φ_j with $\mathrm{supp}\,\varphi_j \subset \mathcal{O}(u_0)$ and consider the solution w_j of (8.16). Let

$$\square(\beta w_j) = -2\langle \nabla\beta, \nabla w_j \rangle - (\Delta\beta)w_j = F_j.$$

The above analysis implies

$$WF(F_j) \cap \{\mu \in \tilde{T}^*(\mathbf{R} \times \Omega) : \mu = \overbrace{(t, x, 1, -\theta)}, t \leq T_1\} = \emptyset. \quad (8.24)$$

Let $u_j(\lambda, x)$ be the solution of the problem

$$\begin{cases} (\Delta + \lambda^2)u_j = 0 & \text{in } \Omega^\circ, \\ u_j + \mathcal{F}_{t \to \lambda}(h_j) = 0 & \text{on } \partial\Omega, \\ u_j \text{ is } (i\lambda)\text{-outgoing,} \end{cases}$$

$\mathcal{F}_{t \to \lambda}$ being the Fourier transform with respect to t. It is easy to check that $u_j(\lambda, x)$ is a tempered distribution with respect to λ for $|x| \leq R$, hence the Fourier transform

$$\tilde{w}_j(\lambda, x) = \mathcal{F}_{t \to \lambda}(\beta w_j) = \beta u_j$$

exists. Setting $\tilde{F}_j(\lambda, x) = \mathcal{F}_{t \to \lambda}(F_j)$, as in the proof of Proposition 8.3.1, we obtain

$$\int_{\partial\Omega} e^{i\lambda\langle x, \theta\rangle} \left(\frac{\partial \tilde{w}_j}{\partial \nu}(\lambda, x)^\bullet - i\lambda\langle \nu, \theta\rangle \tilde{w}_j(\lambda, x) \right) dS_x$$

$$= \int_\Omega e^{i\lambda\langle x, \theta\rangle} \tilde{F}_j(\lambda, x) dx.$$

Taking the inverse Fourier transform, we have

$$\int_{\partial\Omega} \left(\frac{\partial \tilde{w}_j}{\partial \nu} - \langle \nu, \theta\rangle \frac{\partial w_j}{\partial t} \right) (t + \langle x, \theta\rangle, x) dS_x$$

$$= \int_\Omega F_j(t + \langle x, \theta\rangle, x) dx,$$

and applying (8.24) we deduce

$$I_{j,\delta}(\lambda) = \int_{\mathbf{R}} \int_\Omega e^{i\lambda t} \rho_\delta(-t + t_0) F_j(t + \langle x, \theta\rangle, x) \, dt dx$$

$$= \int_{\mathbf{R}} \int_\Omega e^{i\lambda(t - \langle x, \theta\rangle)} \rho_\delta(\langle x, \theta\rangle - t + t_0) F_j(t, x) \, dt dx$$

$$= O(|\lambda|^{-m}), \quad m \in \mathbf{N}. \quad (8.25)$$

Case B. There exists $\sigma > T_1$ with

$$C_\sigma(u_0) \cap \{(\sigma, x, 1, -\theta) \in T^*(\mathbf{R} \times \Omega) : \rho_0 \leq |x| \leq \tau_1 + \sigma + 1\} \neq \emptyset.$$

Then there is a generalized bicharacteristic γ of $\Box$ issued from μ_{u_0} and passing for $t = \sigma$ over some point y, $|y| \geq \rho_0$ with direction θ. This means that $\pi \circ \gamma$ is a (ω, θ)-ray. The assumption $(U_{\omega,\theta})$ implies that γ is uniquely extendible, hence $C_t(u_0) = \gamma(t)$. Let T_γ be the sojourn time of γ and let

$$\text{Im}\,\gamma = \{(t, x(t), 1, -\xi(t)) \in T^*(\mathbf{R} \times \Omega) : |\xi(t)| = 1, \quad t \geq \tau\},$$

where for $\gamma(t) \in H$ instead of $\xi(t)$ we determine $\xi(t+0)$ and $\xi(t-0)$. Recall that H is the set of hyperbolic points introduced in Section 1.2.

Introduce the numbers

$$T_2 = \inf\{\sigma : \sigma \geq \tau, \ \xi(t) = \theta \quad \text{for } t > \sigma\},$$
$$T_3 = \inf\{\sigma : \sigma \geq \tau, \ x(t) \notin \partial\Omega \quad \text{for } t > \sigma\}.$$

Clearly, $T_2 \leq T_3$. It is easy to see that

$$t - \langle x(t), \theta \rangle = T_\gamma \quad \text{for } T_2 \leq t \leq T_3.$$

For $t = T_3$ this follows from the definition of T_γ and the parametrization of $\gamma(t)$ by t. For $T_2 \leq t \leq T_3$ we use the equality $\langle x(T_3) - x(t), \theta \rangle = T_3 - t$. Taking into account (8.10), we obtain

$$|\langle x(t), \theta \rangle - t + t_0| \geq \epsilon_0, \quad T_2 \leq t \leq T_3.$$

Now choose $\mathcal{O}(u_0)$ small enough and assume supp $\varphi_j \subset \mathcal{O}(u_0)$. Then for $\tau \leq t \leq T_3 + 1$ the singularities of w_j will be contained in a small neighbourhood of $\gamma(t)$. This makes it possible to choose $s < T_2$ sufficiently close to T_2 and to arrange

$$\gamma(s) \notin H, \quad \xi(s) \neq \theta, \tag{8.26}$$
$$|\langle x, \theta \rangle - t + t_0| \geq \frac{\epsilon_0}{2} \tag{8.27}$$

for $t \geq s$ and

$$(t, x) \in \text{sing supp}(w_{j|\mathbf{R} \times \partial\Omega}) \cup \text{sing supp}\left(\left(\frac{\partial w_j}{\partial \nu}\right)_{|\mathbf{R} \times \partial\Omega}\right).$$

It is necessary to satisfy (8.27) for $s \leq t \leq T_3 + \epsilon_1$, where $\epsilon_1 > 0$ is chosen so that the singularities of w_j for $t \geq T_3 + \epsilon_1$ lie in the interior of Ω. Moreover, if $\gamma(s)$ is a glancing point, (8.26) yields

$$\xi(s) \neq \theta_{|T_{x(s)}(\partial\Omega)} \tag{8.28}$$

because $|\xi(s)| = |\theta| = 1$.

For $\mathcal{O}(u_0)$ small enough (8.26) and (8.28) imply

$$WF_b(w_j)\cap\{\mu\in\tilde{T}^*(\mathbf{R}\times\Omega) : \mu = \overbrace{(s,x,1,-\theta)}\} = \emptyset.$$

Since $WF_b(w_j)$ is closed, for small $\epsilon > 0$ we have

$$WF_b(w_j)\cap\{\mu\in\tilde{T}^*(\mathbf{R}\times\Omega) : \mu = \overbrace{(t,x,1,-\theta)}, s \le t \le s+\epsilon\} = \emptyset.$$
$$(8.29)$$

Choose a function $\alpha_2 \in C_0^\infty(\mathbf{R})$ such that

$$\alpha_2(t) = \begin{cases} 1 & \text{for } t \le s, \\ 0 & \text{for } t \ge s+\epsilon. \end{cases}$$

Write

$$I_{j,\delta}(\lambda) = I'_{j,\delta}(\lambda) + I''_{j,\delta}(\lambda),$$

where $I'_{j,\delta}(\lambda)$ (resp. $I''_{j,\delta}(\lambda)$) is obtained from $I_{j,\delta}(\lambda)$ replacing w_j by $\alpha_2 w_j$ (resp. by $(1-\alpha_2)(w_j)$). We may assume that the inequality (8.27) holds for (t,x) in some neighbourhood of

$$\text{sing supp}\left(\left(\frac{\partial}{\partial\nu} - \langle\nu,\theta\rangle\frac{\partial}{\partial t}\right)(1-\alpha_2)w_{j|\mathbf{R}\times\partial\Omega}\right).$$

Then $\delta \le \epsilon_0/2$ implies $\rho_\delta(\langle x,\theta\rangle - t + t_0) = 0$ for such (t,x) and

$$I''_{j,\delta}(\lambda) = O(|\lambda|^{-m}), \quad m \in \mathbf{N}.$$

To examine $I'_{j,\delta}(\lambda)$, set $F_j = \Box(\alpha_2 w_j)$. It follows from (8.29) that

$$WF_b(F_j)\cap\{\mu\in\tilde{T}^*(\mathbf{R}\times\Omega) : \mu = \overbrace{(t,x,1,-\theta)}\} = \emptyset.$$

Then for $I'_{j,\delta}(\lambda)$ we can apply the argument from the proof of Proposition 8.3.1. Thus, (8.25) holds in the case B.

In both cases A and B we have chosen a neighbourhood $\mathcal{O}(u_0)$ of each $u_0 \in F_0$. There exists a finite set $\{u_0^{(j)} : 1 \le j \le M\} \subset F_0$ such that

$$F_0 \subset \cup_{j=1}^M \mathcal{O}(u_0^{(j)}).$$

Suppose that the points $u_0^{(j)}$, $j \le N$, $N \le M$, satisfy condition (i) or (ii) of Section 8.2. Choose a partition of unity $\{\varphi_j(x')\}_{j=1}^\infty$ of $Z(\tau)$ so that supp $\varphi_j \subset \mathcal{O}(u_o^{(t)})$, $j = 1,\ldots,N$, supp $\varphi_j \cap F_0' = \emptyset$ for $j > M$. Set

$$\tilde{w} = \sum_{j=1}^N (w_j + \tilde{v}_j) + \sum_{j>N} W_j,$$

where w_j, W_j, $\tilde{v}_j$ are introduced in the previous section. Clearly

$$\begin{cases} \Box\tilde{w} = 0 \quad \text{in } \mathbf{R}_\tau^+ \times \Omega^\circ, \\ \tilde{w} = 0 \quad \text{on } \mathbf{R}_\tau^+ \times \partial\Omega, \\ \tilde{w}_{|t=\tau} = \delta(\tau - x_n), \quad \frac{\partial\tilde{w}}{\partial t}_{|t=\tau} = \delta'(\tau - x_n). \end{cases}$$

This implies $w = \tilde{w}$ in $\mathbf{R}_\tau^+ \times \Omega$. Choosing $\tau < -T_1$, we can replace w by $\tilde{w}$ in $J(\lambda)$. Then (8.14), (8.22) and (8.25) show that $J(\lambda)$ is rapidly decreasing and we conclude that $-t_0 \notin \text{sing supp } s(t, \theta, \omega)$.

Thus, we have proved the following.

Theorem 8.3.2: *Let $\theta \neq \omega$ be fixed and let the condition $(U_{\omega,\theta})$ be fulfilled. Then*

$$\text{sing supp } s(t, \theta, \omega) \subset \{-T_\gamma : \gamma \in \mathcal{L}_{\omega,\theta}(\Omega)\}. \tag{8.30}$$

The inclusion (8.30) is called the Poisson relation for the scattering kernel by analogy with the relation (5.29) in Chapter 5.

The condition $(U_{\omega,\theta})$ has been used only for the analysis of $C_t(u)$ in the case B. If $C_t(u)$ contains many generalized bicharacteristics, scattering with different directions, a localization of $C_t(u)$ might be done to eliminate the contributions connected with the rays having outgoing directions $\eta \neq \theta$.

8.4. Notes

The representation (8.3) of $s(t, \theta, \omega)$ was obtained in [Ma2] (see also [LPh] and [P5] for related results). The outgoing solutions of the reduced wave equation and the outgoing Green function are examined in more detail in [LPh]. The Poisson relation (8.30) for non-convex domains K has been studied in [P1] under some geometric restrictions concerning the rays incoming with directions $\pm\omega$. For several strictly convex obstacles (8.30) has been proved in [PS 5] (see also [Na1, Na2, NS] for some partial results). Theorem 8.3.2 is obtained in [CPS]. For the description of the uniquely extendible bicharacteristics we can apply Corollary 1.2.4. In particular, for generic domains Ω the condition $(U_{\omega,\theta})$ is satisfied for all $\omega, \theta \in S^{n-1}$.

9 SINGULARITIES OF THE SCATTERING KERNEL FOR GENERIC DOMAINS

In this chapter a formula is proved for the main singularity of the scattering kernel at $-T_\gamma$, γ being an ordinary non-degenerate reflecting (ω, θ)-ray satisfying an additional assumption. As a consequence, according also to some results from Chapter 3, certain information on the singular set of the scattering kernel is obtained. A special emphasis is given to three-dimensional generic domains. We prove that for such domains the (ω, θ)-rays of mixed type disappear and any singularity of the scattering kernel has the form $-T_\gamma$ for some reflecting (ω, θ)-ray γ.

9.1. Singularity of the scattering kernel for a non-degenerate (ω, θ)-ray

Let Ω be a connected closed domain in $\mathbf{R}^n$, $n \geq 3$, n odd with smooth boundary $\partial\Omega$ and bounded complement, and let $\omega \neq \theta$ be fixed unit vectors. Throughout we use the notation from the previous chapter (see also Section 2.4).

In what follows γ will be a fixed ordinary non-degenerate reflecting (ω, θ)-ray satisfying the following additional assumption:

(I) $T_\delta \neq T_\gamma$ for every $\delta \in \mathcal{L}_{\omega,\theta}(\Omega)\backslash\{\gamma\}$.

As in Section 6.3 one checks easily that for sufficiently small $\epsilon > 0$,

$$(T_\gamma - \epsilon, \ T_\gamma + \epsilon) \cap \operatorname{sing\,supp} s(t, \theta, \omega) = \{T_\gamma\}.$$

Fix such an $\epsilon > 0$, and let $\rho_\delta(t) \in C_0^\infty(\mathbf{R})$, $0 < \delta \leq \epsilon$, be the function defined in Section 8.2. Following the localization argument from Section 8.2, determine the distribution $u(t, x, \omega)$ as the solution of the problem:

$$\begin{cases} (\partial_t^2 - \Delta_x)u = 0 & \text{in } \mathbf{R} \times \Omega^\circ, \\ u + \varphi(x)\delta(t - \langle x, \omega\rangle) & \text{on } \mathbf{R} \times \partial\Omega, \\ u_{|t < -\rho_0} = 0. \end{cases} \tag{9.1}$$

Here $\varphi \in C_0^\infty(\mathbf{R})$ is chosen in such a way that $\varphi(q_1) = 1$ and the wave front set

$$\Sigma_1 = WF(\varphi(x)\delta(t - \langle x, \omega \rangle))_{|\mathbf{R} \times \partial\Omega}$$

contains only hyperbolic points of $\square$ with respect to $\partial\Omega$.

In what follows $\delta \in (0, \epsilon]$ will be fixed, and for sake of brevity we set $\rho = \rho_\delta$. Our aim is to examine the integral

$$I(\lambda) = \int_\mathbf{R} \int_{\partial\Omega} e^{i\lambda(t - \langle x, \theta \rangle)} \rho(\langle x, \theta \rangle - t + T_\gamma) \left(\frac{\partial}{\partial \nu} - \langle \nu, \theta \rangle \frac{\partial}{\partial t} \right) u \, dt \, dS_x.$$

(9.2)

Let $q_1, \ldots, q_m$ be the successive reflection points of γ and let $\tilde{\gamma}$ be a generalized bicharacteristic of $\square$ such that $\pi(\tilde{\gamma}) = \gamma$. Denote by t_1 the time of the first reflection of γ. Set

$$v_0 = \varphi(x)\delta(t - \langle x, \omega \rangle),$$

with supp φ small enough, and extend v_0 for $t > t_1 + \epsilon_1$, $\epsilon_1 > 0$, so that v_0 is smooth for such t. Next, we apply the construction from Section 6.1 to the mixed problem (9.1), replacing $i^*R_B^+$ by i^*v_0. Using the notation of Section 6.1, define

$$V_k^+ = R_k^+(Id - M_{k-1})i^*V_{k-1}^+, \quad k \geq 1, \quad V_0^+ = v_0,$$

and set

$$U_m = \sum_{k=0}^m (-1)^{k-1}V_k^+.$$

Then $u - U_m \in C^\infty(\mathbf{R} \times \Omega)$, and we can study (9.2) replacing u by U_m. Now we need a representation for

$$i^* \left(\frac{\partial}{\partial \nu} - \langle \nu, \theta \rangle \frac{\partial}{\partial t} \right) R_m^+(I - M_{m-1})i^*V_{m-1}^+.$$

The operator $(I - M_{m-1})i^*V_{m-1}^+$ is related to the graph of the canonical transformation σ_{m-1} defined in Section 6.2. Let (y', y_n) be local coordinates in a small neighbourhood V_1 of q_1 in $\partial\Omega$ such that in these coordinates $\partial\Omega$ is given by $y_n = g(y')$ and $q_1 = (y_0', 0)$, $dg(y_0') = 0$. Similarly, let V_m be a small neighbourhood of q_m in $\mathbf{R}^n$ and let $W_m = \partial\Omega \cap V_m$ be given by $x_n = h(x')$ with $q_m = (x_0', 0)$, $dh(x_0') = 0$. In local coordinates (s, y', σ, η') and (t, x', τ, ξ') in Σ_1 and Σ_m, respectively, the canonical transformation

$$\sigma_{m-1} : \Sigma_1 \to \sigma_{m-1}(\Sigma_1) = \Sigma_m$$

can be expressed by a generating function $\chi(t, x', \sigma, \eta')$ such that

$$\det \begin{pmatrix} \chi_{t\sigma} & \chi_{t\eta'} \\ \chi_{x'\sigma} & \chi_{x'\eta'} \end{pmatrix} \neq 0.$$

Since τ is constant with respect to the action of σ_{m-1}, we have

$$\chi_{t\sigma} = 1, \quad \chi_{tt} = \chi_{t\eta'} = \chi_{tx'} = 0. \qquad (9.3)$$

Consequently, $\det \chi_{x'\eta'} \neq 0$.

Let t_m be the time of the m-th reflection of γ. For t sufficiently close to t_m we have the representation

$$J_m f = (-1)^{m-1} i^*_{W_m} V^+_{m-1}$$

$$= (2\pi)^{-n} \int e^{i\chi(t,x',\sigma,\eta')-i s\sigma - i\langle y',\eta'\rangle} b(t,x',\sigma,\eta') f(s,y') \; ds \; dy \; d\sigma \; d\eta'.$$

Here

$$b \sim \sum_{k=0}^{\infty} b_k(t,x',\sigma,\eta'),$$

b_k being homogeneous of order $-k$ with respect to (σ, η'). Repeating the argument from Section 6.2, for the principal symbol of J_m we get

$$b_0(t,x',\sigma,\eta') = (-1)^{m-1} i^{\sigma_\gamma} \sqrt{|\det \chi x',\eta'|}, \qquad (9.4)$$

$\sigma_\gamma \in \mathbf{N}$ being a Maslov index.

To describe R^+_m, let $q(x,\tau,\xi)$ be the principal symbol of $\Box$ in local coordinates (x',x_n), and let $\xi^\pm_n(x,\tau,\xi')$ be the roots of the equation

$$q(x,\tau,\xi',\xi_n) = 0$$

with respect to ξ_n. Consider the phase function $\varphi^+(t,x,\tau,\xi')$ obtained as a solution of the problem

$$\begin{cases} \dfrac{\partial \varphi^+}{\partial x_n} = \xi^+_n(x,\varphi^+_t,\varphi^+_{x'}), \\ \varphi^+|_{x_n=h(x')} = t\tau + \langle x', \xi' \rangle, \end{cases}$$

where ξ^+_n is determined by the outgoing condition $-\xi^+_n/\tau > 0$. Thus, for t close to t_m we have

$$R^+_m f = (2\pi)^{-n} \int e^{i\varphi(t,x,\tau,\xi')-i s\tau - i\langle y',\xi'\rangle} \cdot a(t,x,\tau,\xi') f(s,y') \; ds \; dy' \; d\tau \; d\xi'$$

with

$$a \sim \sum_{k=0}^{\infty} a_k(t,x,\tau,\xi'),$$

a_k being homogeneous of order $-k$ with respect to (τ, ξ'). Moreover,

$$a_{0|x_n=h(x')} = 1, \quad a_{k|x_n=h(x')} = 0, \quad k \geq 1.$$

Therefore

$$i_{W_m}^* \left(\frac{\partial}{\partial \nu} - \langle \nu, \theta \rangle \frac{\partial}{\partial t} \right) R_m^+ f = B_m f,$$

where B_m is the first-order pseudo-differential operator with principal symbol

$$\beta_1 = i[(1+|dh|^2)^{1/2}\xi_n^+(x', h(x'), \tau, \xi') - \langle \xi', h_{x'} \rangle (1+|dh|^2)^{-1/2} - \langle \nu, \theta \rangle \tau]. \tag{9.5}$$

Denote by B_m^* and J_m^* the operators formally adjoint to B_m and J_m, respectively. For $\psi = t - \langle x', \theta' \rangle - \theta_n h(x')$ we have the asymptotic expansion

$$e^{-\lambda\psi} B_m^*(e^{i\lambda\psi}f) \sim \sum_{j=0}^{\infty} A_j(t, x', d\psi; f)\lambda^{1-j}$$

with

$$A_0 = \bar{\beta}_1(t, x', \psi_t, \psi_{x'})f.$$

Going back to the integral $I(\lambda)$, we get

$$(B_m J_m f_0, e^{i\lambda\psi}\rho) \sim \sum_{j=0}^{\infty}(f_0, J_m^*(A_j e^{i\lambda\psi}))\lambda^{1-j},$$

where

$$f_0(t, y') = \varphi(y)\delta(t - \langle y, \omega \rangle)|_{y_n=g(y')}.$$

Below we treat only the term A_0. Consider

$$I_0(\lambda) = (f_0, J_m^*(\bar{\beta}_1(t, x', \psi_t, \psi_{x'})\rho e^{i\lambda\psi}))$$
$$= \left(\frac{\lambda}{2\pi} \right)^n \int e^{i\lambda\Phi(t,x',y',\sigma,\eta')} \cdot \bar{b}_0(t, x', \sigma, \eta')\bar{\beta}_1(t, x', \psi_t, \psi_{x'})$$
$$\times \rho(\langle x', \theta' \rangle + \theta_n h(x') - t + T_\gamma)(1+|dg(y')|^2)^{1/2} \, dt \, dx' \, dy' \, d\sigma \, d\eta'$$
$$+ \text{ lower terms.} \tag{9.6}$$

The phase function Φ has the form

$$\Phi(t, x', y', \sigma, \eta') = t - \langle x', \theta' \rangle - h(x')\theta_n - \chi(t, x', \sigma, \eta')$$
$$+ (\langle y', \omega' \rangle + g(y')\omega_n)\sigma + \langle y', \eta' \rangle,$$

so the critical points of Φ satisfy the equations:

$$\begin{cases} \chi_t(t, x', \sigma, \eta') = 1, \\ \chi_{x'}(t, x', \sigma, \eta') = -(\theta' + \theta_n \, dh(x')), \\ \eta' = -(\omega' + \omega_n \, dg(y'))\sigma, \\ \chi_\sigma(t, x', \sigma, \eta') = \langle y', \omega' \rangle + \omega_n g(y'), \\ \chi_{\eta'}(t, x', \sigma, \eta') = y'. \end{cases}$$

Let $(\hat{t}, \hat{x}', \hat{y}', \hat{\sigma}, \hat{\eta}')$ be an arbitrary critical point of Φ. Since τ is constant along the generalized bicharacteristics of $\Box$, we get $\hat{\sigma} = 1$, hence we may parametrize the bicharacteristics by the time t. Therefore

$$\hat{\eta}' = -(\omega' + \omega_n \, dg(\hat{x}')).$$

Set $\hat{y} = (\hat{y}', g(\hat{y}'))$ and let $p(y, \sigma, \eta)$ be the principal symbol of $\Box$ in the coordinates (y', y_n). Denote by $\eta_n^+(y', \eta')$ the outgoing root of the equation $p(y', g(y'), 1, \eta', \eta_n) = 0$ with respect to η_n. Then for $\hat{y}'$ close to $\hat{y}'_0, -\hat{\eta} = -(\hat{\eta}', \hat{\eta}_n^+(\hat{y}', 1, \hat{\eta}'))$ is close to the reflected direction

$$\omega - 2\langle \nu(y_0), \omega \rangle \nu(y_0).$$

Let $\hat{T}$ be the length of the generalized geodesic $\hat{l}$ issued from $\hat{y}$ with direction $\hat{\eta}$ and joining $\hat{y}$ and $\hat{x} = (\hat{x}', h(\hat{x}'))$. Then

$$\langle \hat{y}, \omega \rangle = \chi_\tau(\hat{t}, \hat{x}', 1, \hat{\eta}') = \hat{t} - \hat{T},$$

which shows that the reflected direction of $\hat{l}$ at $\hat{x}$ is close to θ. Consequently, the sojourn time of the reflected ray issued from $(\hat{y}, \hat{\eta})$ is close to T_γ. Taking supp φ small enough and using (*I*) and the non-degeneracy of γ, we find

$$\hat{x}' = x'_0, \quad \hat{y}' = y'_0, \quad \hat{t} - \langle \hat{x}_0, \theta \rangle = T_\gamma, \quad \hat{\eta}' = -\omega'.$$

Notice that at the critical points of Φ we have

$$\Phi_{tt} = \Phi_{tx'} = \Phi_{ty'} = \Phi_{t\eta'} = 0, \quad \Phi_{t\sigma} = 1.$$

Put

$$G = \langle \nu(y_0), \omega \rangle g_{y'y'}(y'_0), \quad H = \langle \nu(x_0), \theta \rangle h_{x'x'}(x'_0),$$

and consider the matrix

$$\Delta = \begin{pmatrix} G & 0 & I \\ 0 & -H - \chi_{x'x'} & -\chi_{x'\eta'} \\ I & -\chi_{\eta'x'} & -\chi_{\eta'\eta'} \end{pmatrix}.$$

We have

$$\det \Delta = (\det \chi_{x'\eta'})^2 \det \begin{pmatrix} G & 0 & I \\ 0 & \chi_{x'\eta'}^{-1}(\chi_{x'x'}+H)\chi_{\eta'x'}^{-1} & I \\ -I & I & \chi_{\eta'\eta'} \end{pmatrix},$$

and to find $\det \Delta$ we apply the following.

Lemma 9.1.1: *Let F, M, L be $n \times n$ matrices. Then*

$$\det \begin{pmatrix} F & 0 & I \\ 0 & M & I \\ -I & I & L \end{pmatrix} = \det(FLM + M - F). \tag{9.7}$$

Proof: First assume that F is invertible. Then

$$\det \begin{pmatrix} F & 0 & I \\ 0 & M & I \\ -I & I & L \end{pmatrix} = (\det F) \det \begin{pmatrix} M & I \\ I & L+F^{-1} \end{pmatrix}.$$

One checks easily that

$$\det \begin{pmatrix} M & I \\ I & L+F^{-1} \end{pmatrix} = \det((L+F^{-1})M - I),$$

from which (9.7) follows directly. In the general case replace F by $F_\epsilon = F + \epsilon I$, where $\epsilon > 0$ is taken such that F_ϵ is invertible. Applying (9.7) for F_ϵ and letting $\epsilon \to 0$, we complete the proof. ♠

The symmetry of the matrices $\chi_{y'y'}$, $\chi_{\eta'\eta'}$, H, G and Lemma 9.1.1 yield

$$(\det \chi_{x'\eta'})^{-1} \det \Delta = \det((H + \chi_{x'x'})\chi_{\eta'x'}^{-1}(I + \chi_{\eta'\eta'}G) - \chi_{x'\eta'}G). \tag{9.8}$$

Using the non-degeneracy of γ, we shall show that the right-hand side of (9.8) is not zero. To do this we are going to use the hyperplane Z_ω, orthogonal to ω and the orthogonal projection π_ω of $\mathbf{R}^n$ onto Z_ω (cf. Section 2.4 for this notation). Introduce

$$\Sigma_b = \{(x,\tilde{\xi}) \in T^*(\partial\Omega) : |\tilde{\xi}| \leq 1\}.$$

Consider the diagram

$$
\begin{array}{ccccccccc}
\partial\Omega & \xrightarrow{\mu_1} & \Sigma_b & \xrightarrow{r} & \Sigma_b & \xrightarrow{\mu_2} & T_x^*(\partial\Omega) \\
j_1 \uparrow & & j_1^* \downarrow & & j_2^* \uparrow & & j_2 \downarrow \\
\mathbf{R}^{n-1} & \xrightarrow{\tilde{\mu}_1} & T^*(\mathbf{R}^{n-1}) & \xrightarrow{\tilde{r}} & T^*(\mathbf{R}^{n-1}) & \xrightarrow{\tilde{\mu}_2} & \mathbf{R}^{n-1}
\end{array}
$$

the maps in it being defined as follows. First, μ_1 has the form

$$\mu_1(y) = (y, \langle \nu(y), \omega \rangle \nu(y) - \omega) \in \Sigma_b.$$

Here $\eta(\omega) = -\omega + 2\langle \nu(y), \omega \rangle \nu(y) \in S^{n-1}$ is the reflected direction associated to $-\omega$. To define r, we use the canonical transformation $\sigma_{m-1} : \Sigma_1 \to \Sigma_m$. According to (9.3), the equality $(t, x, 1, \tilde{\xi}) = \sigma_{m-1}(s, y, 1, \tilde{\eta})$ determines two functions $x(y, \tilde{\eta})$ and $\tilde{\xi}(y, \tilde{\eta})$. Define

$$r : \Sigma_b \ni (y, \tilde{\eta}) \mapsto (x(y, \tilde{\eta}), \tilde{\xi}(y, \tilde{\eta})) \in \Sigma_b.$$

The map μ_2, related to the reflection on $\partial\Omega$ in a neighbourhood of q_m, has the form

$$\mu_2(x, \tilde{\xi}) = -\xi + \langle \nu(x), \xi \rangle \nu(x) \in T_x^*(\partial\Omega),$$

where $\xi \in T_x^*(\Omega)$ is the unique unit vector with $\langle \nu(x), \xi \rangle < 0$ the orthogonal projection of which on $T_x^*(\partial\Omega)$ coincides with $\tilde{\xi}$. The maps j_1, j_2, j_1^*, j_2^* are the diffeomorphisms associated to the local coordinates x', y' chosen above. For example,

$$j_1(y') = (y', g(y')), \quad j_2(\xi', \langle dh(x'), \xi' \rangle) = \xi'.$$

Finally, the maps $\tilde{\mu}_1$, $\tilde{r}$, $\tilde{\mu}_2$ are defined in such a way that the considered diagram is commutative.

Clearly, $dj_1(y_0') = \mathrm{Id}$, $dj_2(\theta', 0) = \mathrm{Id}$, and

$$d(\mu_2 \circ r \circ \mu_1)(q_1) = d(\tilde{\mu}_2 \circ \tilde{r} \circ \tilde{\mu}_1)(y_0').$$

In local coordinates y' we have

$$\tilde{\mu}_1(y') = (y', -(1 + |dg|^2)(\omega_n - \langle dg(y'), \omega' \rangle) dg(y')),$$

therefore

$$d\tilde{\mu}_1(y_0') = \begin{pmatrix} I \\ -G \end{pmatrix}.$$

Similarly,

$$d\tilde{\mu}_2(x_0', \theta') = (-H, -I).$$

To find $d\tilde{r}$, observe that the map $\tilde{r}$ is given by the generating function $\chi(t, x', 1, \eta')$, that is

$$\tilde{r} : \begin{pmatrix} \chi_{\eta'} \\ \eta' \end{pmatrix} \mapsto \begin{pmatrix} x' \\ \chi_{x'} \end{pmatrix}.$$

A simple calculation shows that

$$d\tilde{r}(\chi_{\eta'}, \eta') = \begin{pmatrix} \chi_{\eta'x'}^{-1} & -\chi_{\eta'x'}^{-1}\chi_{\eta'\eta'} \\ \chi_{x'x'}\chi_{\eta'x'}^{-1} & \chi_{x'\eta'} - \chi_{x'x'}\chi_{\eta'x'}^{-1}\chi_{\eta'\eta'} \end{pmatrix}.$$

Now, using the form of $d\tilde{\mu}_1(y_0')$ and $d\tilde{\mu}_2(x_0', \theta')$, we find

$$d(\tilde{\mu}_2 \circ \tilde{r} \circ \tilde{\mu}_1)(y_0') = -(H + \chi_{x'x'})\chi_{\eta'x'}^{-1}(I + \chi_{\eta'\eta'}G) + \chi_{x'\eta'}G.$$

Introduce the projection

$$p_x : S^{n-1} \rightarrow T_x^*(\partial\Omega)$$

by $p_x(\zeta) = \zeta - \langle\nu(x), \zeta\rangle\nu(x) \in T_x^*(\partial\Omega)$. Then we can consider the local inverse

$$p_x^{-1} : T_x^*(\partial\Omega) \rightarrow S^{n-1}.$$

In the same way we can consider the local inverse

$$\pi_w^{-1} : Z_w \rightarrow \partial\Omega$$

around q_1. It is easy to see that $\det d(\pi_w^{-1})(u_\gamma) = -\langle\nu(q_1), w\rangle^{-1}$, $\det d(p_x^{-1})(\theta) = -\langle\nu(q_m), \theta\rangle^{-1}$.

There exists a nighbourhood W_γ of $u_\gamma = \pi_w(q_1)$ in Z_w such that the map

$$J_\gamma : W_\gamma \rightarrow S^{n-1}$$

is well defined and smooth. Moreover, we have

$$J_\gamma = p_x^{-1} \circ \mu_2 \circ r \circ \mu_1 \circ \pi_w^{-1}.$$

Since γ is non-degenerate, we have $\det dJ_\gamma(u_\gamma) \neq 0$. On the other hand (9.8) implies

$$\begin{aligned}
\left|\det dJ_\gamma(u_\gamma)\right| &= |\langle\nu(q_1), w\rangle\langle\nu(q_m), \theta\rangle|^{-1} \\
&\quad \times v \left|\det[(H + \chi_{x'x'})\chi_{\eta'x'}^{-1}(I + \chi_{\eta'\eta'}G) - \chi_{x'\eta'}G]\right| \\
&= |\langle\nu(q_1), w\rangle\langle\nu(q_m), \theta\rangle(\det\chi_{x'\eta'})|^{-1} |\det\Delta|.
\end{aligned}$$

Therefore $\det\Delta \neq 0$ which allows us to apply the stationary phase method to the integral (9.6). It follows from the Euler equality for $\chi(t, x', \sigma, \eta')$ that

$$\Phi(\langle x_0, \theta\rangle + T_\gamma, x_0', y_0', 1, -w') = T_\gamma.$$

By (9.5) for $t = \langle x_0, \theta\rangle + T_\gamma$ we have

$$\bar{\beta}_1(t, x_0', \psi_t, \psi_{x'}) = -i(\xi_n^+(x_0', 0, 1, -\theta') - \langle\nu(x_0), \theta\rangle) = 2i\langle\nu(x_0), \theta\rangle,$$

because the outgoing condition implies

$$\xi_n^+(x_0', 0, 1, -\theta') = -\sqrt{1 - |\theta'|^2} = -\theta_n.$$

Taking into account the contributions from $\bar{b}_0$, $\bar{\beta}_1$ and ρ, we obtain

$$
I_0(\lambda) = 2\mathrm{i} \left(\frac{2\pi}{\lambda}\right)^{(n-1)/2} (-1)^{m-1} \exp\left(\mathrm{i}\frac{\pi}{2}\beta_\gamma\right) \mathrm{e}^{\mathrm{i}\lambda T_\gamma}
$$

$$
\cdot \left| \frac{\det \mathrm{d}J_\gamma(u_\gamma)\langle\nu(q_1),\omega\rangle}{\langle\nu(q_m),\theta\rangle} \right|^{-1/2} + O(\lambda^{-(n+1)/2}),
$$

where

$$
\beta_\gamma = -\sigma_\gamma + \frac{1}{2}\operatorname{sgn}\Delta \in \mathbf{N},
$$

σ_γ being the Maslov index from (9.4). Recall that

$$
(s(t,\theta,\omega),\rho(t+T_\gamma)\,\mathrm{e}^{-\mathrm{i}\lambda t}) = \sum_{k=0}^{n-2} c_k(-\mathrm{i}\lambda)^{n-2-k}
$$

$$
\times \int_{\mathbf{R}}\int_{\partial\Omega} \mathrm{e}^{\mathrm{i}\lambda(t-\langle x,\theta\rangle)}\cdot\rho^{(k)}(\langle x,\theta\rangle - t + T_\gamma)\left(\frac{\partial}{\partial\nu} - \langle\nu,\theta\rangle\frac{\partial}{\partial t}\right)w\,\mathrm{d}t\mathrm{d}S_x
$$

with

$$
c_0 = \frac{1}{2}(-1)^{(n+1)/2}(2\pi)^{1-n}.
$$

Thus, we obtain the following results.

Theorem 9.1.2: *Let γ be an ordinary non-degenerate reflecting (ω,θ)-ray satisfying condition (I). Then*

$$
-T_\gamma \in \operatorname{sing\,supp}(t,\theta,\omega),
$$

and for t near $-T_\gamma$ the kernel has the form

$$
s(t,\theta,\omega) = \left(\frac{1}{2\pi\,\mathrm{i}}\right)^{(n-1)/2}(-1)^{m_\gamma-1}\exp\left(\mathrm{i}\frac{\pi}{2}\beta_\gamma\right)
$$

$$
\times \left| \frac{\det \mathrm{d}J_\gamma(u_\gamma)\langle\nu(q_1),\omega\rangle}{\langle\nu(q_m),\theta\rangle} \right|^{-1/2}\delta^{(n-1)/2}(t+T_\gamma)
$$

$$
+ \text{ lower order singularities.} \tag{9.9}
$$

Here $m = m_\gamma$ is the number of reflections of γ, while q_1 and q_m are the first and the last reflection points of γ, respectively. ♠

In some cases we can find the asymptotic of the *scattering amplitude* $a(\lambda,\theta,\omega)$ defined by

$$
\overline{a(\lambda,\theta,\omega)} = \left(\frac{2\pi}{\mathrm{i}\lambda}\right)^{(n-1)/2} \mathcal{F}_{t\to\lambda}s(t,\theta,\omega),
$$

where $\mathcal{F}_{t \to \lambda}$ is the Fourier transform with respect to t.

In the particular case when $m_\gamma = 1$, the integral $I_0(\lambda)$ can be written with a phase function

$$\psi(x') = \langle x', \omega' - \theta' \rangle + h(x')(\omega_n - \theta_n).$$

Then

$$\det \Delta(x'_0) = (\langle \nu(q_1), \omega - \theta \rangle)^{n-1} \det h_{x'x'}(x'_0),$$

and $\langle \nu(q_1), \omega + \theta \rangle = 0$. Thus,

$$\left| \det \, dJ_\gamma(u_\gamma) \right|^{-1/2} = 2^{(1-n)/2} \left| \langle \nu(q_1), \omega \rangle \right|^{(3-n)/2} \left| \det h_{x'x'}(x'_0) \right|^{-1/2}.$$

By $\nu(q_1) = (\theta - \omega)/|\theta - \omega|$ we have

$$\langle \nu(q_1), \omega \rangle = \frac{\langle \theta, \omega \rangle - 1}{|\theta - \omega|} = -\frac{1}{2} |\theta - \omega|,$$

and therefore

$$\left| \det \, dJ_\gamma(u_\gamma) \right|^{-1/2} = \frac{1}{2} |\theta - \omega|^{(3-n)/2} |\mathcal{K}(q_1)|^{-1/2},$$

where $\mathcal{K}(q_1)$ is the Gauss curvature of $\partial \Omega$ at q_1. In this way we have established the following.

Corollary 9.1.3:   *Let γ be as in Theorem 9.1.2, and assume in addition that γ has exactly one reflection point. Then for t near $-T_\gamma$ we have*

$$s(t, \theta, \omega) = \frac{1}{2} \left(\frac{1}{2\pi i} \right)^{(n-1)/2} \exp \left(i \frac{\pi}{2} \beta_\gamma \right) |\theta - \omega|^{(3-n)/2}$$
$$\times |\mathcal{K}(q_1)|^{-1/2} \delta^{(n-1)/2}(t + T_\gamma)$$
$$+ \text{ lower order singularities.}$$

Moreover, if $\partial \Omega$ is strictly convex at q_1 with respect to the normal field ν, then $\beta_\gamma = -\frac{n-1}{2}$. ♠.

9.2. Singularities of the scattering kernel for generic domains

Let Ω, ω and θ be as in the previous section. Set $X = \partial \Omega$. In this section we consider an application of Theorem 9.1.2, also according to some of the results in Chapter 3.

Denote by $\mathcal{L}^m_{\omega,\theta}(\Omega)$ the set of all (ω, θ)-rays of mixed type in Ω, then

$$L_{\omega,\theta} = \mathcal{L}_{\omega,\theta}(\Omega) \setminus \mathcal{L}^m_{\omega,\theta}(\Omega)$$

is exactly the set of all reflecting (ω, θ)-rays in Ω. Set

$$\mathcal{G}(\Omega) = \{-T_\gamma : \gamma \in \mathcal{L}^m_{\omega,\theta}(\Omega)\}.$$

Theorem 9.2.1: *There exists a residual subset $\mathcal{R}$ of $\mathbf{C}(X)$ such that for every $f \in \mathcal{R}$ we have*

$$\{-T_\gamma : \gamma \in L_{\omega,\theta}(\Omega_f)\}\backslash\mathcal{G}(\Omega_f) \subset \text{sing supp } s_{\Omega_f}(t, \theta, \omega), \qquad (9.10)$$

and for t close to $-T_\gamma$, (9.9) holds with Ω replaced by Ω_f, provided $-T_\gamma$ belongs to the left-hand side of (9.10).

Proof: It follows by Theorems 3.2.6, 3.3.3, 3.4.6 and Lemma 7.1.2 that there exists a residual subset $\mathcal{R}$ of $\mathbf{C}(X)$ such that every f in it has the following properties:

(i) every reflecting (ω, θ)-ray in Ω_f is ordinary and non-degenerate;
(ii) $T_\gamma \neq T_\delta$ for every two different reflecting (ω, θ)-rays γ and δ in Ω_f;
(iii) the normal curvature of $f(X)$ does not vanish of infinite order.

To check that $\mathcal{R}$ has the desired properties, fix an arbitrary $f \in \mathcal{R}$. We claim that $\mathcal{G}(\Omega_f)$ is closed in $\mathbf{R}$. Indeed, let $\{\gamma_k\} \subset \mathcal{L}^m_{\omega,\theta}(\Omega_f)$ and let $T_{\gamma_k} \underset{k\to\infty}{\to} T$. Using Lemma 1.2.6, we may assume that $\{\gamma_k\}$ converges to some $\gamma \in \mathcal{L}_{\omega,\theta}(\Omega_f)$. Clearly, $T_\gamma = T$. If γ is a reflecting (ω, θ)-ray, then it would be ordinary by $f \in \mathcal{R}$, and therefore for all sufficiently large k, γ_k would be a reflecting (ω, θ)-ray, which is a contradiction with the non-degeneracy of γ (cf. (i)). Thus, γ is an (ω, θ)-ray of mixed type, which shows that $\mathcal{G}(\Omega_f)$ is closed in $\mathbf{R}$.

Now the desired properties of f follow from Theorem 9.1.2 and conditions (i) and (ii). ♠

As an immediate consequence of the above theorem one gets the following.

Corollary 9.2.2: *Assume in addition that*

$$K = \overline{\mathbf{R}^n \backslash \Omega}$$

is a finite disjoint union of strictly convex domains K_i. Let $\mathcal{O}$ be the set of those $f \in \mathbf{C}(X)$ such that $f(K_i)$ is strictly convex for every i. Then there exists a residual subset S of $\mathcal{O}$ such that for every $f \in S$ the relation (8.30) becomes an equality with Ω replaced by Ω_f, and for each $\gamma \in \mathcal{L}_{\omega,\theta}(\Omega_f)$ and t close to $-T_\gamma$ we have (9.9). ♠

9.3. Glancing ω-rays

From now on till the end of this chapter we consider domains Ω in $\mathbf{R}^3$. As in Section 9.2, Ω will be a connected close domain with smooth boundary $\partial\Omega$ and bounded complement.

Let $\omega \in S^2$ be fixed, and let Z_ω and π_ω be as in Section 2.4.

A curve

$$\gamma = \cup_{i=0}^{k} l_i$$

in Ω, consisting of linear segments $l_i = [x_i, x_{i+1}]$, $x_i \in \partial\Omega$, $i = 1, \dots, k$, and an infinite ray l_0, starting from x_1 with direction $-\omega$, will be called a *glancing ω-ray* in Ω if it has the following properties:

(i) l_i and l_{i+1} satisfy the law of reflection at x_{i+1} with respect to $\partial\Omega$ for every $i = 0, 1, \dots, k-1$;

(ii) l_k is tangent to $\partial\Omega$ and the normal curvature of $\partial\Omega$ at x_k vanishes in direction l_k.

The points $x_1, \dots, x_k$ will be called *vertices of γ*.

Our aim in this section is to show that for generic domains Ω, for any $k \geq 1$, the glancing ω-rays with k vertices in Ω form a discrete subset of a certain manifold. This fact will be applied in the next section.

Let γ be a glancing ω-ray in Ω and let $x_1, \dots, x_s$ be all different vertices of it such that x_s is the last one. Then there exists a surjective ns-map (cf. Section 3.2)

$$\alpha : \{1, \dots, k\} \to \{1, \dots, s\} \tag{9.11}$$

such that

$$\alpha(k) = s, \tag{9.12}$$

and $x_{\alpha(1)}, \dots, x_{\alpha(k)}$ are the successive vertices of γ. If $s > 1$ we may assume that

$$\alpha(k-1) = s - 1. \tag{9.13}$$

Fix arbitrary integers $k \geq s \geq 1$ and a surjective ns-map (9.11) with (9.12) and (9.13), the latter provided $s > 1$. Set

$$X = \partial\Omega.$$

Recall that for $f \in C(X)$ by Ω_f we denote the unbounded domain in $\mathbf{R}^3$ with boundary $f(X)$. Denote by $G(\omega, \alpha)$ the set of those $f \in C(X)$ such that there does not exist

$$y = (y_1, \dots, y_s) \in f(X)^{(s)}$$

such that $y_{\alpha(1)}, \dots, y_{\alpha(k)}$ are the successive vertices of a glancing ω-ray in Ω_f.

Lemma 9.3.1: *Let α be non-invertible, i.e. $k > s$. Then $G(\omega, \alpha)$ contains a residual subset of $C(X)$.*

Proof: We shall assume that $s > 1$. The case $s = 1$ can be considered using some part of the reasonings below.

Given $i = 1, \dots, s - 1$, we determine $I_i(\alpha)$ by (3.25). Denote by U_α the set of those $y = (y_1, \dots, y_s) \in (\mathbf{R}^3)^{(s)}$ such that y_i does not belong to the convex hull of the set

$$\{y_j : j \in I_i(\alpha)\}$$

for every $i = 1, \ldots, s - 1$. It is convenient to set

$$y_0 = \pi_\omega(y_1). \tag{9.14}$$

Define $F : U_\alpha \to \mathbf{R}$ by

$$F(y) = \sum_{i=0}^{s-1} \left\| y_{\alpha(i)} - y_{\alpha(i+1)} \right\|.$$

Let γ be a glancing ω-ray of type α in Ω_f, $f \in \mathbf{C}(X)$, i.e. there exists an ordering $y_1, \ldots, y_s$ of the different vertices of γ such that $y_{\alpha(1)}, \ldots, y_{\alpha(k)}$ are the successive vertices of γ. Then $y = (y_1, \ldots, y_s) \in U_\alpha$. Moreover, for $x = (x_1, \ldots, x_s)$, $f(x_i) = y_i$, $x' = (x_1, \ldots x_{s-1})$ we have $\mathrm{grad}_{x'}\, F(f^s(x)) = 0$ and $\langle y_s - y_{s-1}, \nu(y_s) \rangle = 0$ (the fact that the normal curvature of $f(X)$ at y_s vanishes in direction $y_s - y_{s-1}$ is not needed here).

Assuming $k > s$, we use almost the same argument as in the proof of Theorem 3.2.3. After the above preparation, the details are rather standard and we leave them to the reader. ♠

In view of the above lemma we can restrict our attention to glancing ω-rays which pass only once through each of their vertices.

Fix $k \in \mathbf{N}$ and denote by $D(\omega, k)$ the set of those $f \in \mathbf{C}(X)$ such that the elements $y = (y_1, \ldots, y_k)$ of $f(X)^k$ for which $y_1, \ldots, y_k$ are the successive vertices of a glancing ω-ray in Ω_f form a discrete subset of $f(X)^{(k)}$.

Lemma 9.3.2: $D(\omega, k)$ *contains a residual subset of* $\mathbf{C}(X)$.

Proof: We assume again $k > 1$, the case $k = 1$ can be proved using part of the argument below.

Let U_k be the set of those $y = (y_1, \ldots, y_k) \in (\mathbf{R}^3)^{(k)}$ such that $y_i \notin [y_{i-1}, y_{i+1}]$ for every $i = 1, \ldots, s - 1$. Define $H : U_k \to \mathbf{R}$ by

$$H(y) = \sum_{i=0}^{k-1} \left\| y_i - y_{i+1} \right\|.$$

As before, if $y_1, \ldots, y_k$ are the successive vertices of a glancing ω-ray in Ω_f, then $y = (y_1, \ldots, y_k) \in U_k$, and for $x = (x_1, \ldots, x_k)$, $f(x_i) = y_i$, $x' = (x_1, \ldots, x_{k-1})$ we have $\mathrm{grad}_{x'}\, H(f^k(x)) = 0$, $\langle y_k - y_{k-1}, \nu(y_k) \rangle = 0$, and the normal curvature of $Y = f(X)$ at y_k vanishes in direction $w = y_k - y_{k-1}$. The latter condition can be expressed analytically as follows. Let $r : V \to Y$ be a smooth chart, V being an open neighbourhood of 0 in $\mathbf{R}^2$ and $r(V)$ an open neighbourhood of y_k in Y, $r(0) = y_k$. Writing the standard coordinates in V by $v = (v_1, v_2)$, we have

$$w = \lambda \frac{\partial r}{\partial v_1}(0) + \mu \frac{\partial r}{\partial v_2}(0)$$

for some real λ, μ. Recall the coefficients of the second fundamental form of Y at y_k:

$$L = \left\langle \frac{\partial^2 r}{\partial v_1^2}(0), \nu(y_k) \right\rangle, \quad M = \left\langle \frac{\partial^2 r}{\partial v_1 \partial v_2}(0), \nu(y_k) \right\rangle, \quad N = \left\langle \frac{\partial^2 r}{\partial v_2^2}(0), \nu(y_k) \right\rangle.$$

Then the fact that the normal curvature of Y at y_k is zero in direction w is equivalent to

$$L\lambda^2 + 2M\lambda\mu + N\mu^2 = 0. \tag{9.15}$$

In fact,

$$\lambda = \left\langle w, \left(G\frac{\partial r}{\partial v_1} - F\frac{\partial r}{\partial v_2} \right) \bigg/ (EG - F^2) \right\rangle,$$

$$\mu = \left\langle w, \left(E\frac{\partial r}{\partial v_2} - F\frac{\partial r}{\partial v_1} \right) \bigg/ (EG - F^2) \right\rangle,$$

where $E = \|\partial r/\partial v_1\|^2$, $F = \langle \partial r/\partial v_1, \partial r/\partial v_2 \rangle$, $G = \|\partial r/\partial v_2\|^2$ are the coefficients of the first fundamental form. Therefore (9.15) is equivalent to

$$L \left\langle w, \left(G\frac{\partial r}{\partial v_1} - F\frac{\partial r}{\partial v_2} \right) \right\rangle^2 + 2M \left\langle w, \left(G\frac{\partial r}{\partial v_1} - F\frac{\partial r}{\partial v_2} \right) \right\rangle$$

$$\times \left\langle w, \left(E\frac{\partial r}{\partial v_2} - F\frac{\partial r}{\partial v_1} \right) \right\rangle + N \left\langle w, \left(E\frac{\partial r}{\partial v_2} - \frac{\partial r}{\partial v_1} \right) \right\rangle^2 = 0.$$

Next, we proceed as in the proofs of Theorems 3.3.1 and 3.4.1. To this end we need the bundle $J_k^2(X, \mathbf{R}^3)$ of 2-jets. Let M be the set of those

$$\tau = (j^2 f_1(x_1), \ldots, j^2 f_k(x_k)) \in J_k^2(X, \mathbf{R}^3)$$

such that $(x_1, \ldots x_k) \in X^{(k)}$, $(f_1(x_1), \ldots, f_k(x_k)) \in U_k$, rank $df_i(x_i) = 2$ for all $i = 1, \ldots, k$. Clearly, M is an open submanifold of $J_k^2(X, \mathbf{R}^3)$. The singular set Σ is now defined as the set of all $\tau \in M$ such that

$$\mathrm{grad}_{x'} F \circ (f_1 \times \cdots \times f_k)(x) = 0, \quad \langle f_k(x_k) - f_{k-1}(x_{k-1}), \nu \rangle = 0,$$

where $x = (x_1, \ldots x_k)$, ν is a non-zero normal vector to $f_k(X)$ at $f_k(x_k)$, and the normal curvature of $f_k(X)$ at $f_k(x_k)$ vanishes in direction $w = f_k(x_k) - f_{k-1}(x_{k-1})$.

We shall show that Σ is a smooth submanifold of M of codimension $2k$. Fix coordinates neighbourhoods V_i, $i = 1, \ldots, k$ with $V_i \cap V_j = \emptyset$ whenever $i \neq j$, and set

$$V = M \cap \left(\prod_{i=1}^{k} J^2(V_i, \mathbf{R}^3) \right).$$

Consider arbitrary charts $\varphi_i : W_i \to V_i$, $W_i \subset \mathbf{R}^2$ and define the chart

$$\varphi : V \to (\mathbf{R}^2)^{(k)} \times (\mathbf{R}^3)^{(k)} \times (\mathbf{R}^3)^{2k} \times (\mathbf{R}^3)^{4k}$$

by $\varphi(j^2 f_1(x_1), \ldots, j^2 f_k(x_k)) = (u; v; a; b)$, where

$$u = (u_1, \ldots, u_k), \quad v = (v_1, \ldots, v_k),$$

$$a = (a_{ij}^{(t)}), \quad b = (b_{ijl}^{(t)}), \quad \varphi_i(u_i) = x_i, \quad v_i = f_i(x_i),$$

$$a_{ij}^{(t)} = \frac{\partial(f_i^{(t)} \circ \varphi_i)}{\partial u_i^{(j)}}(u_i), \quad b_{ijl}^{(t)} = \frac{\partial^2(f_i^{(t)} \circ \varphi_i)}{\partial u_i^{(j)} \partial u_i^{(l)}}(u_i) \qquad (9.16)$$

for $i = 1, \ldots, k$, $j, l = 1, 2$, $t = 1, 2, 3$. As before, we use the notation $u_i = (u_i^{(1)}, u_i^{(2)}) \in \mathbf{R}^2$, $v_i = (v_i^{(1)}, v_i^{(2)}, v_i^{(3)}) \in \mathbf{R}^3$, $f_i = (f_i^{(1)}, f_i^{(2)}, f_i^{(3)})$.

In what follows we write the elements of $\varphi(v)$ in the form $\xi = (u; v; a; b)$, where u, v, a, b are determined by (9.16). For such a ξ set

$$\nu(\xi) = \det \begin{pmatrix} e_1 & e_2 & e_3 \\ a_{k1}^{(1)} & a_{k1}^{(2)} & a_{k1}^{(3)} \\ a_{k2}^{(1)} & a_{k2}^{(2)} & a_{k2}^{(3)} \end{pmatrix},$$

where $e_1 = (1, 0, 0)$, $e_2 = (0, 1, 0)$, $e_3 = (0, 0, 1)$. Then $\nu(\xi) \neq 0$ is a normal vector to $f_k(X)$ at $f_k(x_k)$.

Our aim is to show that $\varphi(V \cap \Sigma)$ is a smooth submanifold of $\varphi(V)$ of codimension $2k$. For $\xi \in \varphi(V)$ define: $E(\xi) = \|a_{k1}\|^2$, $F(\xi) = \langle a_{k1}, a_{k2} \rangle$, $G(\xi) = \|a_{k2}\|^2$, $L(\xi) = \langle b_{k11}, \nu(\xi) \rangle$, $M(\xi) = \langle b_{k12}, \nu(\xi) \rangle$, $N(\xi) = \langle b_{k22}, \nu(\xi) \rangle$, $\lambda(\xi) = \langle v_k - v_{k-1}, G(\xi)a_{k1} - F(\xi)a_{k2} \rangle$, $\mu(\xi) = \langle v_k - v_{k-1}, E(\xi)a_{k2} - F(\xi)a_{k1} \rangle$, where a_{ij} and b_{ijl} are the vectors in $\mathbf{R}^3$ with components $a_{ij}^{(t)}$ and $b_{ijl}^{(t)}$, respectively.

For $m = 1, 2, 3$ set

$$\mathcal{O}_m = \{\xi \in \varphi(V) : \nu^{(m)}(\xi) \neq 0\}.$$

Clearly, $\mathcal{O}_m$ are open subsets of $\varphi(V)$ which cover $\varphi(V)$. Thus, it is sufficient to show that $\mathcal{O}_m \cap \varphi(V \cap \Sigma)$ is a smooth submanifold of $\mathcal{O}_m$ of codimension $2k$. We shall do this for $m = 1$, the other cases are the same.

Consider the map $K : \mathcal{O}_1 \to \mathbf{R}^{2k}$, defined by

$$K(\xi) = ((d_{pq}(\xi))_{p=1,\ldots,k-1; q=1,2}; (K_i(\xi))_{i=1,2}),$$

where

$$d_{pq}(\xi) = \sum_{t=1}^3 \frac{\partial H}{\partial y_p^{(t)}}(v)a_{pq}^{(t)}, \quad K_1(\xi) = \langle v_k - v_{k-1}, \nu(\xi) \rangle,$$

$$K_2(\xi) = L(\xi)\lambda(\xi)^2 + 2M(\xi)\lambda(\xi)\mu(\xi) + N(\xi)\mu(\xi)^2.$$

The map K is so defined that $\mathcal{O}_1 \cap \varphi(V \cap \Sigma) = K^{-1}(0)$. Therefore it is sufficient to establish that K is a submersion on $\mathcal{O}_1 \cap \varphi(V \cap \Sigma)$.

Fix an arbitrary $\xi \in \mathcal{O}_1 \cap \varphi(V \cap \Sigma)$ and assume that

$$\sum_{p=1}^{k-1}\sum_{q=1}^{2} D_{pq}\,\mathrm{grad}\,d_{pq}(\xi) + \sum_{i=1}^{2} A_i\,\mathrm{grad}\,K_i(\xi) = 0 \tag{9.17}$$

for some real constants D_{pq} and A_i. Since $\xi = (u; v; a; b) \in \varphi(V)$ and
$V \subset M$, we have $v_k - v_{k-1} \neq 0$. It follows by the definitions of $\nu(\xi)$, $\lambda(\xi)$,
$\mu(\xi)$ and by $K_1(\xi) = 0$ that

$$v_k - v_{k-1} = C(\lambda(\xi)a_{k1} + \mu(\xi)a_{k2})$$

with some coefficient $C \neq 0$. Consequently, either $\lambda(\xi) \neq 0$ or $\mu(\xi) \neq 0$.
Let $\lambda(\xi) \neq 0$, the other case is similar. Consider in (9.17) the derivatives with
respect to $b_{k11}^{(1)}$. The only non-zero derivative is

$$\frac{\partial K_2}{\partial b_{k11}^{(1)}} = \lambda(\xi)^2 \nu^{(1)}(\xi) \neq 0,$$

because $\xi \in \mathcal{O}_1$. Hence $A_2 = 0$. Next, considering the derivatives with respect
to $a_{pq}^{(t)}$, as in the proofs of Theorems 3.3.1 and 3.4.1, we obtain $D_{pq} = 0$ for all
p, q. Now $A_1 = 0$ follows trivially. Thus, K is a submersion at ξ.

In this way we have established that Σ is a smooth submanifold of M of
codimension $2k$. Then for $f \in G(\omega, \alpha)$ the condition $j_k^2 f \pitchfork \Sigma$ is equivalent
to the fact that $\{x \in X^{(k)} : j_k^2 f(x) \in \Sigma\}$ is a discrete subset of $X^{(k)}$,
i.e. $f \in D(\omega, k)$. According to the multijet transversality theorem, we see that
$D(\omega, \kappa)$ contains a residual subset of $C(X)$. ♠

9.4. Generic domains in $\mathbf{R}^3$

Let Ω be an arbitrary connected domain with smooth compact boundary $X = \partial\Omega$
and bounded complement in $\mathbf{R}^3$, and let $\omega \neq \theta$ be two fixed unit vectors in $\mathbf{R}^3$.
Our aim in this section is to establish the following result.

Theorem 9.4.1:   *There exists a residual subset $\mathcal{R}$ or $C(X)$ such that for every
$f \in \mathcal{R}$ there are no (ω, θ)-rays of mixed type in Ω_f.*

As an immediate consequence of this theorem and the results in the previous
sections, we obtain the following.

Corollary 9.4.2:   *The generic connected domains Ω in $\mathbf{R}^3$ with smooth boundaries and bounded complements have the following properties:*

  (a) every (ω, θ)-ray in Ω is an ordinary non-degenerate reflecting (ω, θ)-ray;
  (b) $T_\gamma \neq T_\delta$ for every two different $\gamma, \delta \in \mathcal{L}_{\omega,\theta}(\Omega)$;

(c) the relation (8.30) becomes an equality;

(d) for every $\gamma \in \mathcal{L}_{\omega,\theta}(\Omega)$ *and t close to* $-T_\gamma$ *we have (9.9).* ♠

The rest of this section is devoted to the proof of Theorem 9.4.1. We begin with a technical lemma.

Lemma 9.4.3: *Let X be a smooth surface in* $\mathbf{R}^3$ *and let*

$$ c \,:\, [a, b] \;\rightarrow\; X, $$

$b > a$, *be a geodesic on X. Let* $t_0 \in (a, b)$ *be such that* $c(t_0)$ *is not a point of selfintersection of c. For every sufficiently small interval* (α, β) *containing* t_0 *and every sufficiently small open neighbourhood U of* $c(t_0)$ *in X with*

$$ c(\alpha, \beta) \subset U, \quad U \cap \mathrm{Im}\, c \,=\, c([\alpha, \beta]) \tag{9.18} $$

there exists $f \in \mathbf{C}(X)$, *arbitrarily close to* id *with respect to the* C^∞ *topology such that* $f(x) \,=\, x$ *for all* $x \in X \backslash U$, *and if*

$$ \tilde{c} \,:\, [a, b] \;\rightarrow\; \tilde{X} \,=\, f(X) $$

is the geodesic on $\tilde{X}$ *with* $\tilde{c}(t) \,=\, c(t)$ *for* $t \in [a, \alpha]$, *then*

$$ \tilde{c}((\alpha, \beta]) \cap c((\alpha, \beta]) \,=\, \emptyset. \tag{9.19} $$

Proof: We take U and (α, β) so small that there exist local coordinates x_0, x_1 in U, determined by a chart

$$ r \,:\, V \,=\, (\alpha, \beta) \times (-\delta, \delta) \;\rightarrow\; U \subset X $$

with $\delta > 0$, such that the components g_{ij} of the standard metric on X have the form

$$ g_{00}(x) \,=\, 1, \quad g_{01}(x) \,=\, 0, \quad g_{11}(x) \,=\, G(x) > 0 $$

for all $x \,=\, (x_0, x_1) \in V$. Moreover, we may assume that

$$ G(x) \,<\, 1 \quad \text{for every } x \in V. \tag{9.20} $$

Otherwise we can replace r by another chart $\tilde{r} \,:\, V \rightarrow X$, given by $\tilde{r}(x_0, x_1) \,=\, r(x_0, \epsilon x_1)$. If $\epsilon > 0$ is chosen sufficiently small, then

$$ \tilde{g}_{11}(x_0, x_1) \,=\, \epsilon^2 g_{11}(x_0, x_1) \,<\, 1 $$

for all $x \in V$. Clearly, (9.20) holds for sufficiently small intervals (α, β) and $\delta > 0$. Next, we assume that α, β and δ are fixed with these properties. Mention finally, that $r(t, 0) \,=\, c(t)$ for all $t \in [\alpha, \beta]$.

Fix two smooth functions $\lambda, \mu : \mathbf{R} \to [0,1]$ such that

$$\operatorname{supp} \lambda = [\alpha, \beta], \quad p = \mu(0) > 0, \quad q = \mu'(0) > 0. \tag{9.21}$$

For $\epsilon > 0$ define $f_\epsilon(y) = y$ for $y \in X \backslash U$, and

$$f_\epsilon(y) = r(x) + \epsilon\lambda(x_0)\mu(x_1)\frac{\partial r}{\partial x_0}(x)$$

for $y = r(x) \in U$, $x = (x_0, x_1) \in V$. It is easy to see that for all sufficiently small ϵ we have $f_\epsilon \in C(X)$, therefore $X_\epsilon = f_\epsilon(X)$ is a smooth surface. Moreover,

$$\psi(x) = r(x) + \epsilon\lambda(x_0)\mu(x_1)\frac{\partial r}{\partial x_0}(x)$$

determines a chart $\psi : V \to \psi(V) \subset X_\epsilon$. Denote by $g_{ij}(\epsilon; x)$ the components of the standard metric on $\psi(V) \subset X_\epsilon$. We have

$$g_{00}(\epsilon; x) = 1 + 2\epsilon\lambda'(x_0)\mu(x_1) + O(\epsilon^2),$$
$$g_{01}(\epsilon; x) = \epsilon\lambda(x_0)\mu'(x_1) + O(\epsilon^2),$$
$$g_{11}(\epsilon; x) = G(x_0, x_1) + 2\epsilon\lambda(x_0)\mu(x_1)\left\langle \frac{\partial r}{\partial x_1}(x), \frac{\partial^2 r}{\partial x_0 \partial x_1}(x)\right\rangle + O(\epsilon^2),$$

as $\epsilon \to 0$.

Using the coordinates x_0, x_1, introduce the canonical coordinates x_0, x_1, y_0, y_1 in T^*X_ϵ, and consider the Hamiltonian vector field generated by the Hamiltonian

$$H(\epsilon; x, y) = \frac{1}{2}g_{00}(\epsilon; x)y_0^2 + g_{01}(\epsilon; x)y_0 y_1 + \frac{1}{2}g_{11}(\epsilon; x)y_1^2,$$

where $x = (x_0, x_1)$, $y = (y_0, y_1)$. Denote by $c(\epsilon; t)$ the geodesic on X_ϵ such that $c(\epsilon; t) = c(t)$ for $t \in [a, \alpha]$, and let $(x(\epsilon; t), y(\epsilon; t))$ be the corresponding integral curve on T^*X_ϵ. Writing the Hamiltonian equations for this curve, and then the corresponding variational equations for

$$X_i(t) = \frac{d}{dt}x_i(\epsilon; t)_{|\epsilon=0}, \quad Y_i(t) = \frac{d}{dt}y_i(\epsilon; t)_{|\epsilon=0},$$

according to (9.21), we obtain:

$$\begin{cases} X_0'(t) = Y_0(t) + 2p\lambda'(t), \\ X_1'(t) = G(t,0)Y_1(t) + q\lambda(t), \\ Y_0'(t) = -p\lambda''(t), \\ Y_1'(t) = -q\lambda'(t), \\ X_0(\alpha) = X_1(\alpha) = Y_0(\alpha) = Y_1(\alpha) = 0, \end{cases}$$

for $t \in [\alpha, \beta]$. Consequently, $Y_1(t) = -q\lambda(t)$, and so $X'_1(t) = q\lambda(t)(1 - G(t, 0))$. This and (9.20) imply $X'_1(t) > 0$ for all $t \in (\alpha, \beta)$. Therefore for all sufficiently small $\epsilon > 0$ we have

$$\frac{\mathrm{d}}{\mathrm{d}t} x_1(\epsilon; t) > 0, \quad t \in (\alpha, \beta).$$

Fix such an ϵ. Then $x_1(\epsilon; t) > 0$ for all $t \in (\alpha, \beta)$ and $f = f_\epsilon$ has the desired properties. This proves the assertion. ♠

It follows by Lemma 7.1.2 that there exists a residual subset $\mathcal{K}$ of $C(X)$ such that for $f \in \mathcal{K}$ the normal curvature of $f(X)$ can vanish only of finite order. Then by the properties of the generalized bicharacteristics of $\square$ (cf. Section 1.2), we have that if $f \in \mathcal{K}$ and $\gamma : \mathbf{R} \to \Omega_f$ is a (ω, θ)-ray in Ω_f, then Im γ is a finite union of linear segments (two of them are infinite) and geodesic segments on $\partial\Omega_f$. The ends of all these (linear or geodesic) segments will be called *vertices* of γ (or Im γ).

Proof of Theorem 9.4.1: We are going to construct by induction a sequence

$$\mathcal{V}_1 \supset \mathcal{V}_2 \supset \dots \supset \mathcal{V}_k \supset \dots$$

or residual subsets of $\mathcal{K}$ such that for every k and every $f \in \mathcal{V}_k$ there are no (ω, θ)-rays of mixed type in Ω_f with not more than $k + 1$ different (linear or geodesic) segments.

Set $\mathcal{V}_1 = \mathcal{K}$. Clearly, for every $f \in \mathcal{V}_1$ there are no (ω, θ)-rays of mixed type in Ω_f having exactly two segments (and therefore only one vertex).

Let $k > 1$, and assume that $\mathcal{V}_1 \supset \dots \supset \mathcal{V}_{k-1}$ are already constructed and have the desired properties. To construct $\mathcal{V}_k$ we need some technical preparation.

A map of the form

$$\kappa : \{1, 2, \dots, k\} \to \{0, 1\}$$

will be called *k-design*, if κ is not identically zero, $\kappa(k) = 0$ and $\kappa(i)\,\kappa(i+1) = 0$ for all $i = 1, \dots, k - 2$. An (ω, θ)-ray γ will be called (ω, θ)-*ray with design* κ, if Im γ has $k + 1$ segments $l_0, l_1 \dots l_k$ and for every $i = 1, \dots, k - 1$, l_i is a linear segment if and only if $\kappa(i) = 0$.

Fix an arbitrary k-design κ and set

$$q = \max\{i : 1 \le i \le k - 1, \ \kappa(i) = 1\},$$
$$p = \min\{i : 1 \le i \le k - 1, \ \kappa(i) = 1\}.$$

It follows by Lemma 9.3.2 that for every $\eta \in S^2$ and every $m \in \mathbf{N}$ there exists a residual subset $D(\eta, m)$ of $C(X)$ such that for $f \in D(\eta, m)$ the elements $y = (y_1, \dots, y_m)$ of $f(X)^m$ such that $y_1, \dots, y_m$ are the successive vertices of a glancing η-ray in Ω_f form a discrete subset of $f(X)^{(m)}$. On the other hand, Theorem 3.3.3 and Corollary 3.4.7 imply the existence of a residual subset $\mathcal{T}_k$ of $C(X)$ such that for every $f \in \mathcal{T}_k$ there are only finitely many reflecting (ω, θ)-rays in Ω_f with not more than k reflection points, and all of them are ordinary.

We set

$$W = \mathcal{V}_{k-1} \cap D(\omega, p) \cap D(-\theta, q) \cap \mathcal{T}_k. \qquad (9.22)$$

Then W is a residual subset of $\mathbf{C}(X)$ which is clearly contained in $\mathcal{K}$.

Denote by $\mathcal{V}(k, \kappa)$ the set of those $f \in W$ such that there are no (ω, θ)-rays of mixed type with design κ in Ω_f.

We first show that $\mathcal{V}(k, \kappa)$ is dense in W. To this end we may assume id $\in W$ and then prove that there are $f \in \mathcal{V}(k, \kappa)$ arbitrarily close to id. Suppose id $\in W$. We claim that there exist only finitely many (ω, θ)-rays of mixed type with design κ in Ω. Assume the contrary, and let

$$\gamma_1, \dots, \gamma_m, \dots$$

be an infinite sequence of different (ω, θ)-rays of mixed type with design κ in Ω. Denote by $x_1^{(m)}, \dots, x_k^{(m)}$ the successive vertices of Im γ_m. We may assume that for all m, $\gamma_m(0) = x_1^{(m)}$ and $\dot{\gamma}_m(t) = \omega$ for $t < 0$. Moreover, considering an appropriate subsequence, we may assume that there exists $\lim_{m \to \infty} x_i^{(m)} = x_i \in \partial\Omega$ for all $i = 1, \dots, k$. It then follows by the continuity of the broken Hamiltonian flow that for every $t \in \mathbf{R}$ there exists

$$\lim_{m \to \infty} \gamma_m(t) = \gamma(t) \in \Omega,$$

and γ is an (ω, θ)-ray in Ω. Roughly speaking the ith finite segment of Im γ has end points x_i and x_{i+1}. However, this is not precise, since in general some of these segments can vanish. Denote by $l_0^{(m)}, \dots, l_k^{(m)}$ the successive segments of Im γ_m. Set

$$l_i = \lim_m l_i^{(m)}, \quad i = 0, 1, \dots, k,$$

then each l_i is either the one-point set $\{x_i\}$, or a linear segment, or a geodesic segment on $\partial\Omega$. Clearly, the first segment l_0 is the infinite ray starting from x_1 and having direction $-\omega$.

Let $s + 1$ be the number of different segments of γ, then $s \leq k$. There are two cases.

Case 1. $s = k$. Then each l_i is a non-degenerate linear or geodesic segment, and clearly γ is a (ω, θ)-ray of mixed type with design κ in Ω. Consequently, $x_1, \dots, x_p$ are the successive vertices of a glancing ω-ray in Ω. Moreover, for every $m \in \mathbf{N}$, $x_1^{(m)}, \dots, x_p^{(m)}$ are the successive vertices of a glancing ω-ray in Ω. Since $x_i^{(m)} \to_m x_i$ for all i, this is a contradiction with id $\in W \subset D(\omega, p)$.

Case 2. $s < k$. Then $s \leq k - 1$, and id $\in W \subset \mathcal{V}_{k-1}$ implies that γ cannot be a (ω, θ)-ray of mixed type. Therefore l_i is one point for all i with $\kappa(i) = 1$, and γ is a reflecting (ω, θ)-ray with s reflection points. Since l_i vanishes for at least one i, some segment of γ is tangent to $\partial\Omega$, which is a contradiction with id $\in W \subset \mathcal{T}_k$.

In both possible cases we got a contradiction. This shows that there are only finitely many (ω, θ)-rays of mixed type with design κ in Ω. Let $\gamma, \gamma_1, \dots, \gamma_s$ be all of them. Denote by $x_1, \ldots, x_k$ the successive vertices of Im γ and by $l_0, l_1, \ldots, l_k$ its successive segments. Since id $\in D(-\theta, q)$, there exists a neighbourhood V of x_k in X such that $V \cap \text{Im } \gamma_i = \emptyset$ for all $i = 1, \ldots, s$, and if δ is a glancing $-\theta$-ray in Ω with $k - q$ vertices and the first y_1 of them belongs to V, then $y_1 = x_k$ (and therefore δ is a part of Im γ). Let $0 = t_1 < \ldots < t_k$ be the times with $\gamma(t_i) = x_i$. Since l_q is a geodesic segment on X, we can find $t_0 \in (\alpha, \beta) \subset (t_q, t_{q+1})$ and a small coordinate neighbourhood U of $\gamma(t_0)$ in X such that

$$U \cap \left((\cup_{i=1}^s \text{Im } \gamma_i) \cup \bar{V} \cup \cup_{j=0, j \neq q}^k l_j \right) = \emptyset,$$

and (9.18) holds for $c(t) = \gamma(t)$, $a = t_q$, $b = t_{q+1}$, α, β. By Lemma 9.4.1 there exists $f \in C(X)$, arbitrarily close to id, such that $f = $ id on $X \backslash U$ and (9.19) is satisfied for the geodesic $\tilde{c}$ on $\tilde{X} = f(X)$ with $\tilde{c}(t) = c(t)$ for $t \in [t_q, \alpha]$. We claim that if f is chosen in this way and is sufficiently close to id, then the only (ω, θ)-rays of mixed type with design κ in Ω_f are $\gamma_1, \ldots, \gamma_s$. Indeed, assume this is not true. Then we can find a sequence $f_m \to_m$ id of such f so that for every m there exists an (ω, θ)-ray of mixed type δ_m with design κ in Ω_{f_m}, different from $\gamma_1, \ldots, \gamma_s$. Since id $\in D(\omega, p)$, for large m, the first vertex of Im δ_m is necessarily x_1. We may assume $\delta_m(0) = x_1$, and then we obtain $\delta_m(t) = \gamma(t)$ for all $t \leq \alpha$. Now the construction of f_m shows that the last vertex $y_k^{(m)}$ of δ_m cannot be x_k; otherwise we would have $\delta_m(\beta) = \gamma(\beta)$, which would imply $\tilde{c}(\beta) = c(\beta)$ in contradiction with (9.19). So $y_k^{(m)} \neq x_k$, and the choice of V yields $y_k^{(m)} \notin V$. Considering appropriate convergent subsequences, we may assume that there exists $\delta(t) = \lim_m \delta_m(t)$ for every t. Then δ is clearly a (ω, θ)-ray in Ω with first reflection point x_1, therefore δ coincides with γ. On the other hand, the last reflection point of δ is $y_k = \lim_m y_k^{(m)} \notin V$, so δ cannot coincide with γ. This contradiction shows that $\gamma_1, \ldots, \gamma_s$ are the only (ω, θ)-rays of mixed type with design κ in Ω_f, provided f is constructed as above and is sufficiently close to id. Moreover, if f is sufficiently close to id, then $f \in \mathcal{W}$. Repeating this procedure s times, we find $g \in \mathcal{W}$, arbitrarily close to id, such that there are no (ω, θ)-rays of mixed type with design κ in Ω_g. Then $g \in \mathcal{V}(k, \kappa)$, which shows that $\mathcal{V}(k, \kappa)$ is dense in $\mathcal{W}$.

To establish that $\mathcal{V}(k, \kappa)$ is open in $\mathcal{W}$, we may assume that $f_m \to_m$ id $\in \mathcal{W}$ for some sequence $\{f_m\} \subset \mathcal{W} \backslash \mathcal{V}(k, \kappa)$. Then we have to prove that there exists an (ω, θ)-ray of mixed type with design κ in Ω. For every m there exists an (ω, θ)-ray of mixed type δ_m with design κ in Ω_{f_m}. Choosing again appropriate subsequence and repeating a part of the above argument, we see that there exists an (ω, θ)-ray of mixed type in Ω. Thus, $\mathcal{V}(k, \kappa)$ is open in $\mathcal{W}$.

Set $\mathcal{V}_k = \cap_\kappa \mathcal{V}(k, \kappa)$ where κ runs over the set of all k-designs. Since the latter is finite, $\mathcal{V}_k$ is open and dense in $\mathcal{W}$, so it is residual in $C(X)$. Clearly, $\mathcal{V}_k$ has all the desired properties. This completes the construction of the sequence $\{\mathcal{V}_k\}$.

Finally, define $\mathcal{V} = \cap_{k=1}^\infty \mathcal{V}_k$. Then $\mathcal{V}$ is a residual subset of $C(X)$, and for every $f \in \mathcal{V}$ there are no (ω, θ)-rays of mixed type in Ω_f. This concludes the proof of the theorem. ♠

9.5. Notes

Under more restrictive assumptions on γ and Ω, Theorem 9.1.2 was proved in [P1]. The case of a strictly convex obstacle (cf. Corollary 9.1.3.) was studied some time before by Majda [Ma1], see also [Ma2, MaT, So2, Y]. In Section 9.1 we follow the analysis of [P1] based on the construction of global parametrix given in [GM1]. The results of Section 9.2 are proved in [CPS]. The material in Sections 9.3 and 9.4 is taken from [S4]. Note that to prove a result, similar to Corollary 9.4.2 for domains in $\mathbf{R}^n$, $n > 3$, it is desirable to have an analogue of Theorem 9.4.1 for such domains. Except Lemma 9.3.2, all arguments in Sections 9.3 and 9.4 can be modified to cover the general case. We do not know either for Lemma 9.3.2 or for Theorem 9.4.1 whether their analogues are true for $n > 3$.

10 SCATTERING INVARIANTS FOR SEVERAL STRICTLY CONVEX DOMAINS

In this chapter we study scattering rays and related singularities of the scattering kernel in the exterior Ω of an obstacle K which is a disjoint union of a finite number strictly convex compact domains in $\mathbf{R}^n$. It is shown first that if $\omega \in S^{n-1}$ is fixed, then for most of the vectors $\theta \in S^{n-1}$ all (ω, θ)-rays γ are ordinary with distinct sojourn times providing singularities of the scattering kernel $s(t, \theta, \omega)$.

Starting from the second section, we assume that the convex hull of every two connected components of K does not contain points of any other connected component of K. Under this condition, we prove that for a fixed configuration α of connected components of K one can choose appropriate ω and θ so that for every integer $q \leq 1$ there exists a unique (ω, θ)-ray γ_q with q reflection points following the given configuration in a certain way. It turns out that the reflection points of these rays approximate the corresponding reflection points of the unique periodic reflecting ray γ_α in Ω having type α. Moreover, the period of γ_α is an invariant which can be determined from the asymptotic of the sojourn times T_{γ_q} as $q \to \infty$. Another geometric invariant, related to the Poincaré map of γ_α, can be determined by the asymptotic of the coefficients c_q in front of the main singularity of $s(t, \theta, \omega)$ for $t \sim -T_{\gamma_q}$.

10.1. Singularities of the scattering kernel for generic θ

Let $K \subset \mathbf{R}^n$, $n \geq 2$, have the form

$$K = K_1 \cup \ldots \cup K_s, \tag{10.1}$$

where K_i are disjoint strictly convex compact domains in $\mathbf{R}^n$ with smooth boundaries $\Gamma_i = \partial K_i$. As before, we shall use the notation

$$\Omega = \overline{\mathbf{R}^n \backslash K}.$$

Fix an arbitrary vector $\omega \in S^{n-1}$. Our aim in this section is to prove the following.

Theorem 10.1.1: *There exists a residual subset $R(\omega)$ of S^{n-1} such that every $\theta \in R(\omega)$ has the following properties:*

(a) *each (ω, θ)-ray in Ω is ordinary and every two different (ω, θ)-rays in Ω have distinct sojourn times;*

(b) *if n is odd we have* $\text{sing supp } s_\Omega(t, \theta, \omega) = \{-T_\gamma : \gamma \in \mathcal{L}_{\omega,\theta}(\Omega)\}$, *and for every $\gamma \in \mathcal{L}_{\omega,\theta}(\Omega)$ and t close to $-T_\gamma$ we have (9.9).*

Note that property (b) follows immediately from (a) and Theorems 8.3.2 and 9.1.2. Thus, we have to construct $R(\omega)$ in such a way that (a) is satisfied.

Let us recall some notation from Section 2.4, which will be used in the whole chapter. Given $u \in Z_\omega$, denote by $S_t(u)$ the shift of u after time t along the billiard semi-trajectory $\gamma(u)$ in Ω, starting at u in direction ω. Then

$$\gamma(u) = \{S_t(u) : t \geq 0\}.$$

Let $N_t(u)$ be the velocity vector (i.e. the direction) of $\gamma(u)$ at $S_t(u)$. If t is such that $x = S_t(u) \in \partial\Omega$, we shall use the notation

$$N_{-t}(u) = \lim_{\epsilon \searrow 0} N_{t-\epsilon}(u),$$

and $N_{+t}(u) = \sigma_x(N_{-t}(u))$, σ_x being the symmetry with respect to the tangent hyperplane $T_x\partial\Omega$. Note that $S_0(u) = u$ and $N_0(u) = \omega$. By $x_1(u), x_2(u), \ldots$ we denote the *successive reflection points* of $\gamma(u)$ (if any), and by $t_1(u), t_2(u), \ldots$ the corresponding times (moments) of reflection. It is convenient to set $x_0(u) = u$ and $t_0(u) = 0$. Finally, denote by $r(u)$ the number of reflections of $\gamma(u)$, $0 \leq r(u) \leq \infty$.

Further, let $Z_\omega^{(0)}$ be the set of those $u \in Z_\omega$ such that $r(u) < \infty$. Define the map

$$T : Z_\omega^{(0)} \to \mathbf{R}$$

as follows. Given $u \in Z_\omega^{(0)}$, the vector $N_t(u)$ is constant for sufficiently large t, so there exists $\eta = \lim_{t \to +\infty} N_t(u)$. Clearly, there is exactly one $t > 0$ such that $S_t(u) \in Z_{-\eta}$. We set $T(u) = t$. Let us mention that under the latter notation, if γ is the unique (ω, η)-ray in Ω, passing through u, then $T_\gamma = T(u) + 2a$.

Recall the notion of a configuration from Section 2.4. Given an integer $m \geq 1$, by a *configuration of length m* we mean a sequence

$$\alpha = (i_1, \ldots, i_m) \in \{1, 2, \ldots, s\}^m \tag{10.2}$$

such that $i_j \neq i_{j+1}$ for $j = 1, \ldots, m-1$. For such an α we set $|\alpha| = m$, and define the sets $U_\alpha \subset F_\alpha \subset Z_\omega$ and the map

$$J_\alpha : \bar{F}_\alpha \to \mathbf{R}$$

as in Section 2.4.

Lemma 10.1.2: (a) *There exists a sequence*

$$\mathcal{L}_1 \supset \mathcal{L}_2 \supset \ldots \supset \mathcal{L}_m \supset \ldots$$

of open dense subsets of S^{n-1} such that for every $m \in \mathbf{N}$, if α is a configuration of length m, $u \in F_\alpha$ and $J_\alpha(u) \in \mathcal{L}_m$, then $u \in U_\alpha$;
(b) *For every two configurations α and β there exists a residual subset $\mathcal{L}(\alpha, \beta)$ of S^{n-1} such that for $\theta \in \mathcal{L}(\alpha, \beta)$ the conditions*

$$u \in U_\alpha, \quad v \in U_\beta, \quad r(u) = |\alpha|, \quad r(v) = |\beta|, \quad J_\alpha(u) = J_\beta(v) = \theta, \tag{10.3}$$

imply $T(u) \neq T(v)$.

For the proof of this lemma we need a general fact, concerning the propagation of convex wavefronts in Ω.

Let X be a smooth hypersurface lying in the interior of Ω, and let $e(x)$, $x \in X$, be a continuous field of unit normal vectors to X. We assume that X is **convex** at every $x \in X$ with respect to the normal field $e(x)$, i.e. the corresponding second fundamental form of X is non-negative semi-definite everywhere on X. Consider an open coordinate neighbourhood V of some point x_0 in X and a smooth chart

$$U \ni u \mapsto x(u) \in V \subset X$$

of an open neighbourhood U of 0 in $\mathbf{R}^{n-1}$ onto V with $x_0 = x(0)$. Set $e(u) = e(x(u))$. With respect to this parametrization the *second fundamental form* $\mathrm{II}_u^{(X)}$ of X at u (i.e. at $x(u)$) takes the form

$$\mathrm{II}_u^{(X)}(\xi, \eta) = \sum_{ij=1}^{n-1} b_{ij}^{(X)}(u)\xi_i\eta_j,$$

where

$$b_{ij}^{(X)}(u) = \left\langle e(u), \frac{\partial^2 x}{\partial u_i \partial u_j}(u) \right\rangle, \quad i, j = 1, \ldots, n-1,$$

are the coefficients of $\mathrm{II}_u^{(X)}$ at u. The *first fundamental form* $\mathrm{I}_u^{(X)}$ of X at u is given by

$$\mathrm{I}_u^{(X)}(\xi, \eta) = \sum_{i,j=1}^{n-1} g_{ij}^{(X)}(u)\xi_i\eta_j,$$

with

$$g_{ij}^{(X)}(u) = \left\langle \frac{\partial x}{\partial u_i}(u), \frac{\partial x}{\partial u_j}(u) \right\rangle, \quad i,j = 1,\ldots,n-1.$$

The *sectional (normal) curvature* of X at $x(u)$ in direction ξ (more precisely, in direction $\sum_{i=1}^{n-1} \xi_i \frac{\partial x}{\partial u_i}(u) \in T_{x(u)}X$) is defined by

$$\kappa_u^{(X)}(\xi) = -\frac{\mathrm{II}_u^{(X)}(\xi,\xi)}{\mathrm{I}_u^{(X)}(\xi,\xi)}.$$

For $x \in X$ and $t \in \mathbf{R}$ let $Q_t(x)$ be the shift of x along the billiard semi-trajectory $\chi(x)$ in Ω starting at x in direction $e(x)$, and let $M_t(x)$ be the direction of the trajectory at $Q_t(x)$. Then

$$\chi(x) = \{Q_t(x) : t \geq 0\}.$$

Assume that the trajectory $\chi(x_0)$ hits $\partial\Omega = \partial K$ at some point y_0, reflecting transversally on ∂K at y_0. Then for the angle φ_0 between $\nu(y_0)$ and the direction $N_{t_0}(x_0)$ of $\gamma(x_0)$ at y_0, we have $\cos\varphi_0 > 0$. Here

$$t_0 = \|x_0 - y_0\| > 0.$$

Further, take $T > t_0$ such that $Q_t(x_0)$ makes only one reflection when t runs from 0 to T. Restricting our considerations to a small neighbourhood of x_0 in X, we may assume that for every $x \in X$ the trajectory $Q_t(x)$ makes exactly one transversal reflection on ∂K when t runs from 0 to T. Then

$$Z = Z^{(T)} = \{Q_T(x) : x \in X\}$$

is a smooth hypersurface in $\mathbf{R}^n$. This follows easily, for example, from the next considerations and the implicit function theorem. Set $z_0 = Q_T(x(0))$, and denote by $\kappa(T)$ the minimum of the sectional curvatures of Z at z_0 and by κ_0 the same for ∂K at y_0. Finally, set

$$\kappa_+(t_0) = \lim_{T \searrow t_0} \kappa(T).$$

Before going on, let us mention that for $0 < t < t_0$ we have

$$\kappa_u^{(Z)}(\xi) = \frac{\kappa_u^{(X)}(\xi)}{1 + t\kappa_u^{(X)}(\xi)}.$$

This formula can be easily derived from a part of the argument in the proof of the following lemma. The latter shows that the wavefront $Z = Z^{(T)}$ is always strictly convex and gives an estimate for its minimal normal curvature. Actually, we need only the strict convexity, the estimate of the curvature will not be used in our next considerations.

Lemma 10.1.3:   *Under the above assumptions, the hypersurface Z is strictly convex at z_0 and*

$$\kappa(T) \geq \frac{2\kappa_0 \cos \varphi_0}{1 + 2(T - t_0)\kappa_0 \cos \varphi_0}. \tag{10.4}$$

In particular,

$$\kappa_+(t_0) \geq 2\kappa_0 \cos \varphi_0. \tag{10.5}$$

Proof: Note that (10.4) follows from (10.5) and the above remark. Thus, it is sufficient to prove only (10.5).

For $u \in U$ denote by $y(u)$ the reflection point of $\chi(x(u))$ on ∂K, and set $Y = \partial K$,

$$z(u) = z^{(T)}(u) = S_T(x(u)), \quad f(u) = \frac{z(u) - y(u)}{\|z(u) - y(u)\|}.$$

We shall see later that $f(u)$ is the unit normal to Z at $z(u)$. It follows by our assumptions that

$$u \mapsto y(u), \quad u \mapsto z(u)$$

provide smooth parametrizations for ∂K around y_0 and for Z around z_0. Setting

$$t(u) = \|y(u) - x(u)\|,$$

we have

$$y(u) = x(u) + t(u)e(u), \tag{10.6}$$

and

$$z(u) = y(u) + (T - t(u))f(u). \tag{10.7}$$

Differentiating (10.6) and (10.7) with respect to u_i, $i = 1, \ldots, n - 1$, one finds

$$\frac{\partial y}{\partial u_i} = \frac{\partial x}{\partial u_i} + \frac{\partial t}{\partial u_i}e + t\frac{\partial e}{\partial u_i}, \tag{10.8}$$

$$\frac{\partial z}{\partial u_i} = \frac{\partial y}{\partial u_i} - \frac{\partial t}{\partial u_i}f + (T - t(u))\frac{\partial f}{\partial u_i}. \tag{10.9}$$

Taking the inner product of (10.8) with $e(u)$, we get

$$\frac{\partial t}{\partial u_i}(u) = \left\langle e(u), \frac{\partial y}{\partial u_i}(u) \right\rangle.$$

Then (10.9) implies

$$\frac{\partial z}{\partial u_i}(u) = \frac{\partial y}{\partial u_i}(u) - \left\langle \frac{\partial y}{\partial u_i}(u), e(u) \right\rangle f(u) + O(T - t_0), \tag{10.10}$$

where $O(T-t_0)$ denotes a term which tends to 0 as $T \searrow t_0$. Note that $f(u)$ does not depend on T, provided $T - t_0 > 0$ is sufficiently small. By (10.10) we first obtain

$$
\left\langle \frac{\partial z}{\partial u_i}(u), f(u) \right\rangle = \left\langle \frac{\partial y}{\partial u_i}(u), f(u) \right\rangle - \left\langle \frac{\partial y}{\partial u_i}(u), e(u) \right\rangle = 0,
$$

since $e(u)$ and $f(u)$ are symmetric with respect to the tangent hyperplane to ∂K at $y(u)$. This shows that $f(u)$ is the unit normal to Z at $z(u)$.

Next, (10.10) implies

$$
g_{ij}^{(Z)}(u) = g_{ij}^{(Y)}(u) - \left\langle e(u), \frac{\partial y}{\partial u_i}(u) \right\rangle \left\langle e(u), \frac{\partial y}{\partial u_i}(u) \right\rangle.
$$

Consider an arbitrary $\xi = (\xi_1, \ldots, \xi_{n-1}) \in \mathbf{R}^{n-1}$ and set

$$
\eta = \sum_{i=1}^{n-1} \xi_i \frac{\partial y}{\partial u_i}(u) \in T_{y(u)} \partial K.
$$

Then

$$
\mathbf{I}_u^{(Z)}(\xi, \xi) = \sum_{i,j=1}^{n-1} \xi_i \xi_j g_{ij}^{(Z)}(u) = \mathbf{I}_u^{(Y)}(\xi, \xi) - \langle e(u), \eta \rangle^2 + O(T - t_0)
$$
$$
\leq \mathbf{I}_u^{(Y)}(\xi, \xi) + O(T - t_0). \tag{10.11}
$$

We are going to deduce similar relations between the second fundamental forms. Differentiating (10.8) and (10.9) with respect to u_j, we get

$$
\frac{\partial^2 y}{\partial u_i \partial u_j} = \frac{\partial^2 x}{\partial u_i \partial u_j} + \frac{\partial^2 t}{\partial u_i \partial u_j} e + \frac{\partial t}{\partial u_i} \frac{\partial e}{\partial u_j} + \frac{\partial t}{\partial u_j} \frac{\partial e}{\partial u_i} + t \frac{\partial^2 e}{\partial u_i \partial u_j} \tag{10.12}
$$
$$
\frac{\partial^2 z}{\partial u_i \partial u_j} = \frac{\partial^2 y}{\partial u_i \partial u_j} - \frac{\partial^2 t}{\partial u_i \partial u_j} f - \frac{\partial t}{\partial u_i} \frac{\partial f}{\partial u_j} - \frac{\partial t}{\partial u_j} \frac{\partial f}{\partial u_i}
$$
$$
+ (T - t(u)) \frac{\partial^2 f}{\partial u_i \partial u_j}. \tag{10.13}
$$

To find the term $\partial^2 t / \partial u_i \partial u_j$, we take the inner product of (10.12) with $e(u)$ and get

$$
\frac{\partial^2 t}{\partial u_i \partial u_j} = \left\langle \frac{\partial^2 y}{\partial u_i \partial u_j}, e \right\rangle - \left\langle \frac{\partial^2 x}{\partial u_i \partial u_j}, e \right\rangle - t \left\langle \frac{\partial^2 e}{\partial u_i \partial u_j}, e \right\rangle.
$$

Replacing this term in (10.13), we obtain

$$
\left\langle \frac{\partial^2 z}{\partial u_i \partial u_j}(0), e(0) \right\rangle = \left\langle \frac{\partial^2 y}{\partial u_i \partial u_j}(0), f(0) \right\rangle - \frac{\partial^2 t}{\partial u_i \partial u_j}(0) + O(T - t_0)
$$

$$
= \left\langle \frac{\partial^2 y}{\partial u_i \partial u_j}(0), f(0) - e(0) \right\rangle + \left\langle \frac{\partial^2 x}{\partial u_i \partial u_j}(0), e(0) \right\rangle
$$

$$
+ t_0 \left\langle \frac{\partial^2 e}{\partial u_i \partial u_j}(0), e(0) \right\rangle + O(T - t_0)
$$

$$
= 2 \cos \varphi_0 \left\langle \frac{\partial^2 y}{\partial u_i \partial u_j}(0), \nu(0) \right\rangle + \left\langle \frac{\partial^2 x}{\partial u_i \partial u_j}(0), e(0) \right\rangle
$$

$$
- t_0 \left\langle \frac{\partial e}{\partial u_i}(0), \frac{\partial e}{\partial u_j}(0) \right\rangle + O(T - t_0). \tag{10.14}
$$

Here we have taken into account that $f(0) - e(0) = 2 \cos \varphi_0 \cdot \nu(0)$ and

$$
\left\langle \frac{\partial^2 e}{\partial u_i \partial u_j}, e \right\rangle = - \left\langle \frac{\partial e}{\partial u_i}, \frac{\partial e}{\partial u_j} \right\rangle.
$$

It follows immediately from (10.14) that for $u = 0$ and $\zeta = \sum_{i=1}^{n-1} \xi_i \frac{\partial e}{\partial u_i}(0)$ we have

$$
\mathrm{II}_u^{(Z)}(\xi, \xi) = 2 \cos \varphi_0 \mathrm{II}_u^{(Y)}(\xi, \xi) + \mathrm{II}_u^{(Y)}(\xi, \xi)
$$
$$
- t_0 \|\zeta\|^2 + O(T - t_0).
$$

The convexity of X implies $\mathrm{II}_u^{(X)} \leq 0$, so

$$
\mathrm{II}_u^{(Z)}(\xi, \xi) \leq 2 \cos \varphi_0 \mathrm{II}_u^{(Y)}(\xi, \xi) + O(T - t_0).
$$

Finally, combining the latter with (10.11), one gets

$$
- \frac{\mathrm{II}_0^{(Z)}(\xi, \xi)}{\mathrm{I}_0^{(Z)}(\xi, \xi)} \geq - \frac{2 \cos \varphi_0 \mathrm{II}_0^{(Y)}(\xi, \xi) + O(T - t_0)}{\mathrm{I}_0^{(Y)}(\xi, \xi)}
$$
$$
\geq 2 \kappa_0 \cos \varphi_0 + O(T - t_0),
$$

which clearly yields (10.5). ♠

Proof of Lemma 10.1.2: (a) Set $\mathcal{L}_1 = S^{n-1} \setminus \{\omega\}$. Assume that the sets $\mathcal{L}_1 \supset \dots \supset \mathcal{L}_m$ are already constructed and have the desired properties. We are going to construct $\mathcal{L}_{m+1}$.

Fix an arbitrary configuration

$$\alpha = (i_1, \ldots, i_m, i_{m+1})$$

of length $m+1$ and for $k \leq m$ set

$$\alpha_k = (i_1, \ldots, i_k).$$

Since $\mathcal{L}_m$ is open and dense in S^{n-1},

$$F = S^{n-1} \backslash \mathcal{L}_m$$

is a compact subset of S^{n-1} with empty interior. Fix for a moment an arbitrary $k \leq m$ and set $\beta = \alpha_k$. Next, we use the notation (2.39) and (2.40) from Section 2.4, as well as Proposition 2.4.4. First, consider the map $J_\beta : \bar{F}_\beta \to E_\beta$. Since $F \cap E_\beta$ is a compact subset of E_β with empty interior, and

$$J_\beta^{-1}(F) = J_\beta^{-1}(F \cap E_\beta) \subset (J_\beta^{-1}(F \cap E_\beta) \cap M_\alpha) \cup L_\alpha,$$

it follows by Proposition 2.4.4 that $J_\beta^{-1}(F)$ is a compact subset of Z_ω with empty interior. By the definition of β we have $F_\alpha \subset F_\beta$ and $L_\alpha \subset L_\beta$. On the other hand, $\omega \notin \mathcal{L}_1 \supset \mathcal{L}_m$ and the definitions of L_β and F imply $J_\beta^{-1}(F) \cap L_\beta = \emptyset$. Consequently, $J_\beta^{-1}(F) \cap L_\alpha = \emptyset$, which shows that $J_\beta^{-1}(F) \cap \bar{F}_\alpha$ is contained in M_α. Applying again Proposition 2.4.4, we deduce that $J_\alpha(J_\beta^{-1}(F) \cap \bar{F}_\alpha)$ is a compact subset of S^{n-1} with empty interior. In this way we have established that

$$V = \mathcal{L}_m \backslash \cup_{k=1}^m J_\alpha(F_\alpha \cap J_{\alpha_k}^{-1}(F))$$

is an open dense subset of S^{n-1}. Note that if $u \in F_\alpha$ and $J_\alpha(u) \in V$, then the first m reflection points of $\gamma(u)$ are proper (i.e. transversal) ones.

Denote by $\mathcal{L}_{m+1}^{(\alpha)}$ the set of those $\theta \in V$ such that if $J_\alpha(u) = \theta$ for some $u \in F_\alpha$, then the first $m+1$ reflection points of $\gamma(u)$ are proper ones. We claim that $\mathcal{L}_{m+1}^{(\alpha)}$ is open and dense in V. The openness is clear. To establish the density, fix an arbitrary $\theta \in V \backslash \mathcal{L}_{m+1}^{(\alpha)}$. Then $\theta = J_\alpha(u_0)$ for some $u_0 \in F_\alpha$ such that the first m reflection points $x_1(u_0), \ldots, x_m(u_0)$ of $\gamma(u_0)$ are proper ones and $\gamma(u_0)$ is tangent to ∂K at $x_{m+1}(u_0)$. Take an arbitrary t with $t_m(u_0) < t < t_{m+1}(u_0)$, and choose an open $(n-2)$-dimensional ball U in Z_ω with centre u_0 so small that for every $u \in U$ the trajectory $\{S_s(u) : 0 \leq s \leq t\}$ has exactly m reflections, all of them being proper ones, according to Lemma 10.1.3, $Y = S_t(U)$ is a smooth strictly convex hypersurface in $\mathbf{R}^n$ (Figure 10.1). Note that $\{N_t(u) : u \in U\}$ is a normal field for Y. It is clear now that there exists $v \in U$ such that the ray, starting at $S_t(v)$ in direction $N_t(v)$, intersects transversally $\Gamma_{i_{m+1}}$, which means that $v \in \mathcal{L}_{m+1}^{(\alpha)}$. Therefore $\mathcal{L}_{m+1}^{(\alpha)}$ is dense in V. Since V is open and dense in S^{n-1}, we deduce that $\mathcal{L}_{m+1}^{(\alpha)}$ is also open and dense in S^{n-1}.

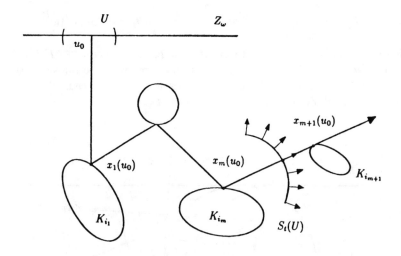

Figure 10.1

Set $\mathcal{L}_{m+1} = \cap_\alpha \mathcal{L}_{m+1}^{(\alpha)}$, where α runs over the (finite) set of all configurations of length $m+1$. Then $\mathcal{L}_{m+1}$ is open and dense in S^{n-1} and has the desired properties. This concludes the proof of (a).

(b) Let $\{\mathcal{L}_m\}$ be the sequence constructed in (a), and set

$$R'(\Omega) = \cap_{m=1}^\infty \mathcal{L}_m.$$

Then $R'(\omega)$ is a residual subset of S^{n-1} such that for every configuration α we have

$$J_\alpha^{-1}(R'(\omega)) \cap F_\alpha \subset U_\alpha.$$

Fix two arbitrary configurations α and β, and set

$$m = \max\{|\alpha|, |\beta|\}.$$

Denote by $\mathcal{L}(\alpha, \beta)$ the set of all $\theta \in R'(\omega)$ such that (10.3) implies $T(u) \neq T(v)$. We are going to show that $\mathcal{L}(\alpha, \beta)$ is open and dense in $R'(\omega)$.

First, we prove that $R'(\omega) \backslash \mathcal{L}(\alpha, \beta)$ is closed in $R'(\omega)$. Consider an arbitrary sequence

$$R'(\omega) \backslash \mathcal{L}(\alpha, \beta) \ni \theta_k \to \theta \in R'(\omega).$$

Then for every k there exist $u_k \in U_\alpha \cap J_\alpha^{-1}(\theta_k)$ and $v_k \in U_\beta \cap J_\beta^{-1}(\theta_k)$ with $r(u_k) = |\alpha|$, $r(v_k) = |\beta|$, $T(u_k) = T(v_k)$. Taking appropriate subsequences, we may assume that $u_k \to u \in \bar{F}_\alpha$, $v_k \to v \in \bar{F}_\beta$. Assume $u \notin F_\alpha$. Then

$u \in F_{\alpha'}$ for some α', and $\gamma(u)$ is tangent to ∂K at some of its points. On the other hand,

$$J_{\alpha'}(u) = J_\alpha(u) = \lim_k J_\alpha(u_k) = \lim_k \theta_k = \theta \in R'(\omega),$$

which is a contradiction with the properties of $R'(\omega)$. Hence $u \in F_\alpha$, and now $J_\alpha(u) = \theta \in R'(\omega)$ implies $u \in U_\alpha$. In the same way one finds $J_\beta(v) = \theta$ and $v \in U_\beta$. Moreover, we have $r(u) = |\alpha|$, $r(v) = |\beta|$. Thus, $\theta \in R'(\omega)\backslash \mathcal{L}(\alpha, \beta)$, which shows that $\mathcal{L}(\alpha, \beta)$ is open in $R'(\omega)$.

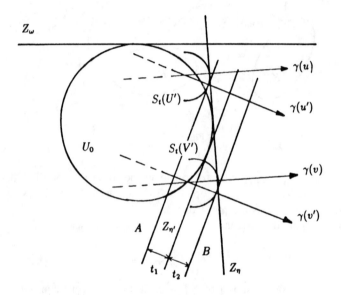

Figure 10.2

To establish the density, fix an arbitrary $\theta \in R'(\omega)\backslash\mathcal{L}(\alpha, \beta)$, and set $\eta = -\theta$. Then there exist u and v having the properties (10.3) with $T(u) = T(v)$. Applying again Lemma 10.1.3, we find open $(n-2)$-dimensional balls $U' \subset U_\alpha$ and $V' \subset U_\beta$, centred at u and v, respectively, such that for $t = T(u) = T(v)$, $S_t(U')$ and $S_t(V')$ are smooth strictly convex hypersurfaces in $\mathbf{R}^n$ (Figure 10.2). Note that these hypersurfaces are tangent to Z_η at $S_t(u)$ and $S_t(v)$, respectively, so there are contained in the closed half-space, determined by Z_η and η. It follows by the strict convexity that for every $\theta' \in S^{n-1}$ close to θ there are unique $u' \in U'$ and $v' \in V'$ with $J_\alpha(u') = J_\beta(v') = \theta'$. Denote by D the set of those $\theta' \in R'(\omega)$ for which there exist $u' \in U'$ and $v' \in V'$ with $J_\alpha(u') = J_\beta(v') = \theta'$ and $S_t(U')$ and $S_t(V')$ have a common tangent hyperplane at $S_t(u')$ and $S_t(v')$, respectively. It is easy to see that D has empty interior in S^{n-1}; in fact, it is contained in a smooth submanifold of S^{n-1} with positive codimension. Since $R'(\omega)$ is dense in S^{n-1}, there exists $\theta' \in R'(\omega) \backslash D$ arbitrarily close to θ. Take such a θ' and denote by A (resp. B) the hyperplane tangent to $S_t(U')$ at $S_t(u')$ (resp. tangent to $S_t(V')$ at $S_t(v')$). Then for $\eta' = -\theta'$, the three hyperplanes A, B and $Z_{\eta'}$ are parallel. Setting $t_1 = T(u') - t, t_2 = T(v') - t$, we see that $|t_1|$ is the distance between A and $Z_{\eta'}$, while $|t_2|$ is the distance between B and $Z_{\eta'}$. Since $A \neq B$, we have $t_1 \neq t_2$, therefore

$$T(u') = t + t_1 \neq t + t_2 = T(v').$$

Moreover, it follows by the properties of the maps J_α and J_β that (10.3) holds if we replace u by u', v by v' and θ by θ'. Thus, $\theta' \in \mathcal{L}(\alpha, \beta)$. This proves the density of $\mathcal{L}(\alpha, \beta)$ in $R'(\omega)$.

Since $R'(\omega)$ is residual in S^{n-1}, and $\mathcal{L}(\alpha, \beta)$ is open and dense in $R'(\omega)$, we get that $\mathcal{L}(\alpha, \beta)$ is residual in S^{n-1}, too. This concludes the proof of (b). ♠

Proof of Theorem 10.1.1: Let $\mathcal{L}(\alpha, \beta)$ be the residual subsets of S^{n-1} from Lemma 10.1.2, chosen as subsets of $R'(\omega)$, the latter being defined in the proof of Lemma 10.1.2 (b). Set

$$R(\omega) = \cap_{\alpha, \beta} \mathcal{L}(\alpha, \beta),$$

where α and β run over the set of all configurations. Then $R(\omega)$ is residual in S^{n-1} and has property (a) from Theorem 10.1.1. As we have already mentioned above, the property (b) follows immediately from (a) and Theorems 8.3.2 and 9.1.2. ♠

10.2. Hyperbolicity of scattering trajectories

Let Ω and K be as in the previous section. From now on we assume that K satisfies the following condition

$$(\mathbf{H}) \begin{cases} \text{for any three distinct connected components,} \\ L, M, N \text{ of } K \text{ the convex hull of} \\ L \cup M \text{ does not contain points of } N. \end{cases}$$

In what follows we use the notation from the beginning of the previous section. Recall that $\kappa_0 > 0$ denotes the minimum of the normal curvatures of ∂K.

Let $m \geq 2$ be an arbitrary integer, and let

$$L', L_1, L_2, \ldots, L_m, L'' \tag{10.15}$$

be an arbitrary sequence of connected components of K such that $L_i \neq L_{i+1}$ for all $i = 1, \ldots, m-1$, $L' \neq L_1, L_m \neq L''$. Note that some of these connected components may coincide. This is clearly the case when m is larger than the number of connected components of K.

Before going on, let us mention that the condition (H) implies the existence of $\varphi_0 \in (0, \pi/2)$ with the following property: if x, y, z belong to the boundaries of connected components L, M and N of K, $L \neq M$, $M \neq N$, the open segments (x, y) and (y, z) have no common points with K and $[x, y]$ and $[y, z]$ satisfy the law of reflection at y with respect to ∂K, then $\varphi < \varphi_0$, where $\varphi \in (0, \pi/2]$ is the angle between $[y, z]$ and the normal $\nu(y)$ to ∂K at y.

Further, it is easy to see that there exists $\psi_0 \in (0, \pi/2)$ with the following property: for every two distinct connected component M and L of K there is a tangent hyperplane H to M such that M and L are contained in one and a same half-space with respect to H and the minimal angle between H and a straight line, having common points with both M and L, is $\geq \psi_0$. Clearly, ψ_0 depends only on K. Let Z' be a hyperplane, tangent to L' and having the properties of H with respect to the pair $M = L'$, $L = L_1$, and Z'' be a hyperplane tangent to L'' and having similar properties with respect to the pair $M = L''$, $L = L_m$. Denote by V be the set of those $x \in Z'$ such that if $y \in \partial L_1$, $z \in \partial L_2$, the open segments (x, y) and (y, z) have no common points with $K \setminus L_1$ and $[x, y]$ and $[y, w]$ satisfy the law of reflection at y with respect to ∂K, then for the angle $\varphi \in (0, \pi/2]$ between $[y, w]$ and the normal $\nu(y)$ we have $\varphi < \varphi_0$. Clearly, V is an open subset of Z'. In a similar way we define an open subset W of Z'', this time the corresponding property concerns points $x \in L_{m-1}$, $y \in L_m$, $z \in W$.

The central moment in this section is the following.

Lemma 10.2.1: *There exist constants $C > 0$ and $\delta \in (0, 1)$, depending only on K, with the following property: if*

$$y_0' \in V, \quad y_1' \in \partial L_1, \ldots, \quad y_m' \in \partial L_m, \quad y_{m+1}' \in W$$

and

$$y_0'' \in V, \quad y_1'' \in \partial L_1, \ldots, \quad y_m'' \in \partial L_m, \quad y_{m+1}'' \in W$$

are two sequences of points such that for every $j = 1, \ldots, m$ the segments $[y_{j-1}', y_j']$ and $[y_j', y_{j+1}']$ satisfy the law of reflection at y_j' with respect to ∂L_j and the segments $[y_{j-1}'', y_j'']$ and $[y_j'', y_{j+1}'']$ satisfy the law of reflection at y_j'' with respect to ∂L_j, then

$$\|y_i' - y_i''\| \leq C(\delta^i + \delta^{m-i}) \tag{10.16}$$

for all $i = 1, \ldots, m$.

To prove this lemma, we need some preparation.

Define the function

$$F : V \times \partial L_1 \times \cdots \times \partial L_m \times W \to \mathbf{R}$$

by

$$F(v; y_1, \ldots, y_m; w) = \|v - y_1\| + \sum_{i=1}^{m-1} \|y_i - y_{i+1}\| + \|y_m - w\|.$$

Applying a standard argument and using the choice of V and W, we see that for every $v \in V$ and every $w \in W$ there exist $y_1(v, w) \in \partial L_1, \ldots, y_m(v, w) \in \partial L_m$ such that

$$F(v; y_1(v, w), \ldots, y_m(v, w); w) = \min\{F(v; y_1 \ldots, y_m; w) :$$
$$(y_1, \ldots, y_m) \in \partial L_1 \times \cdots \times \partial L_m\}$$

(cf. for example the proof of Proposition 10.3.2). Moreover, by an argument, very similar to that in the proof of Proposition 2.4.4, one gets that $y_i(v, w)$ are unique with this property. In fact, $y_i(v, w)$ are the successive reflection points of a billiard trajectory in Ω connecting v and w. This shows that the maps $y_i(v, w)$ depend smoothly on (v, w). The latter can be derived also by the implicit function theorem.

Next, we identify every tangent hyperplane $T_x \partial K$, $x \in \partial K$ (including Z' and Z''), with the $(n-1)$-dimensional linear subspace of $\mathbf{R}^n$ parallel to it, and we measure the lengths of the vectors $\xi \in T_x(\partial K)$ using the standard norm in $\mathbf{R}^n$. In this way we define also the norm of a linear operator between two tangent spaces. That is, we use the standard Riemannian metric on ∂K. We assume also that some basis in Z' is fixed and denote the elements of Z' by $v = (v^{(1)}, \ldots, v^{(n-1)})$. In the same way we fix a basis in Z'' and set $w = (w^{(1)}, \ldots, w^{(n-1)}) \in Z''$. For fixed w let

$$\partial_v y_i(v, w) : Z' \to T_{y_i}(v, w)\partial L_i$$

be the tangential map of the map $V \ni v \mapsto y_i(v, w) \in \partial L_i$ at v. In the same way we define the tangential map

$$\partial_w y_i(v, w) : Z'' \to T_{y_i(v,w)}\partial L_i$$

of the map $W \ni w \mapsto y_i(v, \omega) \in \partial L_i$. We are going to estimate the norms of the linear operators $\partial_v y_i(v, w)$ and $\partial_w y_i(v, w)$.

Lemma 10.2.2: *For all $v \in V$, $w \in W$ and for every $j = 1, \ldots, m$ we have*

$$\|\partial_v y_j(v, \omega)\| \le C' e^{-j\epsilon}, \tag{10.17}$$

and

$$\|\partial_w y_j(v, w)\| \le C' e^{-(m-j)\epsilon}, \tag{10.18}$$

where $C' > 0$ and $\epsilon > 0$ are constants depending only on K.

Proof of Lemma 10.2.2: Fix arbitrary $v_0 \in V$ and $w_0 \in W$. For every $j = 1, \ldots, m$ take a smooth chart

$$\varphi_j : \mathbf{R}^{n-1} \to U_j \subset \partial L_j$$

such that $\varphi_j(0) = y_j(v_0, w_0)$ and $\{\frac{\partial \varphi_j}{\partial u_j^{(p)}}(0)\}_{p=1}^{n-1}$ is an orthonormal basis in $T_{y_j(v_0,w_0)}\partial L_j$. Here $u_j = (u_j^{(1)}, \ldots, u_j^{(n-1)}) \in \mathbf{R}^{n-1}$. As in the proof of Lemma 2.2.6, define the function

$$G : V \times (\mathbf{R}^{n-1})^m \times W \to \mathbf{R}$$

by

$$G(v; u_1, \ldots, u_m; w) = F(v; \varphi_1(u_1), \ldots, \varphi_m(u_m); w).$$

Given $i, j = 1, \ldots, m$, consider the $(n-1) \times (n-1)$ symmetric matrix

$$G_{ij}(v, w) = \left(\frac{\partial^2 G}{\partial u_i^{(p)} \partial u_j^{(q)}} (u(v, w)) \right)_{p,q=1}^{n-1}.$$

Then

$$G_{uu}(v, w) = \begin{pmatrix} G_{11} & G_{12} & \cdots & G_{1m} \\ G_{21} & G_{22} & \cdots & G_{2m} \\ \cdots & \cdots & \cdots & \cdots \\ G_{m1} & G_{m2} & \cdots & G_{mm} \end{pmatrix}$$

is a symmetric $m \times m$ block-matrix. According to our computations in the proof of Lemma 2.2.6, we have

$$G_{ij}(v, w) = 0 \text{ whenever } |i - j| > 1. \tag{10.19}$$

Consequently, there exists a constant $c_0 > 0$, depending only on K such that

$$\|G_{ij}(v, w)\| \leq c_0$$

for all $i, j = 0, 1, \ldots, m - 1$. It then follows by (10.19) that for $C_0 = 6c_0$ we have

$$\|G_{uu}(v, w)\| \leq C_0. \tag{10.20}$$

Moreover, using again the computation in the proof of Lemma 2.2.6, we see that $G_{uu}(v, w)$ is positive definite and

$$2\kappa_0 \cos \varphi_0 I \leq G_{uu}(v, w) \leq C_0.I, \tag{10.21}$$

I being the *identity matrix*.

In what follows denote the matrices $G_{ij}(v_0, w_0)$, $G_{uu}(v_0, w_0)$, etc. briefly by G_{ij}, G_{uu}, etc. We should note that G_{ij} is considered as the matrix of a symmetric linear operator $T_{y_i(v_0, w_0)} \partial L_i \rightarrow T_{y_j(v_0, w_0)} \partial L_j$, so G_{uu} is the matrix of a symmetric positive definite linear operator

$$G_{uu} : \mathcal{T} = \prod_{i=1}^{m} T_{y_i(v_0, w_0)} \partial L_i \rightarrow \prod_{i=1}^{m} T_{y_i(v_0, w_0)} \partial L_i.$$

For $(v, w) \in V \times W$ close to (v_0, w_0) and $j = 1, \ldots, m$ there is a uniquely determined smooth map

$$(v, w) \mapsto u_j(v, w) \in \mathbf{R}^{n-1}$$

such that

$$y_j(v, w) = \varphi_j(u_j(v, w)).$$

Next, consider the maps

$$w \mapsto u_j(v_0, w).$$

For $l = 1, \ldots, n-1$ and $i = 1, \ldots, m$ define the column-vectors

$$\partial_l u_i(w) = \left(\frac{\partial u_i^{(p)}}{\partial w^{(l)}}(v_0, w) \right)_{p=1}^{n-1}, \quad G_i(u_i) = \left(\frac{\partial G}{\partial u_i^{(p)}}(u_i) \right)_{p=1}^{n-1}.$$

For the sake of brevity we set $\partial_l u_i = \partial_l u_i(v_0, w_0)$ and $G_i = G_i(0)$. We shall also consider the elements ξ of $\mathcal{T}$ as column-vectors consisting of m blocks so that the ith block of ξ corresponds to a column-vector from $T_{y_i(v_0, w_0)} \partial L_i$, $i = 1, \ldots, m$.

Consider the linear operator of $\mathcal{T}$ into itself with diagonal block-matrix

$$D = \begin{pmatrix} D_1 & 0 & \cdots & 0 \\ 0 & D_2 & \cdots & 0 \\ \cdots & \cdots & \cdots & \cdots \\ 0 & 0 & \cdots & D_m \end{pmatrix},$$

the ith block D_i being a diagonal $(n-1) \times (n-1)$ matrix

$$D_i = \begin{pmatrix} d_i & 0 & \cdots & 0 \\ 0 & d_i & \cdots & 0 \\ \cdots & \cdots & \cdots & \cdots \\ 0 & 0 & \cdots & d_i \end{pmatrix},$$

with

$$d_i = e^{(m-i)\epsilon}.$$

The constant $\epsilon > 0$ will be chosen in a special way later. Note that $d_m = 1$, so D_m is the identity $(n-1) \times (n-1)$ matrix. For

$$c = e^\epsilon - 1 > 0$$

we have

$$\left| \frac{d_i}{d_{i+1}} - 1 \right| \leq c, \quad \left| \frac{d_{i+1}}{d_i} - 1 \right| \leq c$$

for all $i = 1, \ldots, m-1$. Using these inequalities and according to the choice of C_0 and c_0, by a direct computation we get

$$\left\| G_{uu} - D G_{uu} D^{-1} \right\| \leq 2c C_0.$$

Therefore for $\xi \in \mathcal{T}$ we have

$$\left\| G_{uu} \xi \right\| \leq \left\| D G_{uu} D^{-1} \xi \right\| + 2c C_0 \left\| \xi \right\|.$$

On the other hand, (10.21) implies

$$2\kappa_0 \cos\varphi_0 \, \|\xi\| \le \|G_{uu}\xi\|,$$

which, combined with the previous inequality, gives

$$\|\xi\| \le \frac{1}{2\kappa_0 \cos\varphi_0} \left(\|DG_{uu}D^{-1}\xi\| + 2cC_0 \, \|\xi\|\right).$$

We now choose $\epsilon > 0$ such that

$$c = e^\epsilon - 1 = \frac{\kappa_0 \cos\varphi_0}{2C_0}, \tag{10.22}$$

then $cC_0/\kappa_0 \cos\varphi_0 = \frac{1}{2}$, and therefore

$$\|\xi\| \le \frac{1}{\kappa_0 \cos\varphi_0} \|DG_{uu}D^{-1}\xi\|.$$

Setting

$$b_0 = \frac{1}{\kappa_0 \cos\varphi_0} > 0$$

and $\eta = D^{-1}\xi \in \mathcal{T}$ in the latter inequality, we find

$$\|\eta\| \le b_0 \|DG_{uu}\eta\|, \quad \eta \in \mathcal{T}. \tag{10.23}$$

Fix an arbitrary $i = 1, \ldots, m$. Clearly, for $w \in W$ close to w_0 we have

$$G_i(u_i(v_0, w)) = 0.$$

Differentiating this equality with respect to $w^{(l)}$ and evaluating at $w = w_0$, one gets

$$\sum_{j=1}^{m} G_{ij}\partial_l u_j + \partial_l G_i = 0. \tag{10.24}$$

Here

$$\partial_l G_i = \left(\frac{\partial^2 G}{\partial u_i^{(p)} \partial v^{(l)}}(0)\right)_{p=1}^{n-1}$$

is a column-vector in $T_{y_i(v_0,w_0)}\partial L_i$. According again to the computations in the proof of Lemma 2.2.6, we observe that $\partial_l G_i = 0$ for all $i < m$, while

$$\|\partial_l G_m\| \le c_0',$$

with a constant $c_0' > 0$, depending only on K. Exchanging c_0, we may assume that $c_0' = c_0$. Let $\partial_l G \in \mathcal{T}$ be the vector, consisting of m blocks each of them having $n-1$ entries, such that the first $m-1$ blocks are zero, while the mth block coincides with the column-vector $\partial_l G_m$. Define $\xi \in \mathcal{T}$ so that its ith block coincides with the column-vector $\partial_l u_i$. Using (10.24) for all $i = 1, \ldots, m$, we obtain

$$G_{uu}\eta + \partial_l G = 0,$$

and therefore

$$DG_{uu}\eta + D\partial_l G = 0.$$

Since $D_m = I$ and all blocks of $\partial_l G$, except the last one, are zero, we deduce that

$$DG_{uu}\eta = -\partial_l G,$$

therefore

$$\|DG_{uu}\eta\| \leq c_0.$$

This and (10.23) imply

$$\|D\eta\| \leq b_0 c_0.$$

The ith block of the vector $D\eta \in \mathcal{T}$ has the form $d_i \partial_l u_i$, and according to $d_i = e^{(m-i)\epsilon}$, we obtain

$$\|\partial_l u_i\| \leq b_0 c_0 e^{-(m-i)\epsilon}.$$

This is true for all $i = 1, \ldots, m$ and all $l = 1, \ldots, n-1$. Consequently, for the map $\partial_w u_i(v_0, w_0)$ we have

$$\|\partial_w u_i(v_0, w_0)\| \leq b_0 c_0 \sqrt{n-1}\ e^{-(m-i)\epsilon}.$$

Since $y_i(v, w) = \varphi_i(u_i(v, w))$ and φ_i was chosen in such a way that $\partial_{u_i}\varphi_i(0) = I$, the same estimate holds for $\partial_w y_i(v_0, w_0)$. Therefore (10.18) is satisfied with

$$C' = b_0 c_0 \sqrt{n-1}, \quad \epsilon = \log\left(1 + \frac{\kappa_0 \cos \varphi_0}{2C_0}\right),$$

which clearly depend only on K.

To establish (10.17), we proceed in the same way, considering the maps

$$v \mapsto y_i(v, w_0),$$

and determining the matrix D by $d_j = e^{j\epsilon}$, where ϵ is defined as above. This concludes the proof of the assertion. ♠

Proof of Lemma 10.2.1: Let $\epsilon > 0$ be defined as above and let

$$\delta = e^{-\epsilon}.$$

Then $0 < \delta < 1$ and δ depends only on K.

According to our assumptions, we have

$$y_i(v', \omega') = y_i', \quad y_i(v', w'') = y_i'',$$

for all $i = 1, \ldots, m$, where $v' = y_0'$, $v'' = y_0'' \in V$ and $w' = y_{m+1}'$, $w'' = y_{m+1}'' \in W$. For $s \in [0, 1]$ set $w_s = sw' + (1-s)w''$. Given $i = 1, \ldots, m$, consider the smooth curve

$$c(s) = y_i(v', w_s)$$

on ∂L_i. Since diam $W \leq C_1$, we have $\|w_0 - w_1\| \leq C_1$ with some constant $C_1 > 0$, depending only on K. Combining this with (10.18), we obtain

$$\|\dot{c}(s)\| \leq C' C_1 \delta^{m-i}.$$

Integrating this inequality for s from 0 to 1, we see that the length of the curve c is not greater than $C \cdot \delta^{m-i}$, where

$$C = C' C_1 > 0.$$

Hence there exists a curve with length $\leq C \cdot \delta^{m-i}$ on ∂L_i joining $y_i(v', w')$ and $y_i(v', w'')$. Consequently,

$$\|y_i(v', w') - y_i(v', w'')\| \leq C\delta^{m-i}. \tag{10.25}$$

Now set $v_s = sv' + (1-s)v''$ and consider the curve

$$d(s) = y_i(v_s, w'')$$

on ∂L_i. Applying the same argument and according to (10.17), we get

$$\|y_i(v', w'') - y_i(v'', w'')\| \leq C\delta^i.$$

Combining this with (10.25), one gets

$$\|y_i(v', w') - y_i(v'', w'')\| \leq C(\delta^i + \delta^{m-i}),$$

which proves (10.16). ♠

We conclude this section with another lemma, the proof of which is very similar to that of Lemma 10.2.1.

Fix an arbitrary $w \in S^{n-1}$ and set

$$Z = Z_\omega.$$

Let

$$M_1, M_2, \ldots, M_r, \tag{10.26}$$

$r \geq 2$, be a sequence of connected components of K such that $M_i \neq M_{i+1}$ for all $i = 1, \ldots, r - 1$. We assume that

$$\pi_\omega(M_1) \cap \pi_\omega(M_2) = \emptyset. \tag{10.27}$$

Using this assumption, we may choose φ_0 from the beginning of the section in such a way that if $x \in Z, y \in M_1, z \in M_2$, the open segments (x, y) and (y, z) have no common points with K and $[x, y]$ and $[y, z]$ satisfy the law of reflection at y with respect to ∂K, then $\varphi < \varphi_0$, where $\varphi \in (0, \pi/2]$ is the angle between $[y, z]$ and the normal $\nu(y)$ to ∂K at y. Hereafter we assume that φ_0 is chosen in this way. In this case φ_0 depends on K and the choice of Z (i.e. on the choice of ω).

For $j \leq r(u)$ denote by $\varphi_j(u) \in [0, \pi/2]$ the angle between the normal $\nu(x_j(u))$ and the vector $N_{+t_j(u)}(u)$. It then follows from the choice of φ_0 that

$$\varphi_j(u) < \varphi_0, \quad j < r(u), \tag{10.28}$$

for every $u \in Z$.

We are going to study trajectories $\gamma(u)$ such that $r(u) \geq r$ and $x_j(u) \in \Gamma_j$ for every $j = 1, \ldots, r$. Denote by $\mathcal{U}_r$ the set of all $u \in Z$ with these properties.

Lemma 10.2.3: *For all $u', u'' \in \mathcal{U}_r$ and every $i = 0, 1, \ldots, r$ we have*

$$\|x_i(u') - x_i(u'')\| \leq B\delta^{r-i}, \tag{10.29}$$

where $B > 0$ and $\delta \in (0, 1)$ are constants depending only on K and Z.

Proof: We use almost the same argument as in the proof of Lemma 10.2.1 above. Set $Z' = Z$ and choose Z'' as before, considering the pair $M = M_r$, $L = M_{r-1}$. Define V, W and F in the same way. This time it is more convenient to denote the elements of V by $u_0 = (u_0^{(1)}, \ldots, u_0^{(n-1)})$. Given $w \in W$ there exist uniquely determined $z_0(w) \in Z, z_1(w) \in \partial M_1, \ldots, z_{r-1}(w) \in \partial M_{r-1}$ such that

$$F(z_0(w), z_1(w), \ldots, z_{r-1}(w); w) = \min\{F(z_0, z_1, \ldots, z_{r-1}; w) : \\ (z_0, z_1, \ldots, z_m) \in Z \times \partial M_1 \times \cdots \times \partial M_{r-1}\}.$$

Then for $u = z_0(w)$ we have $z_i(w) = x_i(u)$, $i = 1, \ldots, r - 1$ and w is a point on the ray starting at $x_{r-1}(u)$ in direction $N_{+t_{r-1}(u)}(u)$.

We claim that for every $w \in W$ and every $j = 0, 1, \ldots, r - 1$ we have

$$\|\partial_w z_j(w)\| \leq C' e^{-(r-j-1)\epsilon}, \tag{10.30}$$

where $C' > 0$ and $\epsilon > 0$ are the same constants as in Lemma 10.2.2.

Fix $w_0 \in W$ and take a smooth chart

$$\varphi_j : \mathbf{R}^{n-1} \to U_j \subset \partial M_j,$$

$j = 1, \ldots, r-1$, such that $\varphi_j(0) = z_j(w_0)$ and $\{\frac{\partial \varphi_j}{\partial u_j^{(p)}}(0)\}_{p=1}^{n-1}$ is an orthonormal basis in $T_{z_j(w_0)} \partial M_j$. Take $\varphi_0(u_0) = u_0 + z_0(w_0)$. Define the function

$$G : V \times (\mathbf{R}^{n-1})^{r-1} \times W \to \mathbf{R}$$

by

$$G(u_0, u_1, \ldots, u_{r-1}; w) = F(\varphi_0(u_0), \varphi_1(u_1), \ldots, \varphi_{r-1}(u_{r-1}); w).$$

Next, we repeat the argument from the proof of (10.18) slightly exchanging the notation. After the change of φ_0 the constants c_0, C_0, ϵ, etc. are the same. In this way we establish the inequalities (10.30).

Let $u', u'' \in \mathcal{U}_r$, then $u' = z_0(w')$ and $u'' = z_0(w'')$ for some $w', w'' \in W$. Using an integration as in the proof of Lemma 10.2.1, we see that for every $i = 0, 1, \ldots, r-1$ we have

$$\|z_i(w') - z_i(w'')\| \leq B \cdot \delta^{r-i},$$

where $\delta = e^{-\epsilon}$ is the same as before, and $B = C \cdot e = C'C_1 e$. Since $x_i(u') = z_i(w')$, $x_i(u'') = z_i(w'')$, this implies (10.29). ♠

10.3. Existence of scattering rays and asymptotic of their sojourn times

Let the obstacle K have the form (10.1) and let Ω be the closure of its complement in $\mathbf{R}^n$. We assume again that the condition (H) is satisfied. In this section it is shown that for every configuration α, under a special choice of $\omega, \theta \in S^{n-1}$, there exists an infinite sequence γ_q of (ω, θ)-rays in Ω, following α in a certain way, and the asymptotic of the sojourn times T_{γ_q} as $q \to \infty$ is found.

In what follows we use the notation from Section 10.1. Given $x \in \mathbf{R}^n$, $\eta \in S^{n-1}$, denote by $l(x, \eta)$ the linear ray starting at x with direction η. Sometimes it will be convenient to use the notation

$$\text{dist}\,(x, y) = \|x - y\|\,.$$

Fix an arbitrary configuration α of the form (10.2). We shall say that the pair (ω, θ) of elements of S^{n-1} is α-admissible if the following conditions are satisfied:

(i) every (ω, θ)-ray in Ω is ordinary and any two different (ω, θ)-rays in Ω have distinct sojourn times;

(ii) for every $x \in \Gamma_{i_1}$ the ray $l(x, -\omega)$ (resp. $l(x, \omega)$) has no common points with $K \backslash K_{i_1}$ (resp. with K_{i_2});

(iii) for every $x \in \Gamma_{i_k}$ the ray $l(x, \theta)$ (resp. $l(x, -\theta)$) has no common points with $K \backslash K_{i_k}$ (resp. $K_{i_{k-1}}$).

Lemma 10.3.1: *For every configuration α there exist unit vectors $\omega \neq \theta$ such that (ω, θ) is α-admissible.*

Proof: Set $D_1 = K_{i_1}$, $D_2 = K_{i_2}$. Take an arbitrary hyperplane A, separating D_1 and D_2, such that A is tangent to D_1 at some point x and to D_2 at another point y. Set $\omega' = (x - y)/\|x - y\|$, and consider the convex cone

$$C = \{y + t(u - y) : u \in D_1, t \geq 0\}.$$

It is easy to check that the orthogonal projection of D_1 on the hyperplane $Z_{\omega'}$ is contained in C. We claim that

$$l(u, -\omega') \cap (K \backslash D_1) = \emptyset, \quad u \in D_1. \tag{10.31}$$

To prove this, assume that there exist $u \in D_1$, $t > 0$, $j \neq i_1$ such that $v = u - t\omega' \in K_j$. Then u and v have a common orthogonal projection u' on $Z_{\omega'}$. Since $u, u' \in C$, we have $v \in C$. On the other hand, the definition of C implies that the segment $[y, v]$ contains points of D_1, which is a contradiction of the condition (H). Thus, (10.31) holds. Then by the compactness of $K \backslash D_1$, there exists $\epsilon > 0$ such that

$$l(u, -\omega) \cap (K \backslash D_1) = \emptyset, \quad u \in D_1,$$

holds, provided $\omega \in S^{n-1}$ and $\|\omega - \omega'\| < \epsilon$. Take an arbitrary $\omega \in S^{n-1}$, satisfying the latter inequality and such that $\langle -\omega, \nu(x) \rangle > 0$. Then clearly the condition (ii) is satisfied.

Fix an ω with property (ii) and denote by $R(\omega)$ the residual subset of S^{n-1} from Therorem 10.1.1. Using the density of this set, and applying the above argument for $D_1 = K_{i_k}$ and $D_2 = K_{i_{k-1}}$, we find $\theta \in R(\omega)$ satisfying (iii). Now $\theta \in R(\omega)$ implies that (i) is also satisfied. This proves the assertion. ♠

From now on till the end of the chapter,

$$\alpha = (i_1, \ldots, i_k)$$

will be a *fixed configuration* with $k \geq 2$ and $i_1 \neq i_k$, and l will be a *fixed integer* with $1 \leq l \leq k$. For every integer $q \geq 0$ set

$$\alpha_{q,l} = (i_1, \ldots, i_k; \ldots; i_1, \ldots, i_k; i_1, \ldots, i_l), \tag{10.32}$$

where the block $(i_1, \ldots, i_k)$ is repeated q times. Clearly, $\alpha_{q,l}$ is a configuration of length $qk + l$.

We now fix two arbitrary unit vectors $\omega \neq \theta$ such that the pair (ω, θ) is $\alpha_{1,l}$-admissible. The existence of such vectors is guaranteed by Lemma 10.3.1. The pair (ω, θ) will be also fixed till the end of the chapter. As before we shall use the notation

$$Z = Z_\omega.$$

Proposition 10.3.2: *For every integer $q \geq 0$ there exists a unique (ω, θ)-ray γ_q of type $\alpha_{q,l}$ in Ω.*

Proof: Set for convenience $\eta = -\theta$. Fix an arbitrary integer $q \geq 0$ and set $m = qk + l$ and

$$D = Z \times \Gamma_{i_1} \times \ldots \times \Gamma_{i_m} \times Z_\eta,$$

where i_j are the successive components of $\alpha_{q,l}$. Define the function $F : D \to \mathbf{R}$ by

$$F(\zeta) = \|z_1 - x_1\| + \sum_{j=1}^{m-1} \|x_j - x_{j+1}\| + \|x_m - z_2\|$$

for every $\zeta = (z_1; x_1, \ldots, x_m; z_2) \in D$. Clearly F is continuous and, considering its restriction on an appropriate compact subset of D, we see that there exists $\zeta' = (z_1'; x_1', \ldots, x_m'; z_2') \in D$ with $F(\zeta') = \min F$. Clearly z_1' is the orthogonal projection of x_1' on Z, while z_2' is the projection of x_m' on Z_η. Since (ω, θ) satisfies condition (ii), the segment $[z_1', x_2']$ has no common points with K_{i_1}. For $c > 0$ consider the rotative ellipsoid

$$E_c = \{x \in \mathbf{R}^n : \|z_1' - x\| + \|x - x_2'\| \leq c\}.$$

Let $c > 0$ be the minimal number with $E_c \cap K_{i_1} \neq \emptyset$. Then E_c is tangential to K_{i_1} at some of its points y_1. It is now clear that $y_1 = x_1'$, since F has total minimum at ζ'. Therefore the segments $[z_1', x_1']$ and $[x_1', x_2']$ satisfy the law of reflection at x_1' with respect to Γ_{i_1}.

Repeating this argument several times and using the condition (H), we see that $x_1', \ldots, x_m'$ are the successive reflection points of a (ω, θ)-ray of type of $\alpha_{q,l}$ in Ω. The uniqueness of this ray follows from Corollary 2.4.6. ♠

For every integer $q \geq 0$ set $U_q = U_{\alpha_{q,l}}$. Then U_q is an open subset of $Z = Z_\omega$, and the above proposition implies $U_q \neq \emptyset$. More precisely, there exists a unique $u_q \in U_q$ such that $\gamma_q = \gamma(u_q)$ is an (ω, θ)-ray in Ω of type $\alpha_{q,l}$.

As in Section 10.1, the condition (H) and the choice of Z imply the existence of $\varphi_0 \in (0, \pi/2)$ such that

$$\varphi_j(u) < \varphi_0, \quad j = 1, \ldots, r(u), \tag{10.33}$$

for every $u \in Z$. Then as an immediate consequence of Lemma 10.2.3, we get the following.

Lemma 10.3.3: *There exist constants $C > 0$ and $\delta > 0$, depending only on K and Z, such that*

$$\mathrm{dist}(x_i(u), x_i(v)) \leq C\delta^{qk+l-i}, \quad i = 0, 1, \ldots, qk + l \tag{10.34}$$

for every integer $q \geq 0$ and all $u, v \in U_q$. ♠

We can apply also Lemma 10.2.3, considering the hyperplane $Z_{-\theta}$ instead of $Z = Z_\omega$. Note that γ_q is a (ω, θ)-ray, therefore taking its reflection points in the opposite order, we get a sequence

$$x_{qk+l}(u_q), x_{qk+l-1}(u_q), \ldots, x_2(u_q), x_1(u_q),$$

which is the sequence of the successive reflection points of a $(-\theta, -\omega)$-ray in Ω. In a similar way we can consider also the ray γ_{q+1}. Now, eventually exchanging the constants C and δ (making them dependent on K, Z_ω and $Z_{-\theta}$) to get an analogue of Lemma 10.3.3 with Z replaced by $Z_{-\theta}$, we obtain the inequalities

$$\text{dist}(x_{qk+l-i}(u_q), x_{(q+1)k+l-i}(u_{q+1})) \leq C\delta^{qk+l-i-1}$$

for all $i = 0, 1, \ldots, qk + l$. Setting $j = qk + l - i$, the latter implies

$$\text{dist}(x_j(u_q), x_{j+k}(u_{q+1})) \leq C\delta^{j-1}, \quad j = 0, 1, \ldots, qk + l. \tag{10.35}$$

Next, we assume that C and δ are fixed having the above properties. Using (10.34) for $i = 0$, we see that diam $\overline{U}_q \leq C\delta^{qk+l}$ for every q. Since

$$U_1 \supset \cdots \supset U_q \supset \cdots,$$

we deduce that the intersection of the closures of these sets consists of exactly one point u^∞, i.e.

$$\cap_{q=0}^\infty \overline{U}_q = \{u^\infty\}.$$

It is clear that the trajectory $\gamma^\infty = \gamma(u^\infty)$ has infinitely many reflection points. Moreover, for $x_i^\infty = x_i(u^\infty)$ we have

$$x_{qk+j}^\infty \in \Gamma_{i_j}, \quad q \geq 0, \quad 1 \leq j \leq k.$$

Fix for a moment $q \geq 0$, $r \geq 0$ and $j = 1, \ldots, k$. Take an arbitrary integer $p > q$ and apply Lemma 10.2.1 for $m = pk + j - 2$, the connected components

$$L' = K_{i_1}, \quad L_1 = K_{i_2}, \ldots, \quad L_m = K_{i_{j-1}}, \quad L'' = K_{i_j}$$

of K and the sequences of points

$$x_1^\infty, x_2^\infty, \ldots, x_{pk+j-1}^\infty, x_{pk+j}^\infty$$

and

$$x_{rk+1}^\infty, x_{rk+2}^\infty, \ldots, x_{(p+r)k+j-1}^\infty, x_{(p+r)k+j}^\infty.$$

Then we obtain

$$\text{dist}(x_{qk+j}^\infty, x_{(q+r)k+j}^\infty) \leq C(\delta^{qk+j} + \delta^{(p-q)k-2}). \tag{10.36}$$

Here we have used the inequality (10.16) for $i = qk + j$. Since (10.36) holds for all $p > q$, letting $p \to \infty$, we get

$$\text{dist}(x^\infty_{qk+j}, x^\infty_{(q+r)k+j}) \leq C\delta^{qk+j} \tag{10.37}$$

for all q and r, which shows that the sequence $\{x^\infty_{qk+j}\}_q$ is convergent. Denote by z_j its limit. Then $z_j \in \Gamma_{i_j}$, and (10.37) implies

$$\text{dist}(x^\infty_{qk+j}, z_j) \leq C\delta^{qk+j}, \quad q \geq 0, \quad 1 \leq j \leq k. \tag{10.38}$$

Moreover, it follows from

$$z_{j+k} = \lim_q x^\infty_{(q+1)k+j} = \lim_q x^\infty_{qk+j} = z_j$$

that $z_1, z_2, \ldots, z_k$ are the successive reflection points of a periodic reflecting ray γ_α of type α in Ω. Note that by Corollary 2.2.4, there exists only one such ray.

Remark: One can avoid the use of Lemma 10.2.1 in the above argument, according to the inequalities (10.35), which follow in fact from Lemma 10.2.3. In other words, for our considerations in this section (and the following one as well) only Lemma 10.2.3 from Section 10.2 is necessary. However, Lemma 10.2.1 presents an important property of the billiard trajectories, and since its proof is almost the same as that of Lemma 10.2.3, we include it in this book.

Set

$$d_j = \sum_{p=1}^{j} \|z_p - z_{p+1}\|, \quad 1 \leq j \leq k,$$

$d_\alpha = d_k$, and

$$L^\infty_m = \langle x^\infty_1, \omega \rangle + \sum_{p=1}^{m} \|x^\infty_p - x^\infty_{p+1}\|.$$

Clearly, d_α is the period (length) of γ_α. Using (10.38), for $q \geq 0$ and $r \geq 0$ we find

$$\left| (L^\infty_{(q+r)k+j} - (q+r)d_\alpha - d_j) - (L^\infty_{qk+j} - qd_\alpha - d_j) \right|$$

$$\leq 2C \sum_{p=1}^{rk+1} \delta^{qk+r+j} \leq C_1 \delta^q,$$

where the constant $C_1 > 0$ is determined by C and δ. Hence for every $j \leq k$ there exists

$$L_j = L_{\alpha,\omega,j} = \lim_q (L^\infty_{qk+j} - qd_\alpha - d_j).$$

Moreover, we have the asymptotic

$$L^\infty_{qk+j} = qd_\alpha + d_j + L_j + O(\delta^q) \quad \text{as } q \to \infty. \tag{10.39}$$

In the same way as above for $u^\infty \in Z = Z_\omega$, we find a unique $v^\infty \in Z_\eta$ such that the billiard trajectory $\tilde\gamma^\infty$, starting from v^∞ in direction $\eta = -\theta$ has infinitely many reflection points y^∞_i such that $y^\infty_{qk+r} \in \Gamma_{j_r}$ for all $q \geq 0$, $1 \leq r \leq k$, where

$$(j_1, \ldots, j_k) = (i_l, i_{l-1}, \ldots, i_1; i_k, i_{k-1}, \ldots, i_{l+2}, i_{l+1}).$$

Now for

$$G^\infty_m = -\langle y^\infty_1, \theta \rangle + \sum_{p=1}^{m} \|y^\infty_p - y^\infty_{p+1}\|$$

we get the asymptotic

$$G^\infty_{qk} = qd_\alpha + L_{\alpha,\theta} + O(\delta^q) \quad \text{as } q \to \infty, \tag{10.40}$$

where $L_{\alpha,\theta}$ is a constant, depending only on K, α and θ. This can be proved by the same argument as above, and we omit the details.

Set $T_q = T_{\gamma_q}$,

$$L^{(q)}_p = \langle x_1(u_q), \omega \rangle + \sum_{r=1}^{p} \|x_r(u_q) - x_{r+1}(u_q)\|,$$

and

$$G^{(q)}_p = -\langle x_{qk+l}(u_q), \theta \rangle + \sum_{r=p+1}^{qk+l-1} \|x_r(u_q) - x_{r+1}(u_q)\|.$$

Given $q \geq 0$, we define $p = p(q)$ by

$$p = k\left[\frac{q}{2}\right] + l - 1.$$

Using the choice of the constants C and δ (cf. Lemma 10.3.3), we get the following:

$$\left| L^{(q)}_p - L^\infty_p \right| + \left| G^{(q)}_p - G^\infty_{(q-[q/2])k} \right|$$

$$\leq 2 \sum_{r=1}^{p+1} \|x_r(u_q) - x^\infty_r\| + 2 \sum_{r=1}^{(q-[q/2])k+1} \|x_{qk+l-r+1}(u_q) - y^\infty_r\| \leq C_2\delta^q,$$

where $C_2 > 0$ is a constant, depending only on K, α, ω and θ. Combining this with the asymptotics (10.39) and (10.40), and using the fact that for every q,

$$T_q = L^{(q)}_p + G^{(q)}_p,$$

one obtains the following.

Theorem 10.3.4: *The sojourn times T_q have the asymptotic*

$$T_q = qd_\alpha + L_{\alpha,\omega,\theta} + O(\delta^q) \quad \text{as } q \to \infty, \tag{10.41}$$

where

$$L_{\alpha,\omega,\theta} = L_{l-1} + L_{\alpha,\theta} + d_{l-1},$$

and the constants d_j, L_j and $L_{\alpha,\theta}$, $d_0 = 0$, are determined as above. ♠

In particular, from the sojourn times T_q one can revocer the period d_α of the periodic reflecting ray γ_α.

Corollary 10.3.5: *Let $s = 2$, i.e. $K = K_1 \cup K_2$, and let d be the distance between K_1 and K_2. Let (ω, θ) be α-admissible for $\alpha = (1, 2)$. Then for every $i = 1, 2$ and every integer $m \geq 1$ there exists a unique (ω, θ)-ray in Ω with m reflection points, the first of which belongs to K_i. Let $T_m^{(i)}$ be the sojourn time of this (ω, θ)-ray. There exists a constant $\delta \in (0, 1)$ and for $i = 1, 2$, $j = 0, 1$ a constant $L_{\omega,\theta}^{(i,j)}$, such that*

$$T_{2q+j}^{(i)} = 2qd + L_{\omega,\theta}^{(i,j)} + O(\delta^q) \quad \text{as } q \to \infty.$$

Example 10.3.6: Under the notation in Corollary 10.3.5, assume in addition that $n = 2$ and K_1 and K_2 are discs in $\mathbf{R}^2 = Oxy$, having one and the same radius r and centres $(-a, 0)$ and $(a, 0)$, respectively. We suppose that $a > r > 0$, so K_1 and K_2 are disjoint. Set

$$\omega = (0, -1), \quad \theta = (\cos\theta_0, \sin\theta_0),$$

where $\theta_0 \in (0, \pi/2)$ is taken close to $\pi/2$. By Lemma 10.3.1 we can choose θ_0 in such a way that the pair (ω, θ) is α-admissible for $\alpha = (1, 2)$. In fact, under certain assumptions on r and a, it can be shown that this is true for all θ in a small neighbourhood of $-\omega$ in S^{n-1} (cf. [NS], for example). Using the notation $T_m^{(i)}$ from the above corollary, we then have $T_m^{(i)} \neq T_n^{(j)}$ whenever $(m, i) \neq (n, j)$.

Next, consider the two-dimensional torus K' in $\mathbf{R}^3 = Oxyz$, obtained by rotating K (or K_1 only) about the axis Oy. Let Ω' be the closure of the complement of K' in $\mathbf{R}^3$. We shall consider ω and θ as vectors in $\mathbf{R}^3$ having third component 0. According to our remarks in Example 2.5.2, we have that the scattering length spectrum of Ω' coincides with

$$\{T_m^{(i)} : m \in \mathbf{N}, \quad i = 1, 2\}.$$

Moreover, for $m \in \mathbf{N}$ and $i = 1, 2$ if γ is the (ω, θ)-ray in Ω' with sojourn time $T_m^{(i)}$, then for t close to $-T_\gamma = T_m^{(i)}$ we have (9.9). ♠

10.4. Asymptotic of the coefficients of the main singularity

We continue with the notation and assumptions from the previous section. For the sake of brevity set $\kappa = \cos\varphi_0$.

Given $q \geq 0$, denote by c_q the *coefficient in front of the main singularity* of the scattering kernel $s_\Omega(t, \theta, \omega)$ for t close to $-T_q$. In other words, c_q is the coefficient in front of $\delta^{(n-1)/2}(t + T_q)$ in the formula (9.9) for the (ω, θ)-ray $\gamma = \gamma_q$ and t close to $-T_q$.

Our aim in this section is to find the asymptotic of $|c_q|$ as $q \to \infty$.

Set

$$D = \operatorname{diam} K, \quad d' = \min_{i \neq j} \operatorname{dist}(K_i, K_j), \quad d'' = \frac{1}{d'}.$$

Since the domains K_i are compact and strictly convex, there exist constants $\mu_2 > \mu_1 > 0$ such that

$$\mu_1 \langle v, v \rangle \leq \langle G_x v, v \rangle \leq \mu_2 \langle v, v \rangle, \quad v \in T_x(\partial K),$$

$G_x : T_x(\partial K) \to T_x(\partial K)$ being the differential of the Gauss map of ∂K at x.

Let $x \in \Gamma_i$, $y \in \Gamma_j$, $i \neq j$, and assume that the segment $[x, y]$ is contained in Ω and is transversal to both Γ_i and Γ_j. Denote by Π the hyperplane passing through x and orthogonal to $[x, y]$. Set $e = (y - x)/\|y - x\|$, and denote by π the *projection* $\Pi \to T_x(\partial K)$ along the vector e. As in Section 2.3, define the symmetric linear map $\tilde\psi : \Pi \to \Pi$ by

$$\langle \tilde\psi(u), u \rangle = 2\langle e, \nu(x) \rangle \langle G_x(\pi(u)), \pi(u) \rangle, \quad u \in \Pi. \tag{10.42}$$

We shall say that $\tilde\psi$ is the *operator, determined by the segment* $[x, y]$. A standard exercise shows that

$$\operatorname{spec} \tilde\psi \subset [2\mu_1 \langle \nu(x), e \rangle, 2\mu_2 \langle \nu(x), e \rangle^{-1}]. \tag{10.43}$$

We shall prove two technical lemmas which will be used several times later.

Lemma 10.4.1: *Let* $x, x' \in \Gamma_i$ $y, y' \in \Gamma_j$, $i \neq j$, *and let* $\epsilon > 0$ *be such that* $\operatorname{dist}(x, x') < \epsilon$ *and* $\operatorname{dist}(y, y') < \epsilon$. *Introduce the vectors*

$$e = \frac{y - x}{\|y - x\|}, \quad e' = \frac{y' - x'}{\|y' - x'\|},$$

and assume that $\langle e, \nu(x) \rangle \geq \kappa$, $\langle e', \nu(x') \rangle \geq \kappa$. *Let*

$$\tilde\psi : \Pi \to \Pi, \quad \tilde\psi' : \Pi' \to \Pi'$$

be the operators, determined by the segments $[x, y]$ *and* $[x', y']$, *respectively. Then there exist a linear isometry* $A : \mathbf{R}^n \to \mathbf{R}^n$ *and a constant* $C' > 0$, *de-*

pending only on K and κ, such that $A(\Pi') = \Pi$, $\|A - I\| < C'\epsilon$ and $\|\tilde{\psi} - A\tilde{\psi}'A^{-1}\| < C'\epsilon$.

Proof: Let $A_1 : \mathbf{R}^n \to \mathbf{R}^n$ be the translation determined by the vector $x' - x$. Set $\nu'' = A_1(\nu(x'))$, and denote by A_2 the rotation around a line in $\mathbf{R}^n$ in a angle

$$\varphi = \cos^{-1}\langle\nu(x), \nu(x')\rangle,$$

for which $A_2(\nu'') = \nu(x)$ and such that $A_2 = $ id on $\{\nu(x), \nu''\}^\perp$. Finally, let $e'' = A_2 \circ A_1(e')$ and let A_3 be the rotation with $A_3(e'') = e$ which is identical on $\{e, e''\}^\perp$. We set $A = A_3 \circ A_2 \circ A_1$. Clearly, A is a linear isometry.

It is easy to check that $\|A_i - I\| < \text{const} \cdot \epsilon$ for every $i = 1, 2, 3$. For example,

$$\|A_2 - I\| = \left\|\begin{pmatrix} 1 - \cos\varphi & \sin\varphi \\ -\sin\varphi & 1 - \cos\varphi \end{pmatrix}\right\| = \sqrt{2(1 - \cos\varphi)} = \|\nu(x) - \nu(x')\|,$$

and the smoothness of G_x and the compactness of K imply

$$\|\nu(x) - \nu(x')\| < \text{const} \cdot \|x - x'\| < \text{const} \cdot \epsilon.$$

Similar simple estimates can be written for A_1 and A_3.

Thus, $\|A - I\| < \text{const} \cdot \epsilon$. Next, set $\tilde{\chi} = A\tilde{\psi}'A^{-1}$, $G = G_x$, $G' = G_{x'}$. Take an arbitrary $u \in \Pi$ with $\|u\| = 1$ and set $u' = A^{-1}u$. Then $u' \in \Pi'$ and $\|u'\| = 1$. Let

$$\pi : \Pi \to T_x(\partial K), \quad \pi' : \Pi' \to T_{x'}(\partial K)$$

be the projections along the vectors e and e', respectively, and let $v = \pi(u)$, $v' = \pi'(u')$. We have:

$$\left|\langle\tilde{\psi}(u), u\rangle - \langle\tilde{\chi}(u), u\rangle\right| = \left|\langle\tilde{\psi}(u), u\rangle - \langle\tilde{\psi}'(u'), u'\rangle\right|$$
$$= |2\langle e, \nu(x)\rangle\langle G(\pi(u)), \pi(u)\rangle - 2\langle e', \nu(x')\rangle\langle G'(\pi'(u')), \pi'(u')\rangle|$$
$$\leq 2\left|\langle e, \nu(x)\rangle - \langle e', \nu(x')\rangle\right|\langle G(v), v\rangle + 2\langle e', \nu(x')\rangle\left|\langle G(v), v\rangle - \langle G'(v'), v'\rangle\right|$$
$$< \text{const} \cdot \epsilon + 2\left|\langle G(v), v\rangle - \langle G'(v'), v'\rangle\right|.$$

It follows by the smoothness of the Riemannian metric on ∂K that

$$|\langle G(v), v\rangle - \langle G'(v')v'\rangle| < \text{const} \cdot \|v - v'\|.$$

Now taking into account that $\|\pi\| \leq 1/\kappa$ and $\|\pi'\| \leq 1/\kappa$, we find

$$\|v - v'\| = \|\pi(u) - \pi'(A^{-1}(u))\|$$
$$\leq \|\pi - \pi' \circ A^{-1}\| \leq \|\pi'\| \|A - I\| < \text{const} \cdot \epsilon.$$

Therefore $\left|\langle(\tilde{\psi} - \tilde{\chi})(u), u\rangle\right| < \text{const} \cdot \epsilon$, which implies $\left\|\tilde{\psi} - \tilde{\chi}\right\| < \text{const} \cdot \epsilon$. This proves the assertion. ♠

Further, we are going to apply the above lemma for two sequences of points. Let $x_1, \ldots, x_p$ and $x'_1, \ldots, x'_p$ be points of K such that for every $j = 1, \ldots, p$ the points x_j and x'_j belong to Γ_i for one and the same $i = i(j)$. Assume that

$$\text{dist}(x_j, x'_j) \leq D.a^j, \quad j = 1, \ldots, p \tag{10.44}$$

for some constants $D > 0$ and $a > 0$. Introduce the unit vectors

$$e_j = \frac{x_{j+1} - x_j}{\left\|x_{j+1} - x_j\right\|}, \quad e'_j = \frac{x'_{j+1} - x'_j}{\left\|x'_{j+1} - x'_j\right\|},$$

and assume that

$$\langle e_j, \nu(x_j) \rangle \geq \kappa, \quad \langle e'_j, \nu(x'_j) \rangle \geq \kappa, \quad j = 1, \ldots, p.$$

It follows by the above lemma that for every $j \leq p$ there exists a linear isometry A_j in $\mathbf{R}^n$ such that $A_j(\Pi'_j) = \Pi_j$ and

$$\left\|A_j - I\right\| < C'D(1+a)a^j, \quad \left\|\tilde{\psi}_j - A_j \tilde{\psi}'_j A_j^{-1}\right\| < C'D(1+a)a^j. \tag{10.45}$$

Here $\tilde{\psi}_j : \Pi_j \rightarrow \Pi_j$ and $\tilde{\psi}'_j : \Pi'_j \rightarrow \Pi'_j$ are the operators determined by the segments $[x_j, x_{j+1}]$ and $[x'_j, x'_{j+1}]$, respectively. Let $M_1 : \Pi_1 \rightarrow \Pi_1$ and $M'_1 : \Pi'_1 \rightarrow \Pi'_1$ be arbitrary symmetric non-negatively definite linear operators. Define recursively

$$M_i = \sigma_i M_{i-1}(I + \lambda_i M_{i-1})^{-1}\sigma_i + \tilde{\psi}_i, \quad i = 2, \ldots, p, \tag{10.46}$$

where $\lambda_i = \text{dist}(x_{i-1}, x_i)$ and σ_i is the symmetry with respect to Π_i. We define the maps M'_i, $i = 2, \ldots, p$, in the same way, replacing $\tilde{\psi}_i$, σ_i, λ_i and x_i by $\tilde{\psi}'_i$, σ'_i, λ'_i and x'_i, respectively. Finally, set

$$b = (1 + 2\mu_1 \kappa d')^{-1}, \quad a_1 = \begin{cases} a & \text{if } a \geq 1, \\ \max\{a, b\} & \text{if } a < 1. \end{cases} \tag{10.47}$$

Next, we use the notation $\log = \log_e$.

Lemma 10.4.2: *Under the above assumptions, there exist constants* $E > 0$, $E' > 0$, *depending only on* K, κ *and* a, *such that*

$$\left\|M_j - A_j M'_j A_j^{-1}\right\| < DEa_1^j + b^{2(j-r)}\left\|M_r - A_r M'_r A_r^{-1}\right\|, \tag{10.48}$$

and

$$\left| \log \det \left((I + \lambda_{j+1} M_j)(I + \lambda'_{j+1} M'_{j+1})^{-1} \right) \right|$$
$$< DE' a_1^j + (n-1)db^{2(j-r)+1} \left\| M_r - A_r M'_r A_r^{-1} \right\| \qquad (10.49)$$

for all $1 \le r \le j \le p$.

Proof: First, note that $|\lambda_i - \lambda'_i| < D(1+a)a^i$. Moreover, for every symmetric non-negative definite linear operator M we have

$$\left\| (I + \lambda M)^{-1} \right\| \le (1 + \lambda \sigma)^{-1}, \quad \left\| M(I + \lambda M)^{-1} \right\| \le \frac{1}{\lambda},$$

where $\sigma = \min(\operatorname{spec} M)$. It follows by (10.46) that

$$\min(\operatorname{spec} M_{i-1}) \ge \min(\operatorname{spec} \tilde{\psi}_{i-1}), \quad i \ge 2,$$

therefore

$$\left\| (I + \lambda_i M_{i-1})^{-1} \right\| \le b, \quad \left\| M_{i-1}(I + \lambda_i M_{i-1})^{-1} \right\| \le \frac{1}{\lambda_i} \le d''.$$

Introduce the operator

$$L_i = A_i M'_i A_i^{-1} : \Pi_i \rightarrow \Pi_i .$$

Since $\sigma'_i = A_i^{-1} \sigma_i A_i$, we find

$$L_i = \sigma_i B_i L_{i-1}(I + \lambda'_i L_{i-1})^{-1} B_i^{-1} \sigma_i + A_i \tilde{\psi}'_i A_i^{-1}$$

for $B_i = A_i \circ A_{i-1}^{-1}$. Using (10.46) and the trivial inequality

$$\left\| X - B_i Y B_i^{-1} \right\| \le 2 \left\| X \right\| \left\| I - B_i \right\| + \left\| X - Y \right\|,$$

we have:

$$\left\| M_i - L_i \right\| \le \left\| M_{i-1}(I + \lambda_i M_{i-1})^{-1} - B_i L_{i-1}(I + \lambda'_i L_{i-1})^{-1} B_i^{-1} \right\|$$
$$+ \left\| \tilde{\psi}_i - A_i \tilde{\psi}'_i A_i^{-1} \right\| < C' D(1+a)a^i + 2C'd''D(1+a)^2 a^{i-1}$$
$$+ \left\| M_{i-1}(I + \lambda_i M_{i-1})^{-1} - L_{i-1}(I + \lambda'_i L_{i-1})^{-1} \right\|.$$

The last term can be estimated as follows:

$$\left\| M_{i-1}(I + \lambda_i M_{i-1})^{-1} - L_{i-1}(I + \lambda'_i L_{i-1})^{-1} \right\|$$
$$\le |\lambda_i - \lambda'_i| \left\| (I + \lambda_i M_{i-1})^{-1} M_{i-1} \right\| \left\| L_{i-1}(I + \lambda'_i L_{i-1})^{-1} \right\|$$

$$+ b^2 \left\| M_{i-1} - L_{i-1} \right\| < D d''^2 (1+a) a^i + b^2 \left\| M_{i-1} - L_{i-1} \right\|.$$

Therefore

$$\left\| M_i - L_i \right\| < D E'' a^i + b^2 \left\| M_{i-1} - L_{i-1} \right\|, \quad i = 2, \ldots, p,$$

where $E'' = (1+a)(C' + 2d''(1+a)a^{-1}C' + d''^2) > 0$.
Repeating this procedure $j - r$ times, one gets

$$\left\| M_j - L_j \right\| < D E'' \sum_{t=0}^{j-r-1} a^{j-t} b^{2t} + b^{2(j-r)} \left\| M_r - L_r \right\|, \qquad (10.50)$$

for all $1 \le r \le j \le p$. There are two cases.
Case 1. $a \ge 1$. Then $a > b^2$, and (10.50) implies

$$\left\| M_j - L_j \right\| < D E'' a^j \sum_{t=0}^{j-r-1} (b^2/a)^t + b^{2(j-r)} \left\| M_r - L_r \right\|$$

$$< D E'' a^j \left(1 - \frac{b^2}{a} \right)^{-1} + b^{2(j-r)} \left\| M_r - L_r \right\|.$$

In this case we set $E = E'' a (a - b^2)^{-1}$.
Case 2. $0 < a < 1$. Then $a_1 = \max\{a, b\} < 1$, and (10.50) implies

$$\left\| M_j - L_j \right\| < D E'' \sum_{t=0}^{j-r-1} a_1^{j+t} + b^{2(j-r)} \left\| M_r - L_r \right\|$$

$$< D E'' (1 - a_1)^{-1} a_1^j + b^{2(j-r)} \left\| M_r - L_r \right\|.$$

Now set $E = E''(1 - a_1)^{-1}$.
It follows by the choice of E in both cases that (10.48) holds for $1 \le r \le j \le p$.
Before going on, let us note that if A is an arbitrary linear operator in $\mathbf{R}^k$, then

$$|\det A| \le (1 + \|A - I\|)^k.$$

Using this, we find the following estimates:

$$\det \left((I + \lambda_{i+1} M_i)(I + \lambda'_{i+1} M'_i)^{-1} \right)$$
$$\le \left(1 + \left\| I - (I + \lambda_{i+1} M_i)(I + \lambda'_{i+1} L_i)^{-1} \right\| \right)^{n-1}$$
$$= \left(1 + \left\| (\lambda'_{i+1} L_i - \lambda_{i+1} M_i)(I + \lambda'_{i+1} L_i)^{-1} \right\| \right)^{n-1}$$
$$\le \left(1 + \frac{|\lambda'_{i+1} - \lambda_{i+1}|}{\lambda'_{i+1}} + b \lambda_{i+1} \left\| M_i - L_i \right\| \right)^{n-1}$$

$$< \left(1 + D(1+a)d''a^{i+1} + bd\,\|M_i - L_i\|\right)^{n-1}.$$

In the same way one gets a similar inequality for

$$\det\left((I + \lambda_{i+1}M_i)^{-1}(I + \lambda'_{i+1}M'_i)\right),$$

therefore

$$\left|\log\det(I + \lambda_{i+1}M_i) - \log\det(I + \lambda'_{i+1}M'_i)\right|$$
$$< (n-1)\log(1 + Dd''(1+a)a^{i+1} + bd\,\|M_i - L_i\|)$$
$$< (n-1)\left(Dd''(1+a)a^{i+1} + bd\,\|M_i - L_i\|\right).$$

Finally, applying (10.48) for $j = i$, we get (10.49) with $E' = (n-1)((a_1^2 + a_1)$ $d'' + bdE)$ for all $1 \le r \le j \le p$. This completes the proof of the lemma. ♠

For the reflection points z_i of the unique periodic reflecting ray of type α in Ω (cf. Section 10.3) we define $z_m = z_j$, whenever m has the form $m = qk + j$, $1 \le j \le k$. Let

$$\tilde{\psi}''_j : \Pi''_j \to \Pi''_j$$

be the operator determined by $[z_j, z_{j+1}]$, and let $\mathcal{M}(\Pi''_j)$ be the *space of all symmetric positively definite linear maps* $M : \Pi''_j \to \Pi''_j$. Define $\mathcal{F}_j : \mathcal{M}(\Pi''_j) \to \mathcal{M}(\Pi''_j)$ by

$$\mathcal{F}_j(M) = \sigma''_{j+1}M(I + \lambda''_{j+1}M)^{-1}\sigma''_{j+1} + \tilde{\psi}''_{j+1},$$

where $\lambda''_j = \mathrm{dist}(z_j, z_{j-1})$ and σ''_j is the symmetry with respect to the tangent hyperplane to ∂K at z_j. Using the argument from the proof of Proposition 2.3.2, we deduce that the map

$$\mathcal{F}_k \circ \mathcal{F}_{k-1} \circ \ldots \mathcal{F}_1 : \mathcal{M}(\Pi''_1) \to \mathcal{M}(\Pi''_1)$$

has a unique fixed point M''_1. Then $M''_2 = \mathcal{F}_1(M''_1)$ is the unique fixed point of $\mathcal{F}_1 \circ \mathcal{F}_k \circ \mathcal{F}_{k-1} \circ \ldots \mathcal{F}_2$, etc.

Let $q \ge 0$ be an arbitrary integer. Consider the configuration $\alpha_{q,l}$, and set

$$J_q = J_{\alpha_{q,l}} : F_{\alpha_{q,l}} \to S^{n-1}.$$

Recall that $u_q \in U_q \subset Z$ is the unique point in Z for which there exists a (ω, θ)-ray of type $\alpha_{q,l}$ intersecting Z at u_q (we denote this (ω, θ)-ray by γ_q).

Set

$$m = qk + l$$

and $x_i = x_i(u_q)$ for $i = 1, \ldots, m$ and $x'_i = x_i(u_{q+1})$ for $i = 1, \ldots, m+k$. Define the operators $\tilde{\psi}_i$ and $\tilde{\psi}'_i$ as in the text before Lemma 10.4.2, and set $M_1 =$

$\tilde{\psi}_1, M_1' = \tilde{\psi}_1'$. Next, define M_i recursively by (10.46), and M_i' in a similar way. Finally, set

$$p = \left[\frac{m}{2}\right], \quad t = \left[\frac{p}{2}\right]. \tag{10.51}$$

Clearly, $2p \leq m < 2p+1$, $2t \leq p < 2t+1$, which implies $4t \leq m < 4p+3$. Applying the inequalities (10.34), we get:

$$\text{dist}(x_i, x_i') < C\delta^{p-i}, \quad i = 1,\dots,t, \tag{10.52}$$
$$\text{dist}(x_i, x_i') < C\delta^i, \quad i = t+1,\dots,p. \tag{10.53}$$

Next, (10.35) implies

$$\text{dist}(x_{p+i}, x_{p+k+i}') < C\delta^{p-i-1}, \quad i = 1,\dots,t, \tag{10.54}$$
$$\text{dist}(x_{p+i}, x_{p+k+i}') < C\delta^i, \quad i = t+1,\dots,m-p.$$

Finally, apply Lemma 10.2.1 to the sequences

$$x_1', x_2', \dots, x_m'$$

and

$$z_1, z_2, \dots, z_m$$

(instead of refering to Lemma 10.2.1, one can use the inequalities (10.35) only; cf. the Remark above). Then we find

$$\text{dist}(x_{rk+j}', z_j) \leq C(\delta^{rk+j} + \delta^{m+k-(rk+j)})$$

for $rk+j \leq p+k$. Since $m \geq 2p$, this implies

$$\text{dist}(x_{rk+j}', z_j) < C''\delta^{rk+j}, \quad rk+j \leq p+k,$$

where $C'' = C(1+\delta^{-k}) > C$. Set $D = C'' \cdot \delta^p, a = 1/\delta > 1$, then (10.44) takes the form

$$\text{dist}(x_i, x_i') < Da^i, \quad i = 1,\dots,t.$$

Now we can apply Lemma 10.4.2 to the sequences $x_1,\dots,x_t$ and $x_1',\dots,x_t'$. Since

$$\|M_1 - A_1 M_1' A_1^{-1}\| = \|\tilde{\psi}_1 - A_1 \tilde{\psi}_1' A_1^{-1}\| < C'D(1+a)a < 2CC'\delta^{p-2},$$

it follows by (10.49) for $r = 1$ and $a_1 = 1/\delta$ that

$$|\log \det(I + \lambda_{i+1}M_i) - \log \det(I + \lambda_{i+1}'M_i')|$$
$$< DE'\delta^{-i} + 2(n-1)db^{2i-1}C'C\delta^{p-2}$$
$$< CE'\delta_1^{p-i} + 2(n-1)dCC'\delta_1^{p+2i-3} < F_1\delta_1^{p-i}$$

for all $i = 1, \ldots, t$, where $F_1 = CE' + 2(n-1)dCC'$ and

$$\delta_1 = \max\{\delta, b\} \in (0, 1).$$

Further, observe that (10.52) implies

$$\text{dist}(x_i, x_i') < C\delta^i, \quad i = 1, \ldots, t.$$

This and (10.53) show that Lemma 10.4.2 is applicable with $D = C$, $a = \delta = a_1$, $r = 1$. Then we get

$$\left| \log \det(I + \lambda_{i+1} M_i) - \log \det(I + \lambda_{i+1}' M_i') \right| < F_1 \delta_1^i, \quad i = 1, \ldots, p.$$
(10.55)

Next, we apply Lemma 10.4.2 to the sequences $z_1, \ldots, z_{p+k}$ and $x_1', \ldots, x_{p+k}'$. It follows by (10.43) that

$$\| M_1' \| = \left\| \tilde{\psi}_1' \right\| \le \frac{2\mu_2}{\kappa}.$$

Therefore (10.49) implies

$$\left| \log \det(I + \lambda_{p+j+1} M_{p+j}) - \log \det(I + \lambda_{p+j+1}'' M_{p+j}'') \right|$$
$$< F_2 \delta_1^{p+j}, \quad j = 1, \ldots, \kappa,$$
(10.56)

where $F_2 > 0$ is a constant.

Consider the sequences $x_1, \ldots, x_p$ and $x_{k+1}', \ldots, x_{k+p}'$, and the corresponding isometries

$$A_j' : \Pi_{j+k}' \to \Pi_j$$

(cf. Lemma 10.4.1). Applying again Lemma 10.4.2 and the inequalities (10.35), we find a constant $F_3 > 0$ such that

$$\left\| M_p - A_p' M_{p+k}' A_p'^{-1} \right\| < F_3 \delta_1^p.$$
(10.57)

Now turn to the sequences $x_{p+1}, \ldots, x_{p+t}$ and $x_{p+k+1}', \ldots, x_{r+k+t}'$. It follows by (10.54), (10.57) and Lemma 10.4.2 for $D = C \cdot \delta^p$, $a = 1/\delta = a_1$, $r = p$, that

$$\left| \log \det(I + \lambda_{p+j+1} M_{p+j}) - \log \det(I + \lambda_{p+k+j+1}'' M_{p+k+j}'') \right|$$
$$< F_4 \delta_1^{p-j}, \quad j = 1, \ldots, t,$$
(10.58)

where $F_4 > 0$ is a constant.

Finally, applying Lemma 10.4.2 twice more, we find constants $F_5 > 0$, $F_6 > 0$ such that

$$\left| \log \det(I + \lambda_{p+j+1} M_{p+j}) - \log \det(I + \lambda_{p+k+j+1}'' M_{p+k+j}'') \right|$$

$$< F_5 \delta_1^j, \quad j = t+1, \ldots, m-p, \tag{10.59}$$

$$\left| \log \left| \det M_m \right| - \log \left| \det M'_{m+k} \right| \right| < F_6 \delta_1^p. \tag{10.60}$$

Set $F = \max\{F_1, \ldots, F_6\}$ and

$$\tilde{c} = -\sum_{j=1}^{k} \log \det(I + \lambda''_{j+1} M''_j) < 0. \tag{10.61}$$

Using the matrix representations of $\mathrm{d}J_q(u_q)$ and $\mathrm{d}J_{q+1}(u_{q+1})$ from Proposition 2.4.2, one finds

$$\log \left| \det \mathrm{d}J_{q+1}(u_{q+1}) \right| = \log \left| \det \mathrm{d}J_q(u_q) \right| - \tilde{c} + \epsilon_{q,l}, \tag{10.62}$$

where

$$
\begin{aligned}
\epsilon_{q,l} = & \sum_{i=1}^{p} \left(\log \det(I + \lambda'_{i+1} M'_i) - \log \det(I + \lambda_{i+1} M_i) \right) \\
& + \sum_{i=1}^{k} \left(\log \det(I + \lambda'_{p+i+1} M'_{p+i}) - \log \det(I + \lambda''_{p+i+1} M''_{p+i}) \right) \\
& + \sum_{j=1}^{m-p-1} \left(\log \det(I + \lambda'_{p+k+j+1} M'_{p+k+j}) - \log \det(I + \lambda_{p+j+1} M_{p+j}) \right) \\
& + \left(\log \left| \det M'_{m+k} \right| - \log \left| \det M_m \right| \right).
\end{aligned}
$$

Now combining (10.55), (10.56), (10.58), (10.59), and (10.60), we obtain

$$
\begin{aligned}
\left| \epsilon_{q,l} \right| & < F \left(2 \sum_{i=1}^{t} \delta_1^{p-i} + \sum_{i=t+1}^{p} \delta_1^i + \sum_{i=1}^{k} \delta_1^{p+i} + \sum_{j=t+1}^{m-p-1} \delta_1^j + \delta_1^p \right) \\
& < 6F(1-\delta_1)^{-1} \delta_1^t < F'_0 \delta_0^{kq}, \tag{10.63}
\end{aligned}
$$

with $\delta_0 = \delta_1^{1/4}$ and $F'_0 = 6F(1-\delta_1)^{-1} \delta_1^{-3/4}$.

Recall from the previous section that x_i^∞ are the reflection points of the trajectory $\gamma(u^\infty)$ with $r(u^\infty) = \infty$. Define the operators M_i^∞ in the same way as M_i, replacing the points x_i by x_i^∞. Set $\lambda_i^\infty = \mathrm{dist}(x_{i-1}^\infty, x_i^\infty)$ and

$$c_l = c_l(\omega, \theta) = \log \left| \det M_l^\infty \right| + \sum_{i=1}^{l-1} \log \det(I + \lambda_{i+1}^\infty M_i^\infty) + \sum_{j=1}^{\infty} \epsilon_{j,l}.$$

Then, applying (10.62) q times, we get

$$\log \left| \det \mathrm{d}J_q(u_q) \right| = -q\tilde{c} + c_l + \delta_{q,l},$$

where

$$\delta_{q,l} = -\sum_{j=q}^{\infty} \epsilon_{j,l} + (\log\left|\det dJ_\beta(u_q)\right| - \log\left|\det dJ_\beta(u^\infty)\right|)$$

and $\beta = (i_1, \ldots, i_l)$. The expression in the parentheses on the right-hand side of the latter equality can be estimated from above with $F_0'' \delta_0^{kq}$ for some constant $F_0'' > 0$. Therefore, by (10.63),

$$\left|\delta_{q,l}\right| < F_0' \delta_0^{kq}(1 - \delta_0^k)^{-1} + F_0'' \delta_0^{kq} = F_0 \delta_0^{kq},$$

where $F_0 = F_0'(1 - \delta_0^k)^{-1} + F_0'' > 0$.

Finally, set

$$c_\alpha = \frac{\tilde{c}}{2}.$$

As a consequence of the considerations in this section, we obtain the following.

Theorem 10.4.3: *There exist constants Q_α and δ_0, $0 < \delta_0 < 1$, depending only on K, α, ω and θ, such that*

$$\log\left|c_q\right| = qc_\alpha + Q_\alpha + O(\delta_0^q) \text{ as } q \to \infty. \tag{10.64}$$

Let us note that the constant c_α has a certain geometrical meaning. Namely, we have

$$c_\alpha = -\frac{1}{2}\sum_{j=1}^{n-1} \log\left|\mu_j\right|, \tag{10.65}$$

$\mu_1, \ldots, \mu_{n-1}$ being the eigenvalues of the linear Poincaré map P_{γ_α} outside the unit circle in $\mathbf{C}$. Indeed, the latter eigenvalues are precisely the eigenvalues of the operator S from the proof of Proposition 2.3.2. Using the representation of S found there, we get

$$\log\det S = \log\prod_{i=1}^{k} \det(I + \lambda_{i+1}'' M_i'') = -\tilde{c} = -2c_\alpha,$$

which proves (10.65).

10.5. Notes

Theorem 10.1.1 and Lemma 10.1.2, as well as the whole of Sections 10.3 and 10.4, are taken from [PS5]. Some results related to Theorem 10.3.4 are proved by Nakamura and Soga (cf. [Na1], [Na2], [NS]). Lemma 10.1.3 is well known in the theory of dispersing billiards. It seems that it was first proved for curves in the plane by Sinai [Sin1], see also [Sin2]. The material in Section 10.2 is an adaptation of a part of appendix (b) in [Sjo], perhaps it might be derived also from Section 3 of [I4]. As we have already mentioned in the Remark in Section 10.3, Lemma 10.2.3 is sufficient for our aims in this chapter. Namely, one can slightly exchange the arguments in Sections 10.3 and 10.4 to prove the same results using only Lemma 10.2.3 from Section 10.2.

It can be shown that the set $R(\omega)$ has a full Lebesgue measure in S^{n-1}, i.e. $S^{n-1}\backslash R(\omega)$ has Lebesgue meaures zero. It was proved recently by one of the authors [S5] that for every compact domain K in $\mathbf{R}^n$ with smooth boundary and connected complement Ω, there exists $R_K \subset S^{n-1} \times S^{n-1}$ with full Lebesgue measure in $S^{n-1} \times S^{n-1}$ such that for $(\omega, \theta) \in R_K$ all reflecting (ω, θ)-rays in Ω are ordinary and non-degenerate and have different sojourn times. It is natural to ask if R_K can be chosen in such a way that for $(\omega, \theta) \in R_k$ there are no (ω, θ)-rays of mixed type in Ω. This would imply that (8.30) becomes an equality for all $(\omega, \theta) \in R_K$.

REFERENCES

[Ab] R. Abraham: *Bumpy metrics.* In: Proc. of Symp. in Pure Math. Vol. XIV, A.M.S., Providence, Rhode Island, 1970.

[AbM] R. Abraham, J. Marsden: *Foundations of Mechanics.* New York: Benjamin, 1967.

[AbR] R. Abraham, J. Robbin: *Transversal Mappings and Flows.* New York: Benjamin, 1967.

[AM] K. Anderson, R. Melrose: *The propagation of singularities along gliding rays.* Invent. Math. **41** (1977), 197–232.

[An] D.V. Anosov: *On generic properties of closed geodesics.* Izv. Acad. Nauk SSSR **46**, No. 4 (1982), 675–703 (in Russian); English transl.: Math. USSR Izv. **21** (1983), 1–29.

[Ar] V. I. Arnold: *Mathematical Methods of Classical Mechanics.* Berlin: Springer, 1978.

[ArA] V.I. Arnold, A. Avez: *Problèmes Ergodiques de la Mécanique Classique.* Paris: Gauthier-Villars, 1967.

[BGR] C. Bardos, J. C. Guillot, J. Ralston: *La relation de Poisson pour l'équation des ondes dans un ouvert nonborné. Application à la théorie de la diffusion.* Commun. in Partial Diff. Equations **7** (1982), 905–958.

[BLR] C. Bardos, G. Lebeau, J. Rauch: *Scattering frequences and Gevrey 3 singularities.* Invent. Math. **90** (1987), 77–114.

[Ber] P.H. Bérard: *Spectral Geometry: Direct and Inverse Problems.* Lect. Notes in Math. **1207**, Berlin: Springer, 1986.

[Bir] G.D. Birkhoff: *Dynamical Systems.* New York: AMS, Revised Eds., 1966.

[BunS] L. Bunimovich, Ya.Sinai: *Markov partitions for dispersed billiards.* Commun. Math. Phys. **73** (1980), 247–280.

[BCS] L. Bunimovich, N. Chernov, Ya. Sinai: *Markov partitions for two-dimensional hyperbolic billiards.* Uspehi Mat. Nauk **45**, No. 3 (1990), 97–134 (in Russian); Engl. translation: Russian Math. Surveys **45** (1990), 105–152.

[Car] F. Cardoso: *Propriedades Espectrais do Laplaceano.* 17x Coloquio Brasileiro de Matematica, IMPA, Rio de Janeiro, 1989.

[CPS] F. Cardoso, V. Petkov, L. Stoyanov: *Singularities of the scattering kernel for generic obstacles.* Ann. Inst. Henri Poincaré **53**, No. 4 (1990), 445–466.

[Ch1] J. Chazarain: *Construction de la paramétrix du problème mixte hyperbolique pour l'equation des ondes.* C.R. Acad. Sci. Paris, Ser. A. **276** (1973), 1213–1215.

[Ch2] J. Chazarain: *Formule de Poisson pour les variétés riemanniennes.* Invent. Math. **24** (1974), 65–82.

[Cher1] N. Chernov: *Construction of transversal foliations for multidimensional semi-scattering billiards.* Funct. Anal. Appl. **16** (1982), 33–46 (in Russian); Engl. translation: **16** (1983), 270–280.

[Cher2] N. Chernov: *Topological entropy and periodic points of two-dimensional hyperbolic billiards*. Funct. Anal. and its Appl. **25** (1991), 50–57 (in Russian); Engl. translation: **25** (1991), 39–45.

[C1] Y. Colin de Verdière: *Sur les longeurs des trajectoires périodiques d'un billiard*. In: Géometrie Symplectique et de Contact: Autour du Théorème de Poincaré-Birkhoff, Paris: Hermann, 1984, 122–139.

[C2] Y. Colin de Verdière: *Spectre du laplacien et longuers des géodesiques périodiques* II. Compositio Math. **27** (1973), 159–184.

[CFS] I.P. Cornfeld, S.V. Fomin, Ya. G. Sinai: *Ergodic Theory*. Berlin: Springer, 1982.

[DG] J.J. Duistermaat, V. Guillemin: *The spectrum of positive elliptic operators and periodic geodesics*. Invent. Math. **29** (1975), 39–79.

[ESch] Yu. Egorov, M. Schubin (Ed.): *Differential Equations with Partial Derivatives*. Encyclopaedia of Mathematical Sciences. Vol. **31**, Berlin: Springer, 1992.

[Ger] C. Gérard: *Asymptotique des pôles de la matrice de scattering pour deux obstacles strictement convexes*. Bulletin de la S.M.F., Mémoire no. 31, **116**(1), (1988).

[GG] M. Golubitskii, V. Guillemin: *Stable Mappings and their Singularities*. Berlin: Springer, 1973.

[GKM] D. Gromoll, W. Klingenberg, W. Meyer: *Riemanische Geometrie im Großen*. Lecture Notes in Math. **55**, Berlin: Springer, 1975.

[G1] V. Guillemin: *Sojourn time and asymptotic properties of the scattering matrix*. Publ. RIMS Kyoto Univ. **12** (1977), 69–88.

[G2] V. Guillemin: *Lectures on spectral theory of elliptic operators*. Duke Math. J. **44** (1977), 485–517.

[GM1] V. Guillemin, R. Melrose: *The Poisson sumation formula for manifolds with boundary*. Adv. Math. **32** (1979), 204–232.

[GM2] V. Guillemin, R. Melrose: *An inverse spectral result for elliptic regions in* $\mathbf{R}^2$. Adv. Math. **32** (1979), 128–148.

[GM3] V. Guillemin, R. Melrose: *A cohomological invariant of discrete dynamical systems*. In: Christoffel Centennial Volume, Ed. P.I. Putzer and F. Feher, Basel: Birkhauser, 1981, 672–679.

[HPTL] B. Helffer, Pham The Lai: *Remarque sur la conjecture de Weyl*. Math. Scand. **48** (1981), 39–40.

[Hir] M. Hirsch: *Differential Topology*. Berlin: Springer, 1976.

[H1] L. Hörmander: *The Analysis of Linear Partial Differential Operators*, Vol. I, 2nd Edn, Berlin: Springer, 1983.

[H2] L. Hörmander: *The Analysis of Linear Partial Differential Operators*, Vol. II, Berlin: Springer, 1983.

[H3] L. Hörmander: *The Analysis of Linear Partial Differential Operators*, Vol. III, *Pseudo-differential operators*, Berlin: Springer, 1985.

[H4] L. Hörmander: *The Analysis of Linear Partial Differential Operators*, Vol. IV, Berlin: Springer, 1985.

[I1] M. Ikawa: *Decay of solutions of the wave equation in the exterior of two convex obstacles*. Osaka J. Math. **19** (1982), 459–509.

[I2] M. Ikawa: *On the distribution of poles of the scattering matrix for two convex obstacles*. Hokkaido Math. J. **12** (1983), 343–359; Addendum: **13** (1983) 795–802.

[I3] M. Ikawa: *Precise information on the poles of the scattering matrix for two strictly convex obstacles*. J. Math. Kyoto Univ. **27** (1987), 69–102.

[I4] M. Ikawa: *Decay of solutions of the wave equation in the exterior of several strictly convex bodies*. Ann. Inst. Fourier **38** (1988), 113–146.

[I5] M. Ikawa: *On the existence of the poles of the scattering matrix for several convex bodies.* Proc. Japan Acad. **64**, Ser. A (1988), 91–93.

[I6] M. Ikawa: *Singular perturbation of symbolic flows and poles of the zeta functions.* Osaka J. Math. **27** (1990), 281–300.

[Iv1] V. Ia. Ivrii: *Second term of the spectral asymptotic expansion of the Laplace-Beltrami operator on manifolds with boundary.* Funct. Analysis and its Applications **14**, No. 2 (1980), 25–34 (in Russian); English transl.: **14** (1980), 98–106.

[Iv2] V. Ia. Ivrii: *Sharp spectral asymptotics for the Laplace-Beltrami operator under general elliptic boundary conditions.* Funct. Anal. and its Appl. **15** (1981), 74–75.

[Kac] M. Kac: *Can one hear the shape of a drum?* Amer. Math. Soc. Monthly **73** (1966), 1–23.

[Kat1] A. Katok: *Lyapunov exponents, entropy and periodic orbits for diffeomorphisms.* Publ. Math. I.H.E.S. **51** (1980), 137–173.

[Kat2] A. Katok: *Entropy and closed geodesics.* Ergod. Theory Dynam. Sys. **2** (1982), 339–367.

[Kat3] A. Katok: *The growth rate of the number of singular and periodic orbits for a polygonal billiard.* Commun. Math. Phys. **111** (1987), 151–160.

[KS] A. Katok, J.M. Strelcyn. In collaboration with F. Ledrappier and F. Przytycki: *Smooth Maps with Singularities: Invariant Manifolds, Entropy and Billiards.* Lecture Notes in Math. **1222**, Berlin: Springer, 1986.

[K1] W. Klingenberg: *Riemannian Geometry.* Berlin: Walter de Gruyter, 1982.

[K2] W. Klingenberg: *Lectures in Closed Geodesics.*, Berlin: Springer, 1978.

[KT] W. Klingenberg, F. Takens: *Generic properties of geodesic flows.* Math. Ann. **197** (1972), 323–334.

[Ko] V. Kovachev: *Smoothness of the billiard ball map for strictly convex domains near the boundary.* Proc. Amer. Math. Soc. **103** (1988), 856–860.

[KP] V. Kovachev, G. Popov: *Invariant tori for the billiard ball map.* Trans. Amer. Math. Soc. **317** (1990), 45–81.

[KozT] V. Kozlov, D. Treshchëv: *Billiards: A Genetic Introduction to the Dynamics of Systems with Impact.* Translations of Math. Monographs, Vol. 89, Providence, R.I.: A.M.S., 1991.

[KSS] A. Kramli, N. Simanyi, D. Szasz: *Dispersing billiards without focal points on surfaces are ergodic.* Commun. Math. Phys. **125** (1989), 439–457.

[Lan] E.E. Landis: *Tangential singularities.* Funct. Analysis and its Applications **15** (1981), 36–49 (in Russian); English transl.: **15** (1981), 103–114.

[L1] V.F. Lazutkin: *Convex Billiard and Eigenfunctions of the Laplace Operator.* Leningrad University, 1981 (in Russian).

[L2] V.F. Lazutkin: *The existence of caustics for the billiard problem in a convex domain.* Izv. Akad. Nauk SSSR, Ser. Mat. **37** (1973), 186–216 (in Russian); English transl.: Math. USSR Izv. **7** (1974), 185–214.

[Lang] S. Lang: *Differential Manifolds.* Reading, Mass.: Addison-Wesley, 1972.

[LPh] P. Lax, R. Phillips: *Scattering Theory.* New York: Acadamic, 1967.

[Mag] A. Magnuson: *Symplectic singularities, periodic orbits of the billiard ball map and the obstacle problem.* Thesis, Massachusetts Institute of Technology, 1984.

[Ma1] A. Majda: *High frequency asymptotics for the scattering matrix and the inverse problem of acoustical scattering.* Commun. Pure Appl. Math. **29** (1976), 261–291.

[Ma2] A. Majda: *A representation formula for the scattering operator and the inverse problem for arbitrary bodies.* Commun. Pure Appl. Math. **30** (1977), 165–194.

[MaT] A. Majda, M. Taylor: *The asymptotic behaviour of the diffraction peak in classical scattering.* Commun. Pure Appl. Math. **30** (1977), 639–669.

[Marg] G. Margulis: *Applications of ergodic theory to the investigation of manifolds of negative curvature.* Funct. Anal. Appl. **3** (1969), 335–336 (in Russian).

[MM] S. Marvizi, R. Melrose: *Spectral invariants of convex planar regions.* J. Diff. Geometry **17** (1982), 475–502.

[MSi] H. P. McKean, I. M. Singer: *Curvature and the eigenvalues of the Laplacian.* J. Diff. Geometry **1** (1967), 43–69.

[Me1] R. Melrose: *Equivalence of glancing hypersurfaces.* Invent. Math. **37** (1975), 165–192.

[Me2] R. Melrose: *Singularities and energy decay of acoustical scattering.* Duke Math. J. **46** (1979), 43–59.

[Me3] R. Melrose: *Forward scattering by a convex obstacle.* Commun. Pure Appl. Math. **33** (1980), 461–499.

[MS1] R. Melrose, J. Sjöstrand: *Singularities in boundary value problems*, I. Commun. Pure Appl. Math. **31** (1978), 593–617.

[MS2] R. Melrose, J. Sjöstrand: *Singularities in boundary value problems*, II. Commun. Pure Appl. Math. **35** (1982), 129–168.

[MeyP] K. Meyer, J. Palmore: *A generic phenomenon in conservative Hamiltonian systems.* Proc. Symp. Pure Math. **14**, A.M.S., Providence R.I., 1970, 185–189.

[Mi] J. Milnor: *Eigenvalues of the Laplace operator of certain manifolds.* Proc. Nat. Acad. Sci. U.S.A. **51** (1964), 542.

[MP] S. Minakshisundaram, A. Pleijel: *Some properties of the eigenfunctions of the Laplace operator on Riemannian manifolds.* Canadian J. Math. **1** (1949), 242–256.

[Mor] T. Morita: *The symbolic representation of billiards without boundary conditions.* Trans. Amer. Math. Soc., to appear.

[Na1] S. Nakamura: *Singularities of the scattering kernel for two convex obstacles.* Publ. RIMS Kyoto Univ. **25** (1989), 223–238.

[Na2] S. Nakamura: *Singularities of the scattering kernel for several convex obstacles.* Hokkaido Math. J. **18** (1989), 487–496.

[NS] S. Nakamura, H. Soga: *Singularities of the scattering kernel for two balls.* J. Math. Soc. Japan **40** (1988), 205–220.

[P1] V. Petkov: *High frequency asymptotics of the scattering amplitude for non-convex bodies.* Commun. Partial Diff. Equations **5** (1980), 293–329.

[P2] V. Petkov: *Note on the distribution of poles of the scattering matrix.* J. Math. Anal. Appl. **101** (1984), 582–587.

[P3] V. Petkov: *Singularities of the scattering kernel.* Nonlinear Partial Differential Equations and their Applications. Collège de France Seminar VI (1984), 288–296.

[P4] V. Petkov: *Singularities of the scattering kernel for non-convex obstacles.* Proc. Conf. on Integral Equations and Inverse Problems, Varna, 1984; pp. 200–208 in: *Integral Equations and Inverse Problems.* Pitman Research Notes in Math. Series **235** London: Longman, 1991.

[P5] V. Petkov: *Scattering Theory for Hyperbolic Operators.* Amsterdam: North-Holland, 1989.

[PP] V. Petkov, G. Popov: *Asymptotic behaviour of the scattering phase for non-trapping obstacles.* Ann. Inst. Fourier (Grenoble) **32** (1982), 114–149.

[PS1] V. Petkov, L. Stoyanov: *Periodic geodesics of generic nonconvex domains in* $\mathbf{R}^2$ *and the Poisson relation.* Bull. Amer. Math. Soc. **15** (1986), 88–90.

[PS2] V. Petkov, L. Stoyanov: *Periods of multiple reflecting geodesics and inverse spectral results.* Amer. J. Math. **109** (1987), 619–668.

[PS3] V. Petkov, L. Stoyanov: *Spectrum of the Poincaré map for periodic reflecting rays in generic domains*. Math. Z. **194** (1987), 505–518.

[PS4] V. Petkov, L. Stoyanov: *On the number of periodic reflecting rays in generic domains*. Ergodic Theory Dynam. Sys. **8** (1988), 81–91.

[PS5] V. Petkov, L. Stoyanov: *Singularities of the scattering kernel and scattering invariants for several strictly convex obstacles*. Trans. Amer. Math. Soc. **312** (1989), 203–235.

[PS6] V. Petkov, L. Stoyanov: *Singularities of the scattering kernel for a class of star-shaped non-convex obstacles*. Mat. Appl. Comput. **8** (1989), 167–176.

[PV] V. Petkov, P. Vogel: *La représentation de l'application de Poincaré correspondant aux rayons périodiques réflechissants*. C.R. Acad. Sci. Paris, Ser. A **296** (1983), 633–635.

[Pon] L. Pontriagin: *Ordinary Differential Equations*, Reading, MA: Addison-Wesley, 1962.

[Po1] G. Popov: *Quasimodes for the Laplace operator*. Preprint.

[Po2] G. Popov: *Quasimodes and glancing hypersurfaces*. Proc. Conf. on Microlocal Analysis and Non-linear Waves, Minnesota 1989. Berlin: Springer, 1991.

[Po3] G. Popov: *Length spectrum invariants of an elliptic geodesic*. Preprint, 1990.

[Prot] M. H. Protter: *Can one hear the shape of a drum? Revisited*. SIAM Rev. **29** (1987), 185–197.

[R] R. C. Robinson: *Generic properties of conservative systems. I, II.* Amer. J. Math. **92** (1970), 562–603 and 897–906.

[Se] R. Seeley: *A sharp asymptotic remainder estimate for the eigenvalues of the Laplacian in a domain in* $\mathbf{R}^3$. Adv. Math. **29** (1978), 244–269.

[Sin1] Ya. Sinai: *Dynamical systems with elastic reflections. Ergodic properties of dispersing billiards*. Uspehi Mat. Nauk **25** (1970), 141–192 (in Russian); English transl.: Russian Math. Surv. **25** (1970), 137–189.

[Sin2] Ya. Sinai: *Development of Krylov ideas*. An addendum to the book: N. S. Krylov: *Works on the Foundations of Statistical Physics*. Princeton Univ. Press, 1979.

[Sjö] J. Sjöstrand: *Geometric bounds on the density of resonances for semiclassical problems*. Duke Math. J. **60** (1990), 1–57.

[So1] H. Soga: *Oscillatory integrals with degenerate stationary points and their applications to the scattering theory*. Commun. Partial Diff. Equations **6** (1981), 273–287.

[So2] H. Soga: *Singularities of the scattering kernel for convex obstacles*. J. Math. Kyoto Univ. **22** (1983), 729–765.

[So3] H. Soga: *Conditions against rapid decrease of oscillatory integrals and their applications to inverse scattering problems*. Osaka J. Math. **23** (1986), 441–456.

[S1] L. Stoyanov: *Generic properties of periodic reflecting rays*. Ergodic Theory Dynam. Sys. **7** (1987), 597–609.

[S2] L. Stoyanov: *An estimate from above of the number of periodic orbits for semi-dispersed billiards*. Commun. Math. Phys. **124** (1989), 217–227.

[S3] L. Stoyanov: *A bumpy metric theorem and the Poisson relation for generic strictly convex domains*. Math. Ann. **287** (1990), 675–696.

[S4] L. Stoyanov: *Nonexistence of generalized scattering rays and singularities of the scattering kernel for generic domains in* $\mathbf{R}^3$. Proc. Amer. Math. Soc., **113** (1991), 847–856.

[S5] L. Stoyanov: *On the singularities of the scattering kernel*. Preprint, 1991.

[ST] L. Stoyanov, F. Takens: *Generic properties of closed geodesics on smooth hypersurfaces*. Preprint, 1991.

[T] F. Takens: *Hamiltonian systems: Generic properties of closed orbits and local perturbations*. Math. Ann. **188** (1970), 304–312.

[Tay] M. Taylor: *Grazing rays and reflections of singularities of solutions to wave equations*. Commun. Pure Appl. Math. **29** (1976), 1–38.

[Vi] M. F. Vigneras: *Varietes riemanniennes isospectrales et non isometriques.* Ann. Math. **91** (1980), 21–32.

[Wa] P. Walters: *An Introduction to Ergodic Theory.* Berlin: Springer, 1981.

[W] H. Weyl: *Über die Asymptotische Verteilung der Eigenwerte.* Göttingen Nachr. (1911), 110–117.

[Y] K. Yamamoto: *Characterization of a convex obstacle by singularities of the scattering kernel.* J. Diff. Equations **64** (1986), 283–293.

SYMBOL INDEX

SUBJECT INDEX